COMBINATORIAL MAPS

Efficient Data Structures for Computer Graphics and Image Processing

COMBINATORIAL MAPS

Efficient Data Structures for Computer Graphics and Image Processing

Guillaume Damiand
Pascal Lienhardt

CRC Press
Taylor & Francis Group
Boca Raton London New York

CRC Press is an imprint of the
Taylor & Francis Group, an **informa** business

CRC Press
Taylor & Francis Group
6000 Broken Sound Parkway NW, Suite 300
Boca Raton, FL 33487-2742

First issued in paperback 2019

ISBN-13: 978-1-4822-0652-4 (hbk)
ISBN-13: 978-0-367-37835-6 (pbk)

Visit the Taylor & Francis Web site at
http://www.taylorandfrancis.com

and the CRC Press Web site at
http://www.crcpress.com

to Jean Françon and Yves Bertrand

Contents

Acknowledgements

The authors wish to thank the following persons, who contributed to common works related to combinatorial maps: Ehoud Ahronovitz, Olivier Alata, Sylvie Alayrangues, Eric Andres, Denis Arrivault, Ken Arroyo Ohori, Mehdi Baba-ali, Fabien Baldacci, Antoine Bergey, Yves Bertrand, Philippe Borianne, Sylvain Brandel, Camille Bihoreau, Pascal Bourdon, Achille Braquelaire, Luc Brun, Xiaofeng Chen, David Coeurjolly, Camille Combier, Xavier Daragon, Martine Dexet, Abdoulaye Diakité, Jean-Philippe Domenger, Jean-François Dufourd, Alexandre Dupas, Hervé Elter, Andreas Fabri, Christophe Fiorio, Elsa Fléchon, Laurent Fuchs, Michel Giner, Romain Goffe, Rocio Gonzalez-Diaz, Stéphane Gosselin, Nicolas Guiard, Jean-Paul Gourlot, Michel Habib, Yll Haxhimusa, Colin De La Higuera, Sébastien Horna, Marc Hugon, Adrian Ion, Fabrice Jaillet, Jean-Christophe Janodet, Walter G. Kropatsch, Jacques-Olivier Lachaud, Véronique Lang, Hugo Ledoux, Sébastien Loriot, David Marcheix, Daniel Meneveaux, Christian Olivier, Christophe Paul, Samuel Peltier, Patrick Resch, Jean-Pierre Réveilles, Jarek Rossignac, Tristan Roussillon, Philippe Saade, Emilie Samuel, Francis Sergeraert, Carine Simon, Xavier Skapin, Christine Solnon, Monique Teillaud, Olivier Terraz, Pol Vanhaecke, Frédéric Vidil, Florence Zara.

Many thanks to my friends and my familly for their support: Aline, Benoît, Angèle, Fabien, Anne, Denis, Anne-Laure, Olivier, Benosh, Fred, Delphine, Stéphane, Françoise, Éric, Gene, François, Imna, Daniel, Isabelle, Laurent, Myriam, Olivier, Sandrine, Aurélien, So, Yannick, Sylvia, Stéphan, Élo, Niko, Fab, Véro, Eric, Véronique, Nabila, Christophe, Laure, Jérôme, Céline, Vaghn, Émilie, Jérôme, Rachel, Jérôme, Caroline, Michel, Lucille, Damien, Nathalie, Emma, Bernard, Jeanne, Hélène, Martina, Frédéric, Jon, Pris, Laurent, Vak, Yo, and particularly Christelle, Max and Charlie.

Many thanks to Cécile, Myriam and Michaël.

List of Algorithms

List of Figures

1

Introduction

1.1 Subdivisions of Geometric Objects

Now, everyone knows several applications for which it is necessary to represent geometric objects in a computer, for instance in the fields of architecture, computer-aided design and manufacturing, geology, medical simulation, image synthesis, video games, medical or biological image processing and analysis, etc.

Representing a geometric object is necessary to compute "information":

- *about* the object, for instance computing geometric properties, extracting images, etc. Often, the object is *analyzed* in some way;
- *using* the object, for instance for *simulating* some phenomenon: physical, biological, etc.

So various needs lead to very different methods for representing geometric objects [172, 178, 2]. Among them, many deal with *subdivided* geometric objects.

1.1.1 Subdivided Objects

Look at the cube of Fig. 1.1(a). Different parts can be distinguished, according to:

- geometric properties. A geometric cube can be associated with this cube of the real world, and it is well-known that a full cube is made by (or *subdivided into*) one *volume*, six *faces*, twelve *edges* and eight *vertices*. Each of these parts is a connected set of geometric points: for instance, a face is a part of the *boundary* of the cube, which contains all coplanar points; each edge is the intersection of two faces. Note that all faces are *incident* to four edges and four vertices: all faces have thus the same *structure*.
- color. For instance, the upper face can be itself subdivided into two parts: one corresponds to the letter, the other to the complement of the letter in the upper face of the cube. They are 2-dimensional parts,

(a)

(b)

FIGURE 1.1
(a) A cube. (b) A soccer ball.

sharing curves (i.e. their boundaries), in which edges and vertices can be distinguished, according to geometric linearity. Note that the letter is a connected set of points, but not its complement.

Similarly, the soccer ball depicted in Fig. 1.1(b) can be approximated by a sphere (it is a rough approximation, since the ball is made of volumic pieces of leather, though the sphere is a surface). According to color, pentagonal and hexagonal faces can be distinguished, i.e. faces such that their boundaries contain five or six edges and vertices.

The house depicted in Fig. 1.2(a) can also be assimilated to a 3-dimensional geometric object, in which volumes, faces, edges and vertices can be distinguished, according to the material (wood, stone, glass, etc.), geometric properties, color, or other characteristics. In fact, an *image* of a house is displayed in Fig. 1.2(a); this is a 2-dimensional image, which can subdivided into 2-dimensional regions, according to some property related for instance to color, intensity, etc.

Handling subdivisions is necessary for various application fields, for instance:

- geology: different layers can be distinguished in the subsoil, according to the materials they are made of. Layers can be broken by faults, i.e. they can be made by several blocks. A block can be represented by a volume, associated with some material, and the boundary of a volume can be subdivided into faces, edges and vertices according to the neighbor blocks. A layer corresponds thus to a set of volumes, a fault corresponds to a set of faces, two layers share a set of faces. A computer representation of a subsoil part is depicted in Fig. 1.2(b).
- medicine: everybody now has seen images of the human brain (or can

(a)

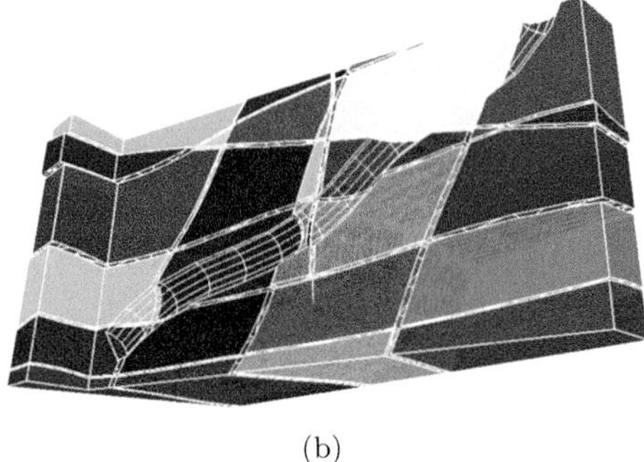

(b)

FIGURE 1.2
(a) A house. (b) Representation of geological layers, in which a drilling is simulated.

easily find such images), and knows that different anatomical parts can be distinguished. In order to examine brain diseases, 3-dimensional brain images are analyzed: an important task consists in identifying the anatomical parts, in order to extract information about them. This task is the image *segmentation*. Then, one can be interested in constructing a representation of the segmented (i.e. subdivided) brain in order, for instance, to simulate a surgical procedure.

- study of spatially distributed data: subdivisions can be constructed in order to study various phenomena. For instance, let $P = \{p_1, \ldots, p_n\}$ be a set of points in the Euclidean plane (these points are called *sites*). The Voronoï face associated to p_i contains any point of the plane whose distance to p_i is less or equal to its distance to any other site (note that a Voronoï face is a convex polygon). A Voronoï edge contains points which are equidistant to their two nearest sites, and a Voronoï vertex is a point which is equidistant to three (or more) nearest sites. This defines a subdivision of the Euclidean plane, the Voronoï diagram (cf. Fig. 1.3(a)), which is a basic tool for many applications, for intense in epidemiology, meteorology, etc.

So, roughly speaking, a subdivided geometric object is composed by *cells* of different dimensions (vertices, edges, faces, volumes, etc.), and *structural relations* exist between these cells, since the *boundary* of a cell is made of cells of lower dimensions. For instance, a cell and a cell of its boundary are said to be *incident*; two cells with the same dimension are *adjacent* if they are both incident to a third cell.

The general framework of this book is the study of *data structures* for the *explicit representation of subdivided geometric objects*, and *operations* for handling these structures. Two fields are mainly addressed: geometric modeling, and image processing. Indeed, different needs emerge from these fields, and it is interesting to illustrate some consequences for the design of data structures and operations. Note that all notions and operations, data structures and algorithms studied in this book are also useful for other fields in which subdivided geometric objects are handled, as computational geometry, discrete geometry, image analysis, etc.

1.1.2 Different Subdivisions

Sometimes, subdivisions satisfy some geometric properties, among which can be distinguished *shape* properties or *structural* properties, which can themselves be more or less *local* or *global.*

For instance:

- local shape properties: the faces of (the representation of) the soccer ball are convex on the sphere, as the faces of any Voronoï diagram in the Euclidean plane. This is not the case for the house, in which concave faces exist, as for the letter of the cube;
- global shape properties: the soccer ball is a sphere, i.e. all points of the surface have the same curvature. This is not the case for the cube, in which sharp edges exist;
- local structural properties: the faces of the soccer ball are pentagons and hexagons, i.e. any face is incident to five or six edges and vertices. There is no such regularity for the faces of the house, nor for the faces of a Voronoï diagram in the Euclidean plane.
- global structural properties: any edge of the soccer ball or of a Voronoï diagram in the Euclidean plane is incident to exactly two faces. More precisely, any *surface* can be constructed by gluing faces along edges, in such a way that at most two faces share an edge. This is not the general case. For instance (cf. Fig. 1.3(b)), when simulating the growing of plants, the structure of the plant is often represented by a tree (in the meaning of graph theory), and the leaves are represented by parts of surfaces: so, the representation of the plant is a 2-*dimensional complex*, i.e. a geometric object which does not satisfy all structural properties of a surface.

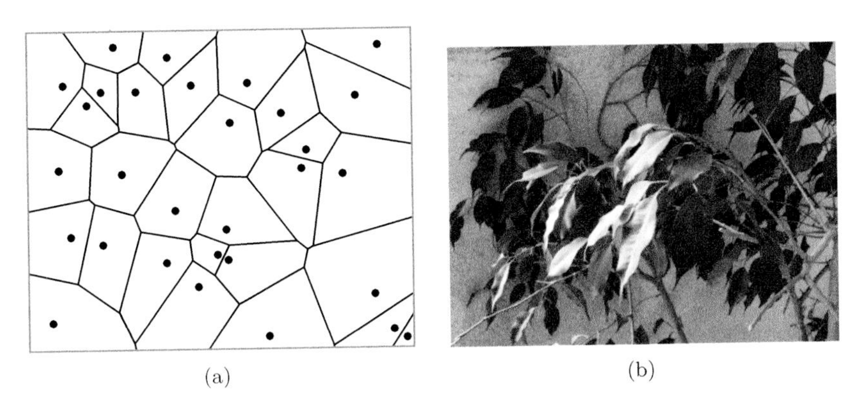

(a) (b)

FIGURE 1.3
(a) A Voronoï diagram. (b) A plant. Usually, the stems are represented by a combinatorial tree, the leaves by parts of surfaces.

1.2 Explicit Representations of Subdivisions

1.2.1 Why Explicit Representations?

Explicit means that the representation of a subdivision is based upon a representation of the cells and their structural relations. More precisely, the structure of a finite subdivision, i.e. a subdivision containing a finite number of cells, can be implemented by a discrete data structure. For instance, the structure of a wire mesh can be represented by a graph: each wire corresponds to an edge, an extremity of a wire corresponds to a vertex.

The shape of the wire mesh is represented by attributes associated with the graph, for instance:

- for a piecewise linear shape: the representation of a geometric point is associated with any vertex; so, each edge is associated with a line segment joining the two points associated with the vertices incident to the edge; in order to avoid degeneracies, two distinct points are associated to two distinct vertices;
- for a piecewise curved shape: a curve (for instance a spline, a Bézier curve, etc.) is associated with any edge, in such a way that the curves associated with adjacent edges share extremity points.

Note that in both cases, the attributes defining the shape have to satisfy some properties in order to fit the structure of the subdivision.

So, the key point for defining data structures for representing subdivisions is to focus on the subdivision structure, i.e. the cells and their structural relations. Often, the data structures are close (in some way) to graphs. So, *structural characteristics* (e.g. connectivity) can be efficiently computed on the data structure, which can be useful in order to *provide information* about the object, or to *control its construction*. The structure of objects can also be used in order to efficiently compare objects or match objects, e.g. for Pattern Recognition applications.

At last, remember that the shape is *locally* defined as attributes of cells: this is often referred to as the *distinction between topology and embedding*[1]. As mentioned above, the shape attributes have to satisfy some properties in order to be coherent according to the structure. Moreover, a relative independence exists between the structure and the shape. For instance, the shape of a wire mesh can be modified although its structure remains invariant, by modifying the points associated with the vertices of the graph representing the wire mesh. Similar modifications can be useful for the animation of articulated mechanical objects or other applications in geometric modeling, computational geometry, discrete geometry, computer graphics, image processing and analysis, etc.

Obviously, shape and structure are not completely independent, and, according to the operations, each one can influence the other. For instance for the simulation of the growing of plants, the creation of a new branch changes the structure of the plant; at the beginning, the branch is so small that it is quite not visible, then its length increases, modifying the shape of the tree; in some way, the structure influences the shape. Conversely, when two branches collide, they can glue to each other: in some way, the shape influences the structure. More formally, these influences will correspond to the parameters and results of operations applied to data structures implementing the subdivisions.

1.2.2 Some Interests of Explicit Representations

Assume a data structure for handling geometric objects has to be conceived for a given application. It can be useful to explicitly represent subdivisions, for instance when the objects are subdivided and when the subdivision itself has a meaning for the application. For instance, explicitly representing all cells of geological layers makes it possible to associate their corresponding material with the volumes. This is similar for all examples discussed above, i.e. it is possible to represent geometric characteristics, color, material, etc. as attributes associated with the cells. In other words, *much information about the object can thus be structured according to the subdivision*, and can be represented as attributes of the cells.

Since a subdivision is represented, *local operations* can be applied to the

[1]Topology denotes the structure, and embedding denotes the shape.

cells. For instance, the rounding operation can be applied to vertices and edges (cf. Fig. 1.4). It consists in:

- first applying *chamfering*: cf. Fig. 1.4(a) and Fig. 1.4(b). This is a structural operation, which consists in "expanding" vertices and edges into faces, the structure of the face depending on the structure of the replaced vertex or edge. For instance, a vertex incident to three edges is basically replaced with a triangular face;
- then computing surfaces patches, which are associated with the chamfered cells: cf. Fig. 1.4(c).

The *structural relations* between cells are also useful or necessary for many *global* operations. For instance, when simulating a traversal of a building, the adjacency relations between the represented rooms make it possible to control the traversal: obviously, it is not possible, when exiting a room, to enter a nonadjacent room.

Subdividing a geometric object corresponds in some way to discretize it, and discretizing a geometric object can be necessary or useful for many processes, for instance:

- simulating some phenomena sometimes involve to solve partial differential equations, for instance simulating fluid flow in geological layers, heat in buildings, etc. Often (cf. finite element method, finite difference method, finite volume method), this requires to discretize the object or the part of geometric space containing the object[2], i.e. to compute a mesh with regular cells (simplices, cubes);
- many processes in computer graphics or in computational geometry handle geometric objects as triangular meshes, since many computations can be efficiently designed for triangles, e.g. computing the intersection of a light ray and a triangle for image synthesis purposes.

1.3 Numerous Structures

Numerous data structures have been conceived in order to represent subdivisions: cf. chapter 9. In fact, according to the *structural properties* of the subdivisions, *optimized data structures* can be conceived. For instance, it is possible to represent the structure of any wire mesh with a graph. If the wire mesh corresponds to a *curve*, i.e. at most two edges are incident to one vertex,

[2] A simple example is the computation of a definite integral using the rectangle method.

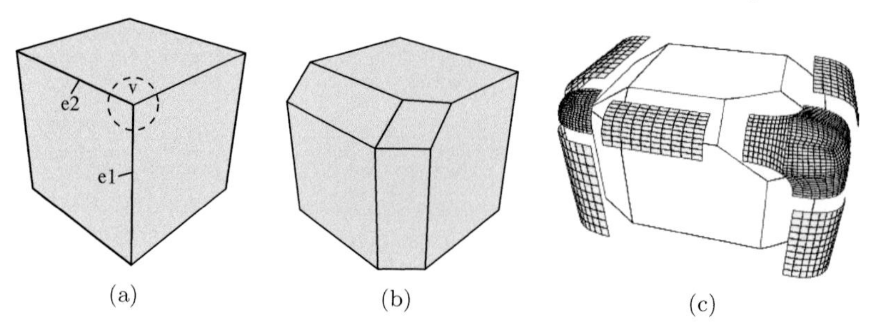

FIGURE 1.4
Chamfering and rounding vertices and edges. (a) A cube. (b) Edges e_1, e_2 and vertex v are chamfered. (c) Rounding, by associating surface patches with chamfered cells (e_1, e_2, v and other cells have been chamfered).

its structure can be represented by a list. This can be generalized for higher dimensional geometric objects.

So, these numerous data structures can be classified according to:

- the local structural properties, i.e. the structural properties of cells. For instance, several *simplicial* structures have been defined in order to represent *triangulations* (for any dimension): cells are vertices, edges, triangles, tetrahedra, etc. Such structures can be generalized for other regular cells, for instance *cubical* structures (cells are vertices, edges, squares, cubes, hypercubes, etc.), *simploidal* structures (cells are products of simplices, as cubes are products of edges). *Cellular* structures have been defined for representing subdivisions with *any* cells: in fact, cells are never "any" cells, they always satisfy some structural properties.

- the global structural properties, related to the way cells are assembled together. For instance, a subdivision can always be constructed by taking cells and gluing them along cells of their boundaries, i.e. by *identifying* cells of their boundaries. Constraints can be added. For instance, *multi-incidence* can be allowed or not, i.e. it is possible, or not, to identify cells which belong to the boundary of the same cell (e.g. a loop is created when two vertices incident to the same edge are identified). Another example is the following. Many structures have been defined for representing *generalizations of surfaces*, i.e. n-dimensional objects which can be constructed by taking n-dimensional cells, and gluing them by identifying $(n-1)-$dimensional cells, in such a way that at most two n-cells are incident to a $(n-1)$-cell. Many structures have also been defined for representing a subclass of these generalizations of surfaces, i.e. for *orientable* such objects *without boundaries*, etc.

1.3.1 Why Numerous Structures?

In fact, there are two questions:

1. *why assuming that objects satisfy some properties?* Mainly for:

    - *coherence of representation.* Representing a volume by its boundary is a well-known method of Geometric Modeling, applied for instance for CAD applications. For instance, take a simple object of our real world, e.g. a wooden cube; it corresponds to a volume, and its boundary is an orientable surface without boundary, made of six faces, twelve edges and eight vertices. But a nonorientable surface does not define the boundary of a volume in the 3-dimensional space. So, when the data structure takes into account that only orientable surfaces without boundaries can be represented, it is impossible to construct an object which cannot be represented in the usual 3-dimensional space;
    - *optimization of operations.* For instance, objects are generally triangulable, and triangulations are well suited for many operations of computer graphics or computational geometry, e.g. as said above, it is easy to detect (and compute) the intersection between a light ray and a triangle;
    - *but it is not always possible to assume that objects still satisfy properties.* Other operations are not so easy to perform on triangulations, for instance boolean operations (union, intersection, difference). Such operations, applied to triangulations, do not directly produce triangulations (cf. Fig. 1.5(a)). A first idea consists in splitting simplices when intersections are processed (cf. Fig. 1.5(b)), but examples exist, showing that the process may be not convergent, since new simplices are added which perhaps produce new intersections; so, it could be convenient here to handle cellular structures instead of simplicial ones;
    - *and don't forget some additional costs.* As said before, it could be efficient to handle triangulations, and thus to triangulate cellular objects; but if some operations are more efficient when applied to a triangular cell, there are more triangular cells than initial ones, so more cells to process. Moreover, information can be lost (or explicit information can become implicit). For instance, assume that the initial structure has a meaning related to the application, e.g. wooden cubes are represented. When triangulated, several triangles correspond to each initial square face: either the information is lost, or it is necessary to compute this information when needed (for instance here by computing the triangles which are coplanar).

2. *why not a general structure?* Everybody knows the answer, which stands for any data structures:

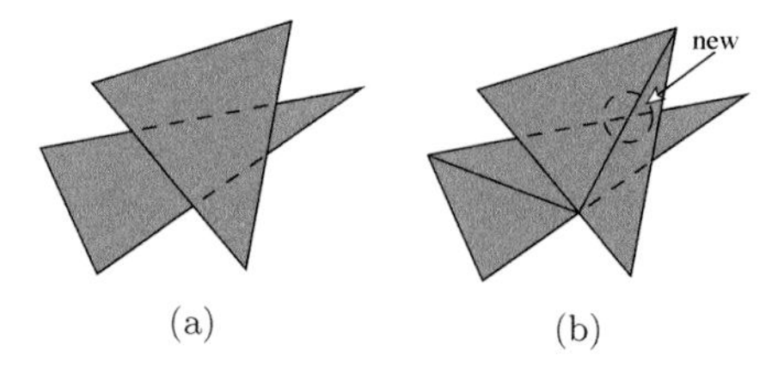

FIGURE 1.5
Boolean operations and simplicial objects. (a) The intersection of two simplicial objects is not a simplicial one. (b) Simplex splitting generates new intersections.

- *optimization of the representation.* The example given above about the representations of wire meshes or wire curves (graphs or lists) can be generalized for any dimension, and thus for all data structures conceived for representing subdivisions. In other words, many structural properties can lead to optimize the representation, by making implicit some information.
- *which can lead to optimize the operations.* Remember that volumes sometimes are represented by their boundaries, which are orientable surfaces without boundary. Intuitively, such a surface has two sides, one directed towards the inside, the other side directed towards the outside. Assume each side corresponds to a given color (e.g. black for inside, white for outside). The surface can be constructed by taking faces having their two sides colored, one black and the other white, and by identifying edges in such a way that adjacent faces are colored coherently, i.e. each side of the surface has one color. Data structures exist, which implement orientable surfaces: so, when identifying edges, there is no choice, faces are glued coherently. If the data structure does not take orientability into account, it is necessary to check how the faces are glued in order to do that in a coherent way; thus the cost of the operation for the general structure is higher than with a specialized structure;
- *or not.* Optimizing the data structure means that no information is lost; but often, some explicit information within the general structure becomes implicit within the optimized structure. When the information is needed for some operations, it has to be computed, leading to an additional cost.

1.3.2 Some Interests of Many Structures

It can be interesting to simultaneously handle several data structures for a given application. For instance for architectural design, a 2-dimensional drawing can be conceived for each floor of the building: a data structure for representing 2-dimensional subdivisions can here be handled. Then, a 3-dimensional representation of each floor is constructed, and it is necessary to handle a data structure for representing 3-dimensional subdivisions. Then, operations are applied in order to join the different floors, producing a 3-dimensional representation of the building. Then, and as said above, the 3-dimensional structure can be discretized (into tetrahedral or cubic meshes) in order to simulate some physical phenomena; the faces of the 3-dimensional structure can also be triangulated in order to efficiently produce images of (parts of) the building.

For geological purposes, surfaces representing faults or frontiers between layers can be reconstructed from 3-dimensional seismic images. Then, 3-dimensional subdivisions can be constructed from these surfaces, representing the different blocks of the layers. For some operations, it can be useful to triangulate the faces of the blocks. For other processes, for instance simulating oil flow, it can be useful to discretize the subdivisions into tetrahedral or cubic meshes.

For some applications, for which huge subdivisions can be represented, it can be useful to simultaneously handle several data structures in order to represent different parts of a subdivision. Indeed, structural irregularities are often localized in a subdivision: so, regular parts are represented using specialized data structures, parts corresponding to irregularities are represented using more general data structures, and correspondences between data structures are maintained in order to put all information together.

1.4 Cellular Structures

This book mainly focusses on cellular structures, i.e. for the representation of subdivisions such that no particular structural property is satisfied by the cells. Mainly two classes of cellular structures can be distinguished:

- *incidence graphs*: the nodes of the graph correspond to the cells, the edges of the graph represent incidence relations between cells. Such graphs represent unambiguously subdivisions in which *no multi-incidence* occurs, i.e. a cell cannot be incident several times to another cell. For instance, it is not possible to unambiguously represent a loop, i.e. an edge which is incident twice to a vertex;

- *ordered models*: such models are based on elements which are more elementary than cells; for instance, it is possible to distinguish the two extremities of an edge, and then to represent unambiguously a loop. Thus, ordered models generalize incidence graphs for the representation of any subdivision, including subdivisions in which multi-incidence occurs. This can be useful for geometric modeling, computational geometry or computer graphics, for instance for handling assemblies of curved geometric objects. In fact, many ordered models have been proposed for geometric modeling, computational geometry, discrete geometry, computer graphics, image processing and analysis purposes.

This book mainly focusses on *combinatorial maps*[3], which make a subset of ordered models. Specialized structures, as *n-Gmaps*, *n-maps*, *chains of maps*, can be distinguished among combinatorial maps, for the representation of subclasses of subdivisions satisfying some structural properties.

The study of combinatorial maps is interesting in order to understand the basics of ordered models:

- only the structural part of a subdivision is described by a combinatorial map; this basic information is the kernel of the whole information which has to be implemented within a data structure. So, it would be possible to answer the question: what is the minimal information which has to be handled in order to represent the structure of a subdivided geometric object? In fact, the basic elements involved in ordered models do not directly correspond to the cells of the subdivisions; so, such models are more difficult to understand than incidence graphs[4]. It is thus important for understanding to focus on the essential information;
- the mathematical foundations of combinatorial maps are sound, and it is possible to formally define the sets of corresponding subdivisions; this is important, since many classical mathematical notions and properties can thus be applied to combinatorial maps. Moreover, since the structure is well-defined, the operations producing combinatorial maps are also well-defined, i.e. they construct the representations of valid geometric objects;
- elementary local operations can be conceived. For instance, two basic operations make it possible to construct any combinatorial map. Such elementary operations make it possible to:
  - carefully control the object during its construction;
  - conceive many high-level operations;

[3]According to the authors, different meanings are related to the terminology "combinatorial maps". Sometimes, it denotes what is called here 2-maps, n-maps, or n-Gmaps. In this book, it is a general term, denoting any structure based on the principles described in the following chapters. So, what is called here n-maps, n-hypermaps, n-Gmaps, n-chains of maps, etc. denote subclasses of combinatorial maps.

[4]The basics of incidence graphs are more related to our usual intuition about subdivisions, although this intuition can be misleading, for instance when multi-incidence occurs.

- efficient data structures have been conceived, based on combinatorial maps, together with operations for handling these structures, for geometric modeling, computational geometry, discrete geometry, computer graphics, image processing and analysis purposes; based on such structures and operations, kernels of geometric modeling softwares can be conceived for handling subdivisions;
- several specialized classes of combinatorial maps can be deduced in order to represent subdivisions satisfying some structural properties, illustrating optimization mechanisms;
- the formal links between combinatorial maps and ordered models, combinatorial maps and incidence graphs, combinatorial maps and simplicial structures, are well-known; it is thus possible to conceive conversion operations between all these structures.

1.4.1 Some Historical Milestones

The concept of combinatorial map has first been defined and studied in mathematics, more precisely in the field of combinatorics. Originally, a map on a surface is a cell decomposition of a surface. In his paper "Combinatorial maps" [212], Vince wrote:

> "The classical approach to maps is by cell decomposition of a surface. A more recent approach, by way of graph embedding, is taken by Edmonds [103], Tutte [207], and others. Our intention is to formulate a purely combinatorial generalization of a map, called a combinatorial map. ..."

So, in the middle of the 1980s, combinatorial maps corresponding to 2-maps, 2-Gmaps and n-Gmaps were known, and also their relations with subdivisions of cellular objects.

Independently, data structures have been defined in geometric modeling, computational geometry, computer graphics, computer vision and image processing for representing cellular objects. The first one is the winged-edge data structure, defined in the middle of the 1970s by Baumgart, for computer vision purposes [17]. This structure has been proved to be equivalent (but not similar in its definition) to 2-maps.

In the middle of the 1980s, several structures were defined, equivalent (and similar in their definition) to 2-maps and 2-maps, 2-Gmaps and 3-Gmaps, since people were interested in handling subdivisions of 2D and 3D cellular objects. At the end of 1980s, structures were proposed, equivalent to n-Gmaps and n-maps, for the representation of generalizations of surfaces for any dimension.

At the end of the 1980s and during the 1990s, many works dealt with the definition of data structures for representing "non-manifold" objects, i.e. complex cellular objects, chains of maps [104] being one representative structure of the results of these works.

As far as we know, no book exists, in the field of geometric modeling, computational geometry, discrete geometry, computer graphics, image processing and analysis, about the representation of subdivided geometric objects. Moreover, many works deal with combinatorial maps, for applications in different fields of mathematics and computer science: many subclasses of combinatorial maps, related notions and operations have been studied. So, a main goal here is to gather important notions related to combinatorial maps in a coherent way, suited for the application of combinatorial maps in geometric modeling, computational geometry, discrete geometry, computer graphics, image processing and analysis, etc.

1.4.2 Outline

Mainly, two subclasses of combinatorial maps, namely n-Gmaps and n-maps, are here studied. The book is organized as follows:

- Chapter 2: first, some basic topological notions and vocabulary are introduced, through a presentation of the classification of surfaces; second, some technical notions are recalled, related to discrete mathematical structures, as permutations and graphs, which are the bases of data structures for representing subdivisions, for instance incidence graphs;

- Chapter 3: starting from subdivided geometric objects, it is shown in an intuitive way how to deduce formal models, i.e. n-Gmaps and n-maps, for representing the structures of the objects;

- Chapters 4 and 5: these chapters are organized in the same way, since they are respectively devoted to n-Gmaps and n-maps. All notions related to these structures are defined; then, elementary operations are defined, for constructing any n-Gmap and n-map. Data structures and iterators are deduced from the mathematical definitions. At last, some useful notions are added;

- Chapter 6: basic operations for handling n-Gmaps and n-maps are defined, such as the closure of a combinatorial map with boundaries, the removal and contraction of cells, and their inverse operations (insertion and expansion); other classical operations are also defined, as chamfering, extrusion and triangulation;

- Chapter 7: combinatorial maps have been implemented and used for Geometric Modeling and Image Processing purposes; the basics of geometric operations and softwares are illustrated for these two application fields;

- Chapter 8: the definitions of simplicial structures are recalled; then cellular objects are defined as simplicial objects structured into cells, and different subclasses are introduced, mainly cellular quasi-manifolds.

The equivalence between n-Gmaps and n-dimensional quasi-manifolds is then stated. The correspondence between n-Gmaps without multi-incidence and n-surfaces, i.e. a subclass of incidence graphs, is also stated; it is shown that other classes of combinatorial maps exist, for instance chains of maps for representing n-dimensional cellular complexes;

- Chapter 9: many ordered models have been defined in order to represent subdivisions. The correspondences between n-Gmaps, n-maps and several ordered models conceived for representing quasi-manifolds, are studied in this chapter;

- Chapter 10: this last chapter contains several (partially) concluding remarks.

2

Preliminary Notions

Two sets of basic notions are studied in this chapter, which is mainly based upon [1, 125]:

- the first ones are related to the objects we are interested in: *subdivisions* of geometric objects. They are illustrated by following Griffith's approach about surface classification [125], and then extended for higher dimensions;
- the second ones are related to the *representations* of these subdivisions. Here we are interested in defining data structures which can be handled in geometric softwares: such representations are algebraic ones, based upon well-known discrete structures equivalent to graphs.

Some notions are not formally defined, since the goal of this chapter is to give an intuition about *(subdivisions of) quasi-manifolds*, and about *identification*, which is the basic construction operation. Quasi-manifolds are formally defined in chapter 8. See [1, 196, 180, 133] for introductions to topology.

2.1 Basic Topological Notions

Some classical vocabulary is introduced here.

2.1.1 Basic Elements

Balls: cf. Figs. 2.1(a), (b) and (c).

Let $\mathbb{R}^n = \{x = (x_1, \ldots, x_n) | x_i \in \mathbb{R}, 1 \leq i \leq n\}$ be the usual $n - dimensional$ space. Let $r \in \mathbb{R}$.

- $B^n_r(x) = \{y \in \mathbb{R}^n | d(x, y) < r\}$, where $d(x, y)$ denotes the usual Euclidean distance between x and y, is the *n-dimensional open ball of radius r around x in* $\mathbb{R}^n$;
- $\bar{B}^n_r(x) = \{y \in \mathbb{R}^n | d(x, y) \leq r\}$ is the n-dimensional *closed* ball of radius r around x in $\mathbb{R}^n$;

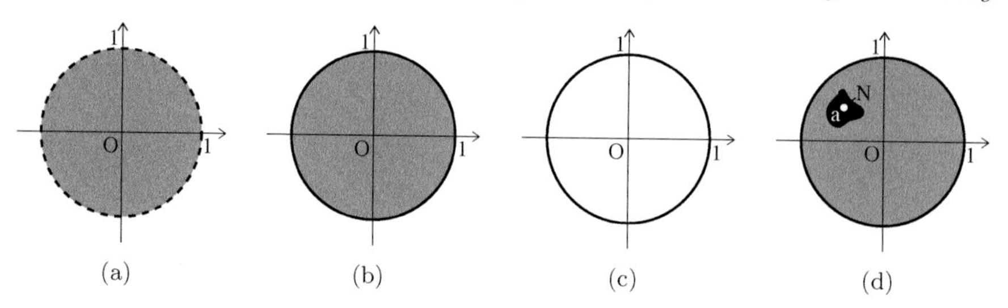

FIGURE 2.1
(a) $B_1^2(O)$, where $O = (0, 0)$.
(b) $\bar{B}_1^2(O)$.
(c) $S_1^1(O)$.
(d) N is a neighborhood of point a in $\bar{B}_1^2(O)$.

- $S_r^{n-1}(x) = \{y \in \mathbb{R}^n | d(x, y) = r\}$ is the $(n-1)$-dimensional *sphere* of radius r around x in $\mathbb{R}^n$.

For instance, $B_1^1(0) = I_1 =]-1, 1[$, $\bar{B}_1^1(0) = \bar{I}_1 = [-1, 1]$ and $S_1^0 = \{-1, 1\}$. Note that it is possible to define n-dimensional balls and $(n-1)$-dimensional spheres in $\mathbb{R}^m$ when $m \geq n$.

Open and closed sets
Let $A \subset \mathbb{R}^n$. A is an *open set* of $\mathbb{R}^n$ if, for every $x \in A$, there is some $\epsilon > 0$ such that $B_\epsilon^n(x) \subset A$; note that $\emptyset$ and $\mathbb{R}^n$ are open sets. Let $B \subset \mathbb{R}^n$. B is a *closed set* of $\mathbb{R}^n$ if $\mathbb{R}^n \setminus B$ is an open set of $\mathbb{R}^n$; note that $\emptyset$ and $\mathbb{R}^n$ are also closed sets.

Arbitrary unions of open sets are open sets; finite intersections of open sets are open sets. Conversely, finite unions of closed sets are closed sets; arbitrary intersections of closed sets are closed sets. For instance, let $B(x)$ be the intersection of all open balls of any radii around x. It is easy to see that x belongs to $B(x)$, but any other point of $\mathbb{R}^n$ does not belong to $B(x)$ (if $y \neq x$, y does not belong to $B_{d(x,y)/2}^n(x)$, so y does not belong to $B(x)$): so $B(x) = \{x\}$ is a closed set, since $\mathbb{R}^n \setminus \{x\}$ is an open set.

Neighborhood: cf. Fig. 2.1(d).
Let $A \subset \mathbb{R}^n$. A subset N of A is a neighborhood of a point $a \in A$ if there is an open set V in A such that $a \in V \subset N$.

Interior, closure, boundary: cf. Fig. 2.2.
Let $A \subset \mathbb{R}^n$. $\breve{A}$, the *interior* of A, is the union of all open sets contained in A: in other words, the interior of A is the biggest open set contained in A. $\bar{A}$, the *closure* of A, is the intersection of all closed sets which contain A: in

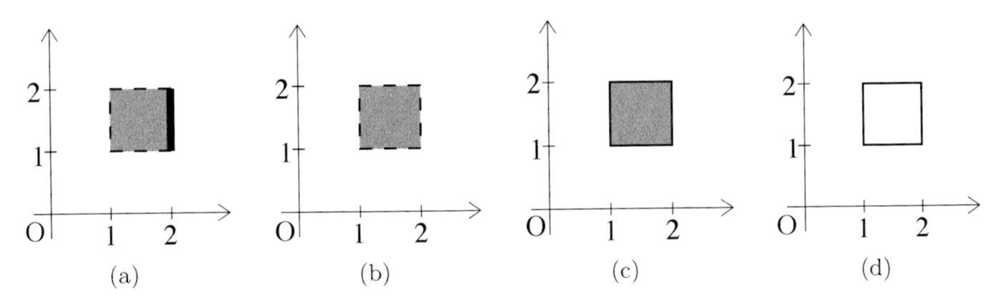

FIGURE 2.2
(a) $A = \{(x, y) | 1 < x \leq 2 \text{ and } 1 < y < 2\}$.
(b) $\breve{A} = \{(x, y) | 1 < x < 2 \text{ and } 1 < y < 2\}$.
(c) $\bar{A} = \{(x, y) | 1 \leq x \leq 2 \text{ and } 1 \leq y \leq 2\}$.
(d) $\partial A = \{(x, y) | ((x = 1 \text{ or } x = 2) \text{ and } 1 \leq y \leq 2) \text{ or } (1 \leq x \leq 2 \text{ and } (y = 1 \text{ or } y = 2))\}$.

other words, the closure of A is the smallest closed set containing A. ∂A, the *boundary* of A, is $\bar{A} \setminus \breve{A}$. For instance, for any x and r, $\bar{B}_r^n(x)$ is the closure of $B_r^n(x)$, $B_r^n(x)$ is the interior of $\bar{B}_r^n(x)$, $S_r^{n-1}(x)$ is the boundary of $B_r^n(x)$; it is also the boundary of $\bar{B}_r^n(x)$. Note also that $\breve{\breve{A}} = \breve{A}$ and $\bar{\bar{A}} = \bar{A}$ for any A.

2.1.2 Continuous Map, Homeomorphism

Let $A \subset \mathbb{R}^n$, and let $f : A \to \mathbb{R}^m$. f is a *continuous* map if for any $x \in A$ and for any $\epsilon > 0$ it exists $\delta > 0$ such that $f(B_\delta^n(x)) \subset B_\epsilon^m(f(x))$. Equivalently, f is continuous if and only if $f^{-1}(V)$ is an open subset of A for every open set V in R^m, where $f^{-1}(V)$ is the set of all points x of A such that $f(x) \in V$. For instance, a continuous map exists between $S_1^1(O)$ and C (depicted in Fig. 2.3), but no continuous map exists between $S_1^1(O)$ and C' (depicted also in Fig. 2.3).

$f : A \to B$ is a *homeomorphism* if it is a continuous one-to-one mapping which has a continuous inverse; in this case, A and B are said *homeomorphic* (or *topologically equivalent*).

For instance (cf. Fig. 2.3), all curves without boundary are homeomorphic, and all are homeomorphic to the unit 1-dimensional sphere $S_1^1(O)$, where $O = (0, 0) \in \mathbb{R}^2$; since any point $x \in S_1^1(O)$ has a neighborhood which is homeomorphic to $I_1 =]-1, 1[\subset \mathbb{R}$, any point of any curve without boundary satisfies the same property. Similarly, all curves with boundary are homeomorphic, and all are homeomorphic to $\bar{I}_1 = [-1, 1] \subset \mathbb{R}$.

Intuitively, A and B are homeomorphic if you can deform continuously A in order to get B. For instance, take a wire (cf. Fig. 2.4(a)) and deform it (cf. Fig. 2.4(b)), it is still the initial wire: nothing but the shape is modified, so

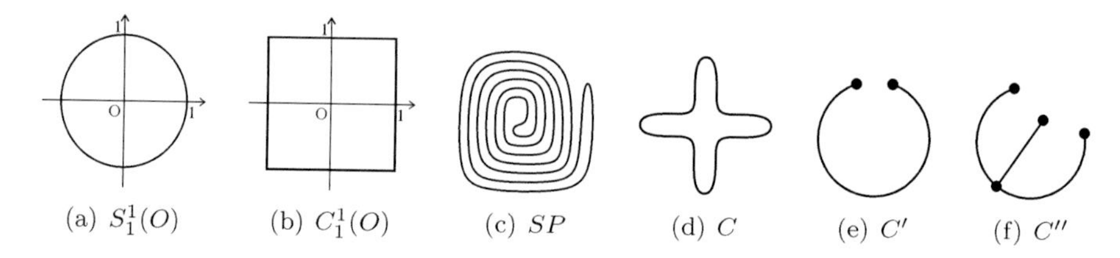

FIGURE 2.3
$S_1^1(O)$, $C_1^1(O)$, SP and C are curves without boundary, and all are homeomorphic; but they are not homeomorphic to C' (which is a curve with boundary), nor to C'' (which is not a curve).

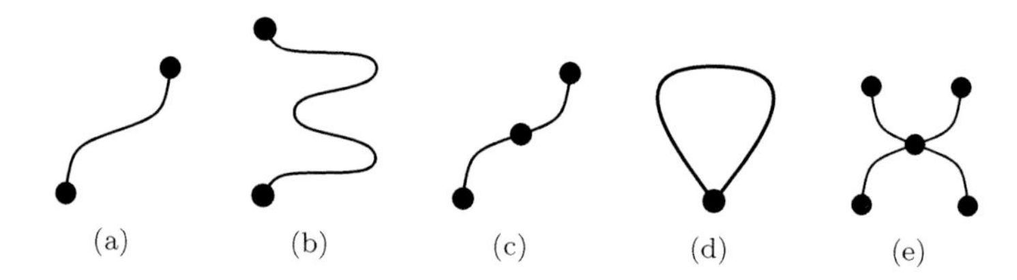

FIGURE 2.4
(a) A wire.
(b) A deformed wire is still a wire.
(c) Two welded wires are equivalent to a wire.
(d) Weld the two extremities of a wire and you get something else.
(e) Weld several wires at the extremity of each wire and you get also something else.

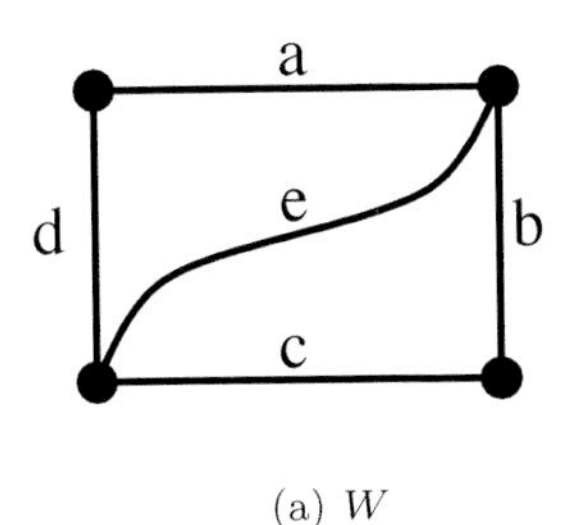

(a) W

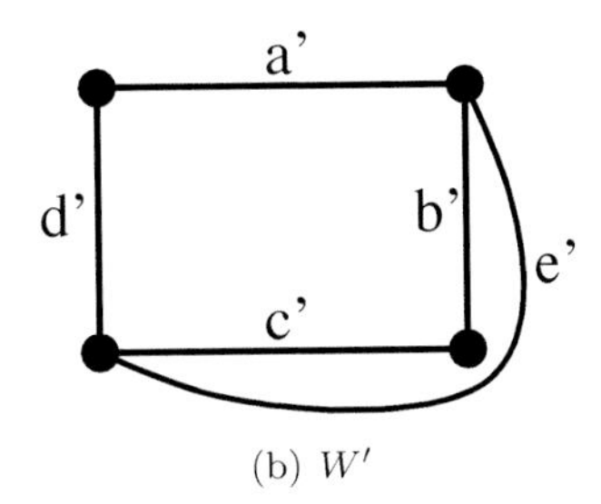

(b) W'

FIGURE 2.5
A homeomorphism exists between W and W', such that a (resp. b, c, d, e) is associated with a' (resp. b', c', d', e'). But no homeomorphism exists which associates the plane containing W with the plane containing W' and which associates W and W' as previously.

the object you get is homeomorphic to the initial wire. More generally, take any two wires, they are homeomorphic. If you weld two wires at an extremity of each wire (cf. Fig. 2.4(c)), you get something which is equivalent to a wire (you can get the resulting object by simply extending a wire), but if you weld the two extremities of a wire (cf. Fig. 2.4(d)), you get something which is not equivalent to the initial wire; similarly, if you weld more than two wires at an extremity of each wire (cf. Fig. 2.4(e)), you get also something which is not equivalent to a wire. So, the notion of homeomorphism is intuitively related to that of continuous deformation: in other words, a continuous deformation does not change the topology, although the notion of topological change is intuitively related to operations as cutting and pasting in their *general* cases, since we have seen in Fig. 2.4(c) that some particular cases of pasting (and conversely of cutting) do not change the topology.

It is important to be careful with these intuitive interpretation of topological equivalence. Take five wires a, b, c, d, e and weld them in order to get object W depicted in Fig. 2.5(a). W is homeomorphic to W' depicted in Fig. 2.5(b), by associating a with a', b with b', c with c', d with d' and e with e'. If you want a proof, take W and deform it in order to get W'! To do that, it is necessary to deform the object in the 3-dimensional space: in the plane, it is impossible to deform W in order to get W' (with the association of wires as described above), since e cannot go through the curve made by a, b, c and d. In other words, a homeomorphism exists between W and W' which associates each wire x with the corresponding x', but no homeomorphism exists between the plane in which W lies and the plane in which W' lies, which associates each wire x with the corresponding wire x'.

Similarly, take a strip of paper (cf. Fig. 2.6(a)) and stick two opposite "edges", you get an annulus: cf. Fig. 2.6(b). If you do the same thing, but you twist twice the strip of paper before sticking, you get an object which

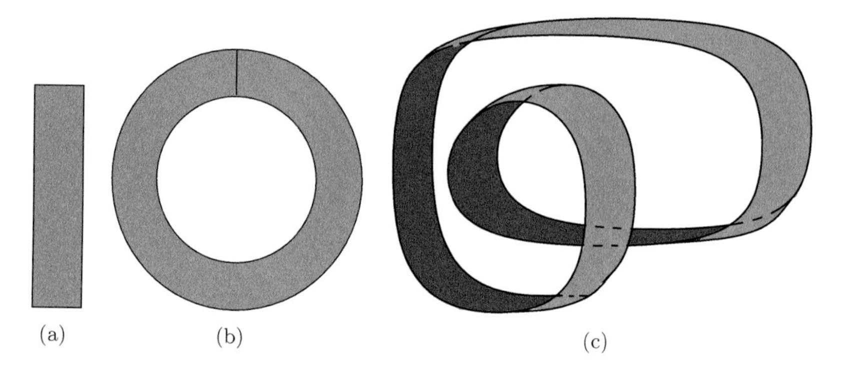

FIGURE 2.6
(a) A strip of paper.
(b) Stick two opposite edges and get an annulus.
(c) Stick two opposite edges after twisting twice the strip, and get another annulus, homeomorphic to the annulus depicted in (b).

is also an annulus (even if it is not obvious): cf. Fig. 2.6(c). Moreover, you cannot deform in the 3-dimensional space the first annulus in order to get the second one (but you could do that in a higher dimensional space). As before, the two annuli are homeomorphic, but no homeomorphism exists between the 3-dimensional space and itself, which associates the two annuli.

2.2 Paper Surfaces

We have seen in the previous section that when the notion of topological equivalence is taken into account, two types of curves can be distinguished: curves without boundary, which are all homeomorphic, and curves with boundary, which are also all homeomorphic. It is also possible to define a categorization of surfaces, but it is more complicated than for curves. This is the goal of this section, in which Griffiths's approach is followed (cf. [125]).

2.2.1 Basic Elements

Assume a sheet of paper is so thin that it has no volume at all: it is called *a face* (i.e. a 2-*dimensional cell*). Moreover, the boundary of the paper sheet is *subdivided* into *edges* and *vertices* (respectively 1-*dimensional* and 0-*dimensional cells*), i.e. for instance, a standard sheet of paper is bounded by four edges (corresponding to the straight lines of its boundary), meeting two by two at the four corners of the paper sheet, which correspond to the four vertices. It

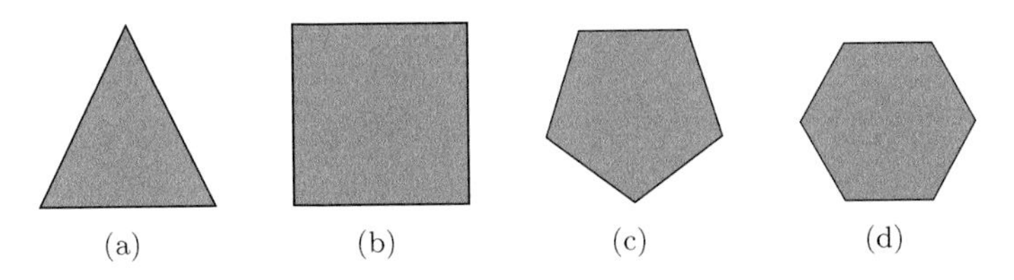

FIGURE 2.7
(a) A triangle. (b) A square. (c) A pentagon. (d) A hexagon.

is of course possible to handle triangular paper sheets (their boundaries are made by three edges and three vertices), or any *polygonal* sheet of paper (cf. Fig. 2.7). Note that the boundary of any face is a cycle of edges and vertices.

More precisely, the face is the interior of the sheet of paper, and each edge is the interior of a straight line of the boundary of the sheet of paper (in other words, a cell does not contain its boundary). So, the set of cells (face, edges, vertices) makes a *partition*[1] of the sheet of paper.

2.2.2 Basic Construction Operation: Identification of Edges

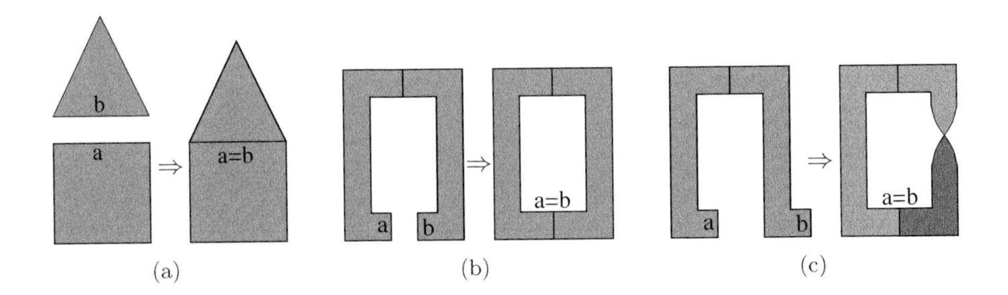

FIGURE 2.8
(a) Gluing a square with a triangle.
(b) Making an annulus.
(c) Making a Möbius strip.

Surfaces can be constructed by pasting such paper sheets along their boundaries, but it is necessary to take the structure of the boundaries into account: more precisely, the basic construction operation consists in gluing

[1]Let S be a set; remember that a partition of S is a set of subsets $\{S_i\}_{i=1,n}$ such that the union of all S_i is equal to S and the intersection of any two distinct subsets is empty. For a sheet of paper, that means that the whole sheet is equal to the union of all its cells, and any two distinct cells have no common points.

two faces by sticking two edges of their boundaries. Once glued, the two edges are *identified* into one edge (cf. Fig. 2.8).

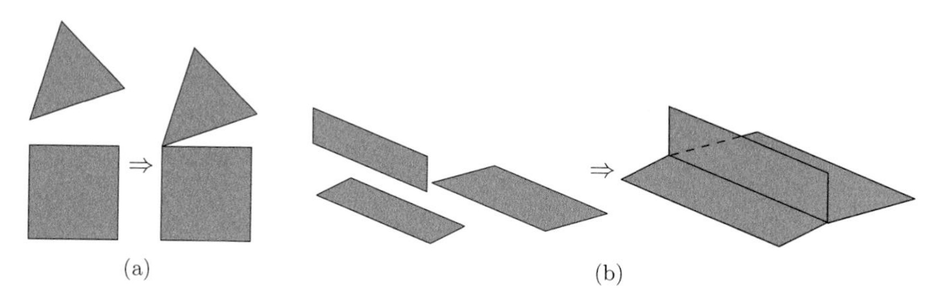

FIGURE 2.9
Forbidden identifications ((a) of vertices, (b) of edges), since the resulting objects are not surfaces.

It is forbidden to glue two faces by simply identifying two vertices; it is also forbidden to glue more than two edges: in both cases, the result is not a surface (cf. Fig. 2.9).

So a paper surface can be made of a single face, or by many faces glued by identifying several pairs of edges. It is thus *subdivided* (i.e. partitioned) into vertices, edges and faces.

Incidence and adjacency relations

The "boundary" relation corresponds to the *incidence* relation: more precisely, an edge (resp. a vertex) which belongs to the boundary of a face is incident to the face, and conversely, the face is incident to the edge (resp. to the vertex). Moreover, if a vertex is part of the boundary of an edge (i.e. the vertex meets the edge), the vertex and the edge are incident to each other. Usually, adjacency is defined in the following way: two i-cells are *adjacent* if they share a common cell in their boundaries. For instance, two faces (resp. two edges) are adjacent if they share an edge or a vertex (resp. a vertex). Adjacency is sometimes defined in a generalized way: two cells are adjacent if they are incident to the same cell: for instance, two vertices are adjacent if they are incident to the same edge or to the same face.

Surface boundary

A *free* edge is incident to a single face; it is *sewn*[2] when it is incident to two faces. The *boundaries* of a subdivided surface are made by the free edges: moreover, they are *cycles* of free edges. When a surface has no free edge, it is *without boundary*, for instance the surface of a cube, made by gluing six square

[2]The use of the word "sew" comes from the fact that, for constructing paper surfaces, instead of gluing faces by sticking two edges into one edge, it is also possible to sew the faces together around the resulting edge.

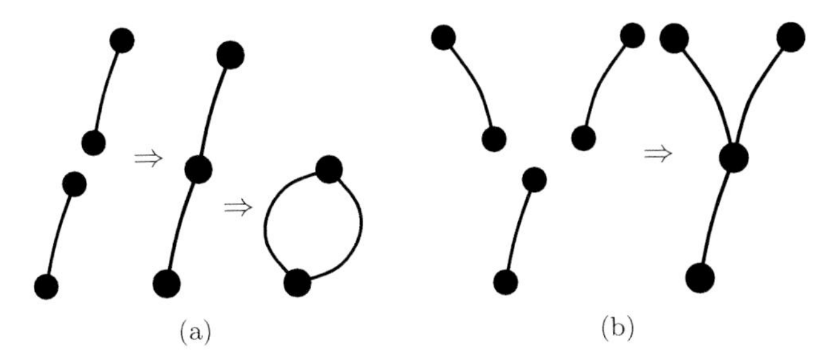

FIGURE 2.10
(a) Correct identifications of vertices (for 1-dimensional objects), producing curves.
(b) Forbidden identifications of vertices, which do not produce a curve.

faces: the initial twenty-four edges are identified into the resulting twelve sewn edges of the cube.

Unnecessary constraints
In order to simplify the study of surface construction, assume the boundary of any face contains at least three edges and three vertices. Similarly, assume that the boundary of an edge contains at least, and thus exactly, two vertices (in other words, no edge is a loop). Moreover, when two edges are identified, they belong to the boundaries of two distinct faces: in other words, it is forbidden to identify two edges incident to a same face.

In fact, these constraints are added in order to reduce the number of possible configurations obtained when gluing together sheets of paper, but it is still possible to construct *any surface*, even when these constraints are satisfied. For instance, if it is not possible to construct an annulus by identifying a pair of edges incident to one face, it is possible to construct an annulus by identifying two pairs of edges of two faces! So, with these "unnecessary constraints", it is not possible to construct any subdivision of any surface, but it is possible to construct subdivisions of any surfaces.

These constraints are very different from the two previous ones, added for constructing only surfaces. This can be observed by considering the following characteristic property of surfaces, informally stated as: any point of any surface, which is not a boundary point, has a *neighborhood* (i.e. a "part around") which is *homeomorphic to an open 2-dimensional ball*[3]. In particular, this is true for any interior point of any face. More generally:

- when two faces are glued by identifying two free edges, any point of the resulting edge has a neighborhood which is homeomorphic to an open 2-dimensional ball (since this point results from the identification of two

[3]A boundary point has a neighborhood homeomorphic to a 2-dimensional *half* ball.

distinct boundary points, its neighborhood results from the union of two 2-dimensional half balls which were the neighborhoods of the two initial points);

- on the contrary, when two faces are glued by identifying two vertices, the neighborhood of the resulting vertex is homeomorphic to two 2-dimensional half balls which share a point, i.e. it is not homeomorphic to a 2-dimensional ball;

- similarly, when three (or more) faces are glued by identifying more than two edges, the neighborhood of any point of the resulting edge is not homeomorphic to a 2-dimensional ball. As seen before, this is similar in dimension 1: assume subdivided curves are constructed by welding pieces of wires. A piece of wire makes a curve; when two pieces of wires are welded, they still make a curve; when three pieces of wires are welded at a same point, they do not make a curve... (cf. Fig. 2.10).

Orientability

The annulus and the Möbius strip (cf. Fig. 2.8) are made similarly (two free edges, which are not adjacent, are identified), but they are clearly different: the annulus has two boundaries, the Möbius strip has only one. Moreover, assume you handle real sheets of paper, and assume the two sides of each sheet of paper have different colours, for instance one side black and the other white: it is possible to get the same colour for each side of the annulus (in fact, the annulus has still two sides), but it is impossible to get the same colour along one side of the Möbius strip, since the Möbius strip has only one side! This well-known difference is related to *orientability*.

An edge is always orientable (cf. Fig. 2.11(a)): that means it is possible to choose a direction for going along this edge, i.e. an origin vertex and an extremity vertex. Note that any edge can be oriented in *two* ways, which are *opposite* (if you have chosen an origin vertex and an extremity vertex, you get the opposite orientation by exchanging the two vertices).

A face is always orientable: that means intuitively that it has two sides, i.e. you can go around all points of the faces by walking and turning always left (or oppositely always right); this corresponds also to the fact that the boundary of the face is a cycle of edges and vertices, which can be oriented into one direction or into the opposite direction (cf. Fig. 2.11(b)). So, orienting a face consists in choosing an orientation of its boundary, and thus each edge of this boundary is itself oriented; but the orientations of the edges are *coherent* along the boundary of the face.

The idea is quite similar for defining orientable surfaces: all faces are oriented, but in a coherent way, meaning that (cf. Figs. 2.11(c) and (d)):

- any sewn edge is oriented in opposite ways for its two incident faces; this is not intuitive, but this corresponds to the fact that if you have chosen to turn left for instance for one face, when the relative orientations of

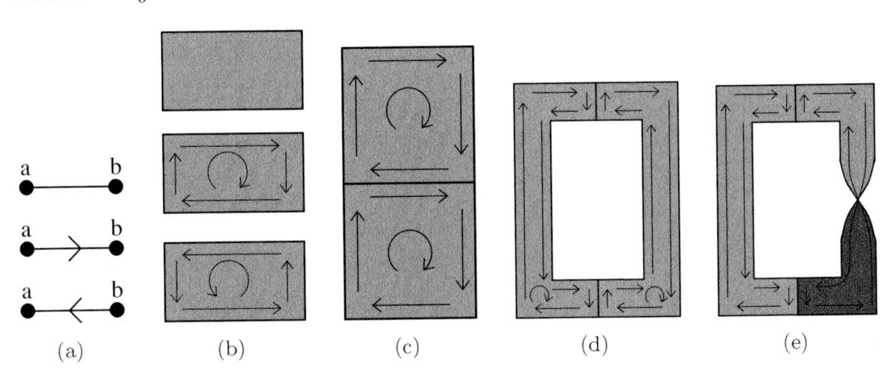

FIGURE 2.11
(a) An edge and its two opposite orientations.
(b) A face and its two opposite orientations.
(c) Coherent orientation of two faces sharing an edge: the boundary of the surface is coherently oriented, and the common edge has opposite orientations with respect to its incident faces.
(d) Coherent orientation of an annulus (meaning the annulus is orientable).
(e) It is impossible to define a coherent orientation of a Möbius strip (meaning the Möbius strip is not orientable).

the common edge are opposite, you have also chosen to turn left for the other face;

- the free edges make oriented boundaries, i.e. oriented cycles of edges.

When a surface is orientable (i.e. it can be coherently oriented), two opposite orientations can be defined (cf. the annulus, for instance). When a surface is not orientable, it is impossible to define a coherent orientation (cf. the Möbius strip, for instance, in Fig. 2.11(e)).

Note also that, when a boundary is coherently oriented, you can close it by gluing along it a face having the same number of edges, in an oriented coherent way. As a consequence, if a surface with boundaries is orientable, you can close it into an orientable surface without boundaries: for instance, you can close a face by pasting another face along its boundary; you can close an annulus by pasting two faces along the two boundaries; in both cases you get a *topological sphere* (if your paper is extensible, you can modify the resulting shape in order to get a geometric sphere). But if a surface with boundaries is nonorientable (i.e. it cannot be oriented coherently in the way described above), then you can try to close the boundaries in any way, it is impossible to get an orientable surface without boundaries.

An important property of orientable surfaces without boundaries (as a sphere, a torus, etc.) is the fact that it can be *embedded* into the usual 3–dimensional space (i.e. you can give it a shape in this space) in such a way that it divides it into three parts: an "internal" (bounded) volume, the surface itself, and an "external" (unbounded) volume. This is similar in dimension 1, but here, all curves are orientable: when embedded into the usual plane, any curve without boundary divides it into three part: an "internal" (bounded) face, the curve itself and an "external" (unbounded) face.

2.3 Classification of Paper Surfaces

2.3.1 Topological Surfaces

Several numbers can be associated with any surface paper S:

- $b(S)$ is the *number of boundaries* of S;
- $c(S)$ is the *Euler characteristic* of S, defined by:

$$c(S) = v(S) - e(S) + f(S)$$

  where $v(S)$, $e(S)$ and $f(S)$ are respectively the numbers of vertices, edges and faces of S.

- $q(S)$ is the *orientability factor* of S, defined by:

$$q(S) = \begin{cases} 0 & \text{if } S \text{ is orientable;} \\ 1 & \text{if } S \text{ is nonorientable and } (b(S) + q(S)) \text{ is odd;} \\ 2 & \text{otherwise.} \end{cases}$$

- $g(S)$ is the *genus* of S, defined by: $g(S) = 1 - (b(S) + c(S) + q(S))/2$.

For instance, all polygons depicted in Fig. 2.7 have one boundary; their Euler characteristic is equal to 1, they are orientable, so their orientability factor is equal to 0, and their genus is null. All polygons have these characteristics (all have one boundary, they are orientable, their Euler characteristic is equal to 1, since they have one face, and vertices and edges in equal numbers), but other surfaces too, for instance the gluing of a square and a triangle, or the gluing of two squares: cf. Fig. 2.8(a) and Fig. 2.11(c). The annulus has two boundaries, its Euler characteristic is equal to 0, it is orientable, so its genus is equal to 0. The Möbius strip has one boundary, its Euler characteristic is equal to 0, it is not orientable, so its orientability factor is equal to 1 and its genus is equal to 0.

All these numbers are integers, and for any paper surface S, $b(S)$, $q(S)$ and $g(S)$ are positive.

A *topological surface* $TS_{b,q,g}$, with $0 \leq q \leq 2$, is the set of all paper surfaces S satisfying $b(S) = b$, $q(S) = q$, $g(S) = g$. Any paper surface belongs to a unique topological surface $TS_{b,q,g}$; and any topological surface contains

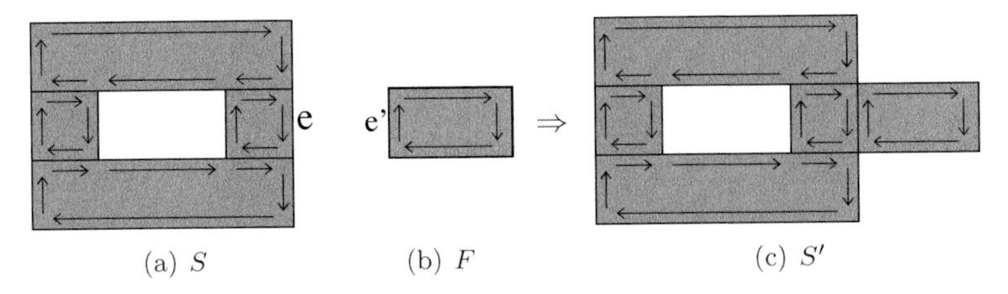

(a) S (b) F (c) S'

FIGURE 2.12
Adding a face.

an infinity of paper surfaces. The important fact here is the *classification theorem*:

all paper surfaces belonging to a same topological surface are homeomorphic, and all homeomorphic paper surfaces belong to the same topological surface.

In order to get a partial intuition of this important property, the construction of paper surfaces will be studied, by distinguishing several cases of the identification operation. Note that when a surface is not orientable, any further identification of two edges cannot produce an orientable surface (this can also be proved by studying all the possibles cases of identification and their impact on the orientation).

Let S be a paper surface of $TS_{b,q,g}$, such that $b \geq 1$, let e be an edge of a boundary of S, and let c be its Euler characteristic. Let S' be the result of an identification involving e and another free edge e', and let b', c', q' and g' be respectively its number of boundaries, Euler characteristic, orientability factor and genus.

Adding a face: Fig. 2.12.

Let F be a paper sheet which does not belong to S, and let e' be an edge of F.

When e is identified with e', the boundary to which e belongs is modified: in S', this boundary is made by the same edges, except e which is replaced by the edges of F, e' excepted. So $b' = b$.

Let x be the number of vertices (or edges) of F. The differences between the numbers of vertices, edges and faces of S' and S are respectively $x - 2$, $x - 1$ and 1 (since four vertices are identified into two vertices, two edges are identified into one edge, and one face is added to S). So $c' = c$.

The orientability is not changed. If S is orientable, it can be oriented, and it is possible to define an orientation of F in such a way that S' is coherently oriented (take for e' the inverse orientation of e), and thus S' is orientable. If S is not orientable, S' is not orientable too. So $q' = q$, since $b' = b$ and $c' = c$.

At last, $g' = g$ since $b' = b$, $c' = c$ and $q' = q$.

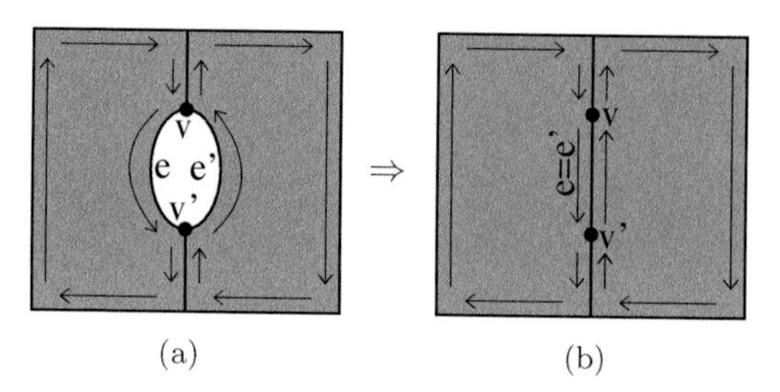

FIGURE 2.13
Removing a boundary.

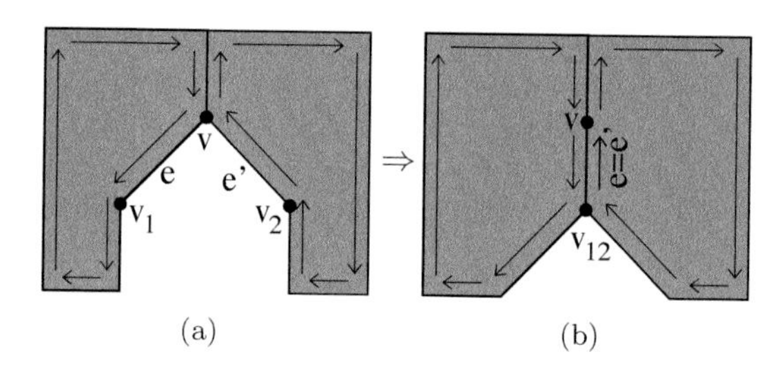

FIGURE 2.14
Shortening a boundary.

Removing a boundary: Fig. 2.13.

Let e and e' be such that they belong to a same boundary, and they share their two extremity vertices v and v': so their incident boundary contains only e and e'. There is only one possible way to identify e and e', since loops are not allowed, and the boundary vanishes. So $b' = b - 1$.

This identification removes one edge (e and e' are identified), so $c' = c+1$.

The orientability is not modified. If S is orientable, it can be coherently oriented: assume e is oriented from v to v'; thus e' is oriented from v' to v, since they are consecutive in their incident boundary, and this boundary is coherently oriented; so the edge resulting from the identification of e and e' is coherently oriented with respect to its incident faces. So $q' = q$ since $b' = b-1$ and $c' = c + 1$.

At last, $g' = g$ since $b' = b - 1$, $c' = c + 1$ and $q' = q$.

Shortening a boundary: Fig. 2.14.

Let e and e' be such that they belong to a same boundary, and they share exactly one vertex v. There is only one possible way to identify e and e', since loops are not allowed: in this case, the two vertices v_1 and v_2 of e and e' which are initially different are identified into one vertex v_{12}, and the resulting edge is incident to two distinct vertices v and v_{12}.

Since e and e' are both incident to v in S, and since $v_1 \neq v_2$, the boundary incident to e and e' contains other edges. The identification of e and e' removes these edges from the boundary, but it still exists, and no new boundary is created. So $b' = b$.

This identification removes one edge (e and e' are identified) and one vertex (v_1 and v_2 are identified), so $c' = c$.

The orientability is not modified. If S is orientable, it can be coherently oriented: assume e is oriented from v to v_1; thus e' is oriented from v_2 to v, since they are consecutive in their incident boundary, and this boundary is coherently oriented; so the edge resulting from the identification of e and e' is coherently oriented with respect to its incident faces, and the boundary remains coherently oriented (the edge before e' is oriented to v_2 and the edge after e is oriented from v_1, and they become consecutive in the resulting boundary, both incident to v_{12}). So $q' = q$ since $b' = b$ and $c' = c$.

At last, $g' = g$ since $b' = b$, $c' = c$ and $q' = q$.

Note that a paper surface which contains only one face belongs to $TS_{1,0,0}$. Any paper surface you can construct by "adding a face" or "shortening a boundary" belongs also to $TS_{1,0,0}$. All these paper surfaces are *discs*, and they are homeomorphic to a 2-dimensional closed ball. If you apply "removing a boundary" to a disc, you get a *sphere*, i.e. an element of $TS_{0,0,0}$, homeomorphic to a 2-dimensional sphere.

Making a boundary, or making a twist: Figs. 2.8(b) and (c)

Let e and e' be such that they belong to a same boundary, and they do not share any vertex. So, they are not consecutive in their incident boundary. Assume the extremity vertices of e (resp. e') are v_1 and v_2 (resp. v'_1 and v'_2). There are two ways for identifying e and e': the first one identifies v_1 and v'_1, v_2 and v'_2, the second one identifies v_1 and v'_2, v_2 and v'_1. In one case ("making a boundary"), a new boundary is created (for instance, an annulus is constructed from a disc); in the other case ("making a twist"), no new boundary is created (for instance, a Möbius strip is constructed from a disc).

1. If the case is "making a boundary":

   - by definition, $b' = b + 1$;
   - four vertices are identified into two vertices, two edges are identified into one edge, so $c' = c - 1$;
   - it is easy to see that the orientability is not changed by this identification (if the surface is orientable, it can be oriented, so e and e' are oriented, and there is only one way to identify e and e' in such a way that the resulting edge is coherently oriented: this way corresponds to "make a boundary" !). So $q' = q$ since $b' = b + 1$ and $c' = c - 1$;
   - $g' = g$ since $b' = b + 1$, $c' = c - 1$ and $q' = q$.

2. If the case is "making a twist":

   - by definition, $b' = b$;
   - four vertices are identified into two vertices, two edges are identified into one edge, so $c' = c - 1$;
   - it is easy to see that the resulting surface is not orientable: in particular, if the surface is orientable before identification, there is only one way to identify e and e' in such a way that no boundary is created, and in this case, the resulting edge is not coherently oriented with respect to its incident faces. *We state* that $q' = q + 1$. Note that you can easily deduce from the identification cases that $b + c$ is even if S is orientable: thus, if S is orientable, $b' + c'$ is odd and $q' = 1$. Note also that if S is not orientable, the parity of $b' + c'$ is the converse of the parity of $b + c$.
   - $g' = g$ since $b' = b$, $c' = c - 1$ and $q' = q + 1$.

By applying b times "making a boundary" to a disc, you get a *sphere with b boundaries* (a disc is thus a sphere with one boundary). If you apply "removing a boundary" to a Möbius strip, you get a *projective plane*. If you apply twice "making a twist" to a disc, you get a *Klein bottle with one boundary*: cf. Fig. 2.15(a) and Fig. 2.16(c). If you apply "removing a boundary" to a Klein bottle with one boundary, you get a *Klein bottle*: cf. Fig. 2.15(b).

Note that if you apply "making a twist" several times, you can get a surface such that its orientability factor is greater than 2: we will see below that, thanks to the *exchange theorem*, there is no contradiction with the surface classification as stated at the beginning of this section.

Making a hole, or making two twists: Fig. 2.16.

Let e and e' be two edges of S such that they are not incident to the same boundary. The identification of e and e' has for consequence that the two boundaries are merged into one boundary. For instance, look at the annulus of Fig. 2.16(a): there are two ways to identify e and e': in the first case (cf. Fig. 2.16(b)), you get a *torus with one boundary*, in the second case you get the object depicted in Fig. 2.16(c), which is, as we will see, a Klein bottle with one boundary. More precisely, two cases are distinguished:

1. "Making a hole". This case is defined by the fact that S' is orientable (and thus S is orientable too);
2. "Making two twists". This case is defined by the fact that S' is not orientable.

In both cases, $b' = b - 1$, since two distinct boundaries are merged into one boundary, and $c' = c - 1$, since four vertices are identified into two vertices, and two edges are identified into one edge. The other characteristics are:

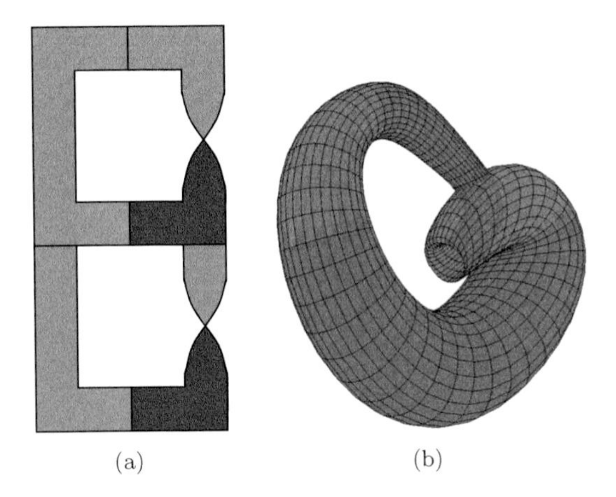

FIGURE 2.15
(a) A Klein bottle with one boundary. (b) A Klein bottle.

1. "Making a hole". By definition, $q' = 0$, thus $g' = g + 1$, since $q' = q$, $b' = b - 1$ and $c' = c - 1$;

2. "Making two twists". Note that you can construct the Klein bottle with one boundary depicted in Fig. 2.16(c)[4], starting from the disc depicted in Fig. 2.16(d) (it is a disc, since it can be constructed by applying twice "adding a face" to a face), by applying twice "making a twist", in order to identify e_1 and e_2, then e_3 and e_4. More precisely:

   - the identification of e_1 and e_2 is clearly a twist;
   - after this identification, e_3 and e_4 belongs to the same boundary; they are not adjacent, and their identification does not create a new boundary: this is thus another twist.

   This can be generalized and we get that $q' = q + 2$ and $g' = g$, since $b' = b - 1$ and $c' = c - 1$.

Exchange theorem: Fig. 2.17.

All possible cases have been studied (note that the conditions defining the cases are exclusive and cover all cases), but the classification described at the beginning of this section is still not complete, since the orientability factor can take any positive value: you can add as many twists as you want. The *exchange theorem* states that $TS_{b,q,g} = TS_{b,q-2,g+1}$ when $q \geq 3$. For instance,

[4]It is a Klein bottle, although the relation with the Klein bottle with one boundary depicted in Fig. 2.15(a) is not obvious! Nevertheless, it is possible to transform the second one to get the first one. Note that in order to get a proof of equivalence, it would be necessary to prove that the transformations do not change the topology...

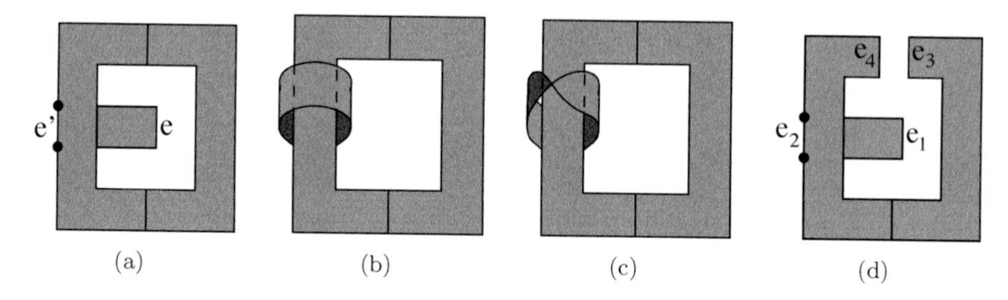

FIGURE 2.16
(a) An annulus.
(b) A torus with one boundary.
(c) A Klein bottle with one boundary.
(d) A disc.

look at the object depicted in Fig. 2.17: it is made of several faces M, V, U, K, B and R. You can construct the surface made by M, K, U and V by applying "adding a face", "shortening a boundary" and once "making a boundary", so it belongs to $TS_{2,0,0}$. B is added by "adding a face" and applying once "making a hole", so the corresponding surface belongs to $TS_{1,0,1}$. At last, add R by applying "adding a face" and once "making a twist", and the resulting surface belongs to $TS_{1,1,1}$. You can construct the same surface by first gluing M, K, B, by applying twice "adding a face". You can add R by applying "adding a face" and once "making a twist", so the corresponding surface belongs to $TS_{1,1,0}$. Add V by applying "adding a face", once "making a twist" and then "shortening a boundary": the surface belongs to $TS_{1,2,0}$. Last, add U by applying "adding a face" and "making a twist", and the resulting surface belongs to $TS_{1,3,0}$. This can be generalized, and we get the *exchange theorem* (see [125] for a complete study and proofs):

Let S be a paper surface, such that the characteristics obtained through its construction are b, q, g. If $q \geq 3$, then the characteristics of S are equivalent to b, $q-2$, $g+1$. At last, S belongs to the topological surface:

- $TS_{b,0,g}$, *if* $q = 0$;
- $TS_{b,1,u}$, *if q is odd; and* $u = g + (q-1)/2$;
- $TS_{b,2,v}$, *if q is even and not null; and* $v = g + q/2 - 1$.

2.3.2 Constructing any Subdivision of Any Surface

Now we can forget the "unnecessary constraints" which were useful in order to simplify the study of surface construction. So, in the general case, an edge

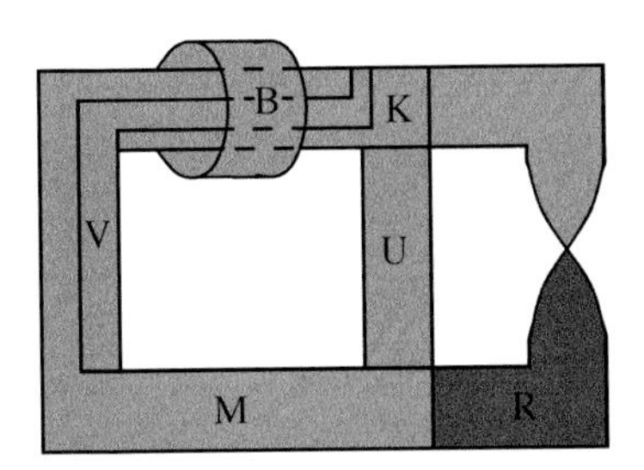

FIGURE 2.17
First construction: $(M, K, U, V) \in TS_{2,0,0}$; $(M, K, U, V, B) \in TS_{1,0,1}$; $(M, K, U, V, B, R) \in TS_{1,1,1}$.
Second construction: $(M, K, B, R) \in TS_{1,1,0}$; $(M, K, B, R, V) \in TS_{1,2,0}$; $(M, K, B, R, V, U) \in TS_{1,3,0}$.

can be a loop (i.e. incident twice to a vertex), two free edges incident to one face can be identified.

So, a face is a disc, its boundary is a cycle of edges and vertices which contains at least one edge and one vertex. Any two free edges can be identified, involving maybe, "by continuity", the identifications of vertices of their boundaries (note that more than two vertices can be identified together).

For instance, two new cases are depicted on Fig. 2.18 (for a complete study, see [164]):

- two edges incident to one face are identified, involving the creation of a new boundary made of one edge and one vertex: Fig. 2.18(a); this case is similar to "making a boundary";
- two edges, respectively incident to vertices v_1 and v, v and v_2 (these three vertices being distinct) are identified[5]. The initial configuration is similar to that of "shortening a boundary" (cf. Fig. 2.14), but the edges are identified in the other way, i.e. the first edge, taken with the orientation v_1v is identified with the other one, taken with the orientation vv_2, inducing thus a twist in the surface and the identification of vertices v, v_1 and v_2;
- often, new cases are "combinations" of the cases studied above. For instance, adding a face incident to one edge and one vertex can remove a boundary made by one edge and one vertex (cf. Fig. 2.18(b)). Another example is the construction of the projective plane: take a face such that its boundary is made of two edges and two vertices; the identification

[5]The identification of two edges corresponds to the existence of a homeomorphism between these two edges. The homeomorphism is "extended" onto the boundaries of the edges by continuity, and remains usually a homeomorphim; the identification corresponds to the fact that any pair of corresponding points is identified into one point.

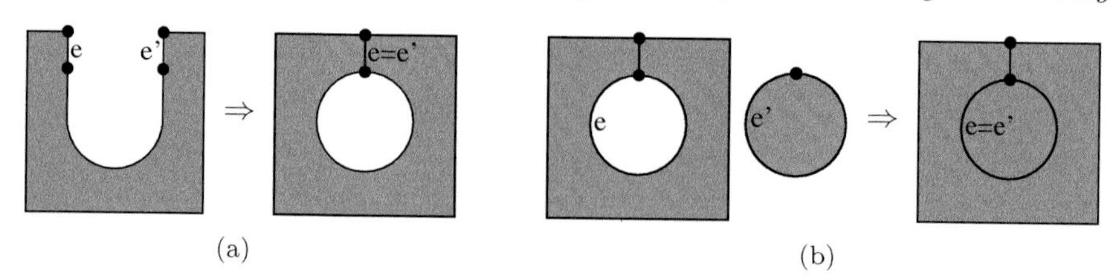

FIGURE 2.18
(a) The identification of edges e and e' involve the creation of a new boundary, which is a loop.
(b) Removing a "loop boundary" by adding a face.

of these two edges remove the boundary. Note that the edges can be identified in two ways: a sphere can be created, but also a projective plane, according to the way the edges are identified. In the case of the projective plane, a new twist is added.

Connectivity

Until now, all considered surfaces are *connected*, i.e. informally, made of one piece: intuitively, you can walk on the surface between any two of its points. More formally, a *path* between two points x and y in $X \subset \mathbb{R}^n$, where $x, y \in X$, means a continuous map $\gamma : [0,1] \rightarrow X$, such that $\gamma(0) = x$ and $\gamma(1) = y$. x and y are said *connected* by the path γ. The path is *closed* when $\gamma(0) = \gamma(1)$.

Let $X \subset \mathbb{R}^n$, and let R be the relation defined on points of X, such that xRy if and only if x and y are connected by a path in X. R is an equivalence relation, and its equivalence classes are the *connected components* of X: in other words, a path in X exists between any two points of a connected component, and no path exists in X between two points of two distinct connected components.

We can generalize our presentation of surfaces in order to take into account nonconnected surfaces, i.e. surfaces made of several connected components; the definition of the characteristic numbers (number of boundaries, orientability factor, genus) could be adapted, but it is more interesting (since more precise) to look at the characteristics of each connected component.

2.4 Manifolds, Quasi-manifolds, Pseudo-manifolds, Complexes

2-dimensional manifolds

Let us define a *polygon of degree* k as a closed disc (i.e. a 2-dimensional closed ball) the boundary of which is a cycle of k edges and k vertices, $k \geq 1$. A polygon contains one face (i.e. a 2-dimensional cell), which is the interior of the polygon: note that each point of the face has a neighborhood which is homeomorphic to an open disc (i.e. a 2-dimensional open ball). Each edge (i.e. a 1-dimensional cell) is homeomorphic to a 1-dimensional open ball (and so is the neighborhood of any point of the edge), and each vertex (i.e. a 0-dimensional cell) is (homeomorphic to) a point. Any subdivision of surface can be constructed by identifying at most two free edges (and thus the boundaries of these edges, by continuity): so, each point of a sewn edge has also a neighborhood which is homeomorphic to an open disc (and so is any point of a surface without boundary). This property: *each point has a neighborhood homeomorphic to a disc* characterizes surfaces without boundary, also called 2-*dimensional manifold without boundary.* This property can be extended in order to characterize any surface (i.e. any 2-*dimensional manifold*): each interior point has a neighbordhood homeomorphic to a disc, each point of a boundary has a neighborhood homeomorphic to a "*half* disc".

Note that we could define a 1-*dimensional manifold* in a similar way. The basic element is here the 1-dimensional closed ball, made of one edge (homeomorphic to a 1-dimensional open ball, as the neighborhood of any of its points) and its two extremity vertices. Any subdivision of a curve can be constructed by identifying at most two free vertices: so each sewn vertex has also a neighborhood which is homeomorphic to an open 1-dimensional ball, and so is the neighborhood of any point of any curve without boundary (also called 1-dimensional manifold without boundary). A curve with boundary can be characterized by the fact that any interior point has a neighborhood homeomorphic to a 1-dimensional ball, and a boundary point has a neighborhood homeomorphic to a "*half* 1-dimensional ball".

n-dimensional simplices

In order to construct n-dimensional "objects", we need n-dimensional cells. For this section, we choose *n-dimensional simplices*, thus some usual definitions are needed. The points $v_0, \ldots, v_k$ of $\mathbb{R}^n$ are *linearly independent* if the vectors $v_1 - v_0, \ldots, v_k - v_0$ are linearly independent. The *line segment* between two points x and y of $\mathbb{R}^n$ is the set $\{z = tx + (1-t)y, t \in [0,1]\}$. A subset A is *convex* if $x, y \in A$ implies that the line segment between x and y is contained in A. The *convex hull* of a set A is the intersection of all convex sets containing A.

Let $k \geq 0$. A *k-dimensional simplex* σ, or *k-simplex*, is the convex hull of $k+1$ linearly independent points $v_0, \ldots, v_k \in \mathbb{R}^n$; σ is denoted $v_0 ... v_k$. The points v_i are the *vertices* of σ. Let $\{w_0, \ldots, w_m\}$ be a subset of the vertices of σ, then $w_0 ... w_m$ is a *m-dimensional face*, or *m-face* of σ: cf. Fig. 2.19. When $m < k$, the face is *proper* else it is *principal*. Note that a simplex contains its boundary, which is made by all its proper faces: so a simplex, or any of its face, is homeomorphic to a closed ball (note also that we could define the faces, or cells, so that they are homeomorphic to open balls). So a 0-simplex

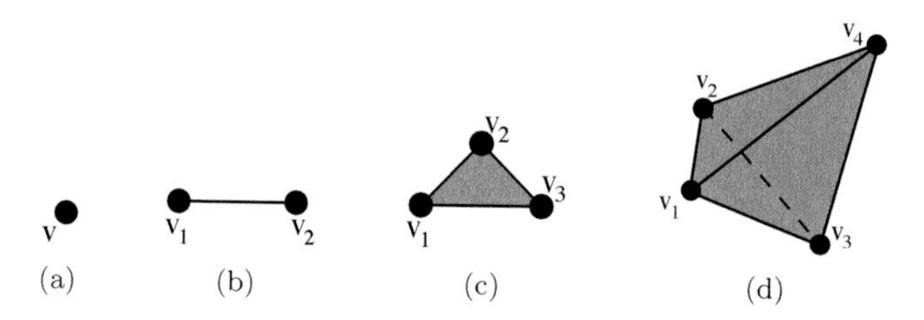

FIGURE 2.19
(a) A 0-simplex v.
(b) A 1-simplex v_1v_2 has two 0-faces v_1 and v_2, and a 1-face v_1v_2.
(c) A 2-simplex has three 0-faces, three 1-faces and one 2-face.
(d) A 3-simplex has four 0-faces, six 1-faces, four 2-faces and one 3-face.

is a point, a 1-simplex is a line segment, a 2-simplex is a triangle, a 3-simplex is a tetrahedron, etc.

Quasi-manifolds, manifolds, pseudo-manifolds and orientability

Let *closed n-cells* be homeomorphic to n-simplices (cf. Figs. 2.20(a), (b) and (c)), such that the homeomorphism propagates the structure of simplices, i.e. a closed n-cell contains an n-cell and its boundary, made of k-cells ($0 \leq k < n$). We can construct *n-dimensional quasi-manifolds* by taking such closed n-cells and by identifying pairs of free $(n-1)$-cells (and their boundaries). The identification operation can be defined by considering a homeomorphism between the two free $(n-1)$-cells, extending it on the boundaries of the $(n-1)$-cells by continuity, and then by identifying any points associated by the "extended homeomorphism": cf. Figs. 2.20(d), (e) and (f). Note that we can define several notions as that of free cells, sewn cells, boundaries of quasi-manifolds, as for the 2-dimensional case.

This construction of quasi-manifolds can be generalized for taking into account more general cells: cf. chapter 8. Anyway, note that 1- and 2-dimensional quasi-manifolds are manifolds, i.e. curves and surfaces: there is no difference between these two notions for dimensions up to 2.

Let us now define a *n-dimensional manifold without boundary* as an n-dimensional quasi-manifold without boundary, such that each point has a neighborhood homeomorphic to an n-dimensional open ball (this definition can be extended for manifolds with boundaries). These notions are different for dimensions higher than two, since quasi-manifolds exist, which are not manifolds:

- a closed n-cell, as defined above (i.e. being homeomorphic to an n-simplex), is homeomorphic to a closed n-ball, so it is a manifold with one boundary. We will see later that other definitions of n-cells do not involve this property (cf. chapter 8);

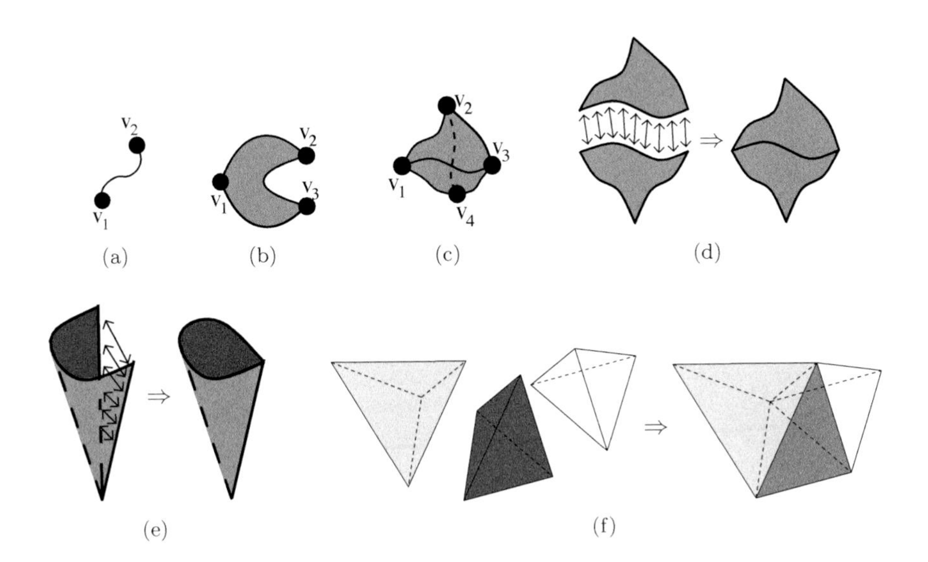

FIGURE 2.20
(a) A closed 1-cell.
(b) A closed 2-cell.
(c) A closed 3-cell.
(d) Two closed 2-cells; arrows at left denote the homeomorphism between two 1-cells, leading to their identification.
(e) An identification of two edges incident to a closed 2-cell.
(f) Construction of a 3-dimensional quasi-manifold (which is a manifold).

- other quasi-manifolds can be manifolds: for instance, look at Fig. 2.20(f). The neighborhood of any point at the boundary of a free 2-cell is a half 3-ball. When two 2-cells are identified, the two half-balls corresponding to two identified points become one 3-ball. For points of identified edges or vertices, half-balls are glued and make half-balls (since all edges and vertices belong to the boundary of the resulting object). Instead of looking at neighborhoods, you can also look at *links* of points, i.e. the boundary of neighborhoods, which can also be defined for points of 3-cells as the intersection of a small 2-sphere centered at the point and the 3-cell. Such links are discs for boundary points, spheres for interior points;

- look now at the object depicted in Fig. 2.21(a): it is a cube, made by gluing twenty-four tetrahedra (four tetrahedra are glued for making a pyramid with a square basis, this square will be a face of the cube, and six such pyramids are glued together). The link of point v (at the center of the upper face of the cube) is made of a half-sphere inside the cube, closed by a disc at the surface of the cube. The boundary of this disc is made of four parts $l_1 \subset f_1$, $l_2 \subset f_2$, $l_3 \subset f_3$, $l_4 \subset f_4$. If face f_1 is identified with face f_2 and face f_3 is identified with face f_4 in such a way that edges e_1 and e_3 (resp. e_5 and e_6, e_7 and e_8) are identified, and vertices v_1 and v_3 are identified, the resulting object is still a manifold (and you can construct the resulting object in the usual 3-dimensional space). For instance, the link of point v becomes a sphere contained in the resulting object, since l_1 (resp. l_4) is identified with l_2 (resp. l_3). If face f_1 is identified with face f_3 and face f_2 is identified with face f_4 in such a way that edges e_1, e_2, e_3 and e_4 are identified together, edge e_5 (resp. e_6) is identified with edge e_7 (resp. e_8) and vertices v_1, v_2, v_3 and v_4 are identified together, the resulting object is still a quasi-manifold, but it is not a manifold (and you cannot construct the resulting object in the usual 3-dimensional space): for instance, the link of vertex v is now obtained by identifying l_1 with l_3, l_2 with l_4, producing a torus with one handle (i.e. its genus is equal to 1), which is not the boundary of a 3-ball; so, the neighborhood of v is not homeomorphic to a 3-ball.

We will see that it is possible to define data structures such that one can be sure that any instance corresponds to a quasi-manifold; but as far as we know, no one knows how to define data structures such that one can be sure that any instance corresponds to a manifold. This corresponds to the fact that (the structure of) quasi-manifolds can be defined combinatorially, i.e. as discrete structures (the related mathematical field was named "combinatorial topology"). This is not the case for manifolds, since, as far as we know, no one knows how to *combinatorially characterize balls*: in other word, there is no discrete computation which makes it possible to know whether an "object" is a ball or not.

Another aspect is the following: as far as we know, there is *no classi-*

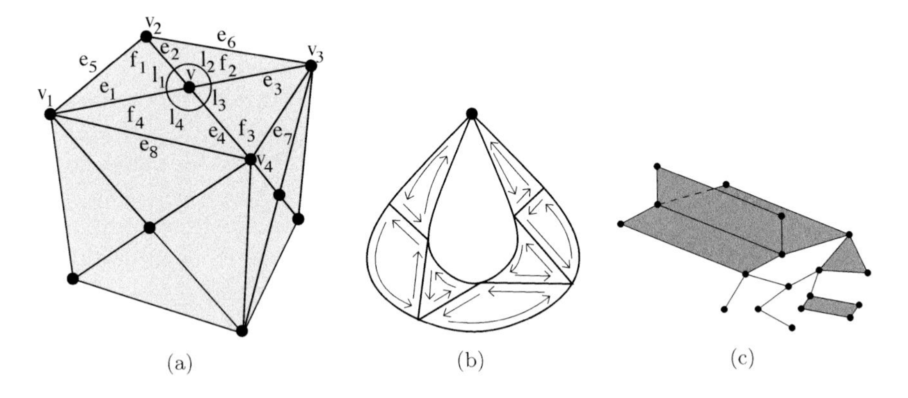

FIGURE 2.21
(a) A cube made by twenty-four tetrahedra.
(b) An orientable 2-dimensional pseudo-manifold.
(c) A 2-dimensional complex.

fication for n-dimensional manifolds, similar to that in the 2-dimensional case, although several notions can be generalized. For instance, *orientability* can be defined for quasi-manifolds, and thus for manifolds, and more generally for *pseudo-manifolds*. Informally, an n-dimensional pseudo-manifold (cf. Fig. 2.21(b)) can be constructed by gluing closed n-cells:

- by identifying at most two free $(n-1)$-cells and their boundaries (as for quasi-manifolds);
- by identifying k-dimensional cells, $0 \leq k < n-1$;
- in such a way that, given any two n-cells σ and σ', a sequence $\sigma_0, \cdots, \sigma_p$ exists, such that $\sigma = \sigma_0$, $\sigma' = \sigma_p$, and σ_j meets σ_{j+1} in a common $(n-1)$-dimensional cell, $0 \leq j < p$.

The definition of pseudo-manifolds is (structurally) a combinatorial one, as the definition of quasi-manifold. Moreover, let SM^n (resp. SQM^n, SPM^n) be the set of n-dimensional manifolds (resp. quasi-manifolds, pseudo-manifolds): then $SM^n \subset SQM^n \subset SPM^n$. In other words, any manifold is a quasi-manifold, which is itself a pseudo-manifold. But pseudo-manifolds exist, which are not quasi-manifolds (cf. for instance Fig. 2.21(b)), as quasi-manifolds exist, which are not manifolds.

A pseudo-manifold without boundary is such that no free $(n-1)$-cell exists. A pseudo-manifold without boundary is *orientable* if its n-cells can be oriented coherently, i.e. any two n-cells which share a common $(n-1)$-cell induce opposite orientations on this $(n-1)$-cell. Note that this definition is consistent if the definition of n-cell orientation is known. In fact, the boundary of a closed

n-cell is a manifold (i.e. a pseudo-manifold) of lower dimension. Let us define the orientation of a closed n-cell (and thus of the corresponding n-cell) as the orientation of its boundary: since we know how to define an orientation for $n = 1$, it is thus possible to define an orientation for any n. It is also possible to define the orientability of pseudo-manifolds with boundaries, and thus the orientability of any (quasi-)manifold.

Other notions can be extended for any dimension, for instance the Euler characteristic, which is equal to the alternate sum of the number of cells. These notions will be studied for quasi-manifolds in following chapters.

At last, note that other sets of objects can be defined. For instance, *complexes* can be constructively defined in the following way: take a set of closed cells of any dimensions up to n, and identify cells of the boundaries of the closed cells; you get an n-dimensional complex (cf. Fig. 2.21(c)).

2.5 Discrete Structures

It is necessary to define data structures in order to handle such quasi-manifolds (and more generally complexes) in computers. We will distinguish between the representation of the structure of the object (its "topology") and the representation of its shape (its "embedding", usually in some $\mathbb{R}^n$).

For instance (cf. Fig. 2.22), assume a triangulated surface embedded in $\mathbb{R}^3$ is described by a set of triangles, each triangle being defined by three points. A possible data structure is a *list* of *Triangle*, each *Triangle* being a tuple of nine floating numbers, defining the coordinates of three points of $\mathbb{R}^3$. This structure does not clearly distinguish between the representations of the topology and of the shape of the surface. In order to know that two triangles are adjacent, it is necessary to compare the coordinate values, i.e. to take the shape into account. Moreover, it is well-known that imprecisions or errors about floating numbers (as consequences of some computations, for instance) can lead to inexact representations of the surface. So it may be impossible to check whether two triangles are adjacent or not.

An alternative representation is the following. A surface still corresponds to a list of *Triangles*, but a *Triangle* is now a tuple of three pointers to a *Point*, each *Point* being a tuple of three floating numbers corresponding to the three coordinates of a point in $\mathbb{R}^3$. The precondition upon the data structure is the fact that a point (resp. a triangle) is represented only once (two different *Points* correspond to two distinct points of $\mathbb{R}^3$, two different *Triangle* correspond to two distinct triangles). Now, the structure of the triangulation is represented, "independently" of its shape. For instance, in order to check whether two triangles are adjacent, pointers to points are compared, producing an exact information; the coordinates of the points are not involved here.

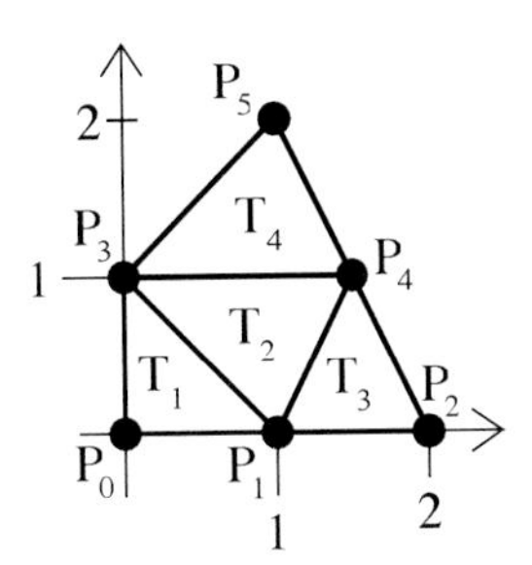

FIGURE 2.22
A triangulation S.
A first representation of S is: $S = (T_1, T_2, T_3, T_4)$, $T_1 = ((0,0),(1,0),(0,1))$, $T_2 = ((1,0),(0,1),(1.5,1))$, $T_3 = ((1,0),(2,0),(1.5,1))$, $T_4 = ((0,1),(1.5,1),(1,2))$.
An alternative representation is: $S = (T_1, T_2, T_3, T_4)$, $T_1 = (\&P_0, \&P_1, \&P_3)$, $T_2 = (\&P_1, \&P_3, \&P_4)$, $T_3 = (\&P_1, \&P_4, \&P_2)$, $T_4 = (\&P_3, \&P_4, \&P_5)$, $P_0 = (0,0)$, $P_1 = (1,0)$, $P_2 = (2,0)$, $P_3 = (0,1)$, $P_4 = (1.5,1)$, $P_5 = (1,2)$. $\&P$ denotes a reference to point P.

Note that more accurate data structures have been conceived in order to handle triangulated surfaces (cf. [183] for instance).

The *structure* of quasi-manifolds can be combinatorially described, i.e. by a discrete structure. It is also the case for particular subsets of quasi-manifolds, as orientable quasi-manifolds, or for more general objects as complexes (cf. the following chapters). But as said before, this is not possible for any subset of complexes, for instance for manifolds.

2.5.1 Discrete Mappings

The discrete structures we will define are algebras, i.e. they are defined by discrete objects on which functions are defined. Such algebras can be equivalently defined as graphs and can be easily implemented[6]. The following definitions will state the vocabulary.

(Multi)sets, (multi)graphs

An element appears once in a *set*, it can appear several times in a *multiset*. From now on, all (multi)sets are discrete ones, and they are usually finite ones, i.e. any (multi)set contains a *finite* number of elements.

An *oriented multigraph* (resp. *graph*) is a pair (V, E), where V is a set of

---

[6]Graphs (and thus discrete objects on which relations, functions, permutations, etc. are defined) can be implemented using arrays, or records and pointers, etc. For more details about possible implementations, see [56].

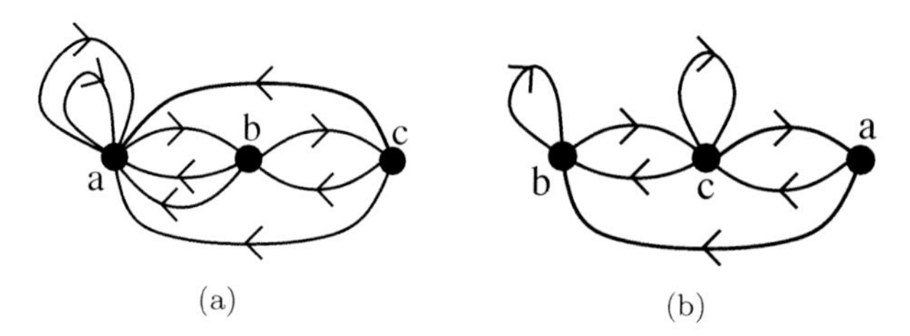

FIGURE 2.23
(a) A multigraph: $V = \{a, b, c\}$, $E = ((a, a), (a, a), (a, b), (b, a), (b, a), (b, c), (c, b), (c, a), (c, a))$.
(b) A graph: $V = \{a, b, c\}$, $E = \{(a, b), (b, b), (b, c), (c, b), (c, c), (c, a), (a, c)\}$.

vertices and E is a multiset (resp. *set*) of *edges*, an edge being a pair of vertices of V (cf. Fig. 2.23). Edge (x, y), where $x, y \in V$, is *oriented* from x to y, x is the *origin* of the edge and y is its extremity; the edge is incident to x and y, and conversely; when $x = y$, the edge is a *loop*. A *bipartite* oriented graph is such that its set of vertices can be partitioned into two subsets V_1 and V_2, such that each edge is incident to a vertex of V_1 and to a vertex of V_2. All classical notions related to graphs, as paths, connected components, etc, are supposed to be known.

Relations, functions, permutations, involutions, compositions

Let F and G be two sets. A *relation* R between F and G is a set of pairs (x, y) such that $x \in F$ and $y \in G$, i.e. it is a subset of $F \times G$, the *cartesian product* of F and G, which is $\{(x, y)|x \in F, y \in G\}$. An oriented graph can be associated with any relation (and conversely), i.e. the graph $(F \cup G, R)$: cf. Fig. 2.24(a). The graph is bipartite when F and G are disjoint.

Relation R is *symmetric* when $xRy \implies yRx$: cf. Figs. 2.24(b) and (c); in other words, when two vertices x and y are connected by (x, y) in the associated graph, then they are also connected by (y, x). A *nonoriented edge* $\{x, y\}$ can be defined by a set of two opposite edges (x, y) and (y, x), and a *nonoriented graph* can be defined as the graph associated with a symmetric relation.

A *function* (or a *map*) ϕ is a relation such that, for any $x \in F$, there is exactly one $y \in G$ such that $(x, y) \in \phi$; y is denoted $\phi(x)$; so, in the corresponding graph, any vertex corresponding to an element of F is the origin of exactly one edge (cf. Fig. 2.24(d)). Let $F' \subset F$, and let $G' \subset G$ be the set of all *images* of elements of F' by ϕ, i.e. $G' = \{y \in G|\exists x \in F', \phi(x) = y\}$. Then $\phi' : F' \to G'$, defined by: $\forall x \in F', \phi'(x) = \phi(x)$, is the *restriction* of ϕ to F', denoted $\phi' = \phi_{|F'}$. The graph corresponding to ϕ' is a *subgraph* of the graph associated with ϕ, which contains vertices corresponding to elements of F',

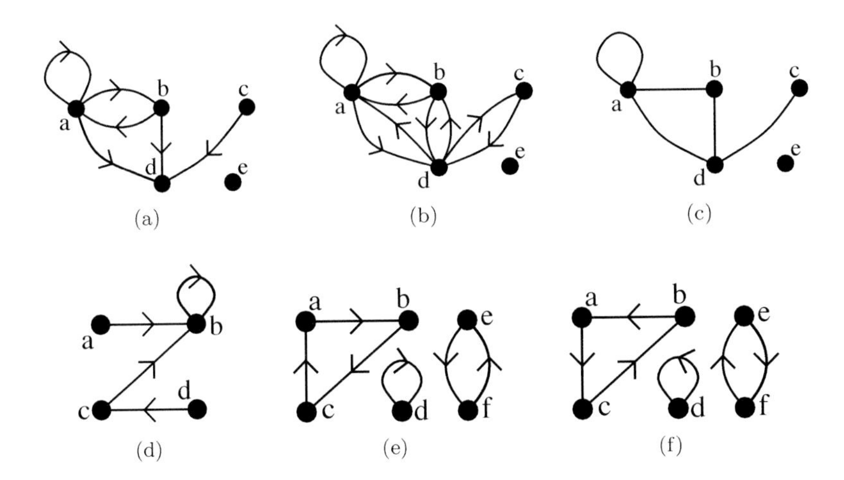

FIGURE 2.24
(a) Graph of a relation R: $F = \{a, b, c\}$, $G = \{a, b, d, e\}$, $R = \{(a, a), (a, b), (a, d), (b, a), (b, d), (c, d)\}$.
(b) Graph of a symmetric relation R': $F' = G' = \{a, b, c, d, e\}$, $R' = \{(a, a), (a, b), (a, d), (b, a), (b, d), (c, d), (d, a), (d, b), (d, c)\}$.
(c) Another representation of the nonoriented graph associated with R'.
(d) Graph of a function ϕ: $F = \{a, b, c, d\}$, $G = \{b, c\}$, $\phi = \{(a, b), (b, b), (c, b), (d, c)\}$.
(e) A permutation P: $F = \{a, b, c, d, e, f\}$, $P = \{(a, b), (b, c), (c, a), (d, d), (e, f), (f, e)\}$.
(f) P^{-1}, the inverse of P, is: $P^{-1} = \{(b, a), (c, b), (a, c), (d, d), (e, f), (f, e)\}$.

all edges which have these vertices as origins, and their extremity vertices.

The function is *injective* (i.e. it is an *injection*) when, for any $x, y \in F$, $x \neq y$ implies $\phi(x) \neq \phi(y)$; so, in the corresponding graph, any vertex corresponding to an element of G is the extremity of at most one edge.

The function is *surjective* (i.e. it is a *surjection*) when, for any $y \in G$, $x \in F$ exists such that $y = \phi(x)$; so, in the corresponding graph, any vertex corresponding to an element of G is the extremity of at least one edge.

A function is *bijective* (i.e. it is a *bijection* or a one-to-one mapping) when it is injective and surjective: so, in the corresponding graph, any vertex corresponding to an element of F is the origin of an edge, and any vertex corresponding to an element of G is the extremity of an edge.

The *inverse* bijection ϕ^{-1} is a bijection between G and F such that its graph is obtained by reversing all edges; in other words, $\phi(x) = y \implies \phi^{-1}(y) = x$.

When $F = G$, the bijection is a *permutation* and the corresponding graph is a set of elementary cycles, since each vertex is the origin and the extremity of one edge; note that the inverse of a permutation is also a permutation (cf. Figs. 2.24(e) and (f)). Let P be a permutation: a *fixed point* for P is any element x such that $P(x) = x$; a fixed point corresponds thus to a vertex incident to a loop in the associated graph.

Id denotes the *identity*, i.e. any permutation such that $Id(x) = x$ for any x. In other words, all elements are fixed points for Id.

An *involution* I is a permutation which is its own inverse, i.e. $I = I^{-1}$, meaning $I(x) = y \implies I(y) = x$. The associated graph is thus a nonoriented graph, and any connected component is made by either one vertex incident twice to an edge (i.e. a loop) or two vertices incident to an edge: cf. Fig. 2.25(a).

Let F, G, H be three sets, f and g be two functions: $f : F \to G, g : G \to H$, such that f is surjective. $g \circ f : F \to H$, the *composition* of f and g, is defined by: $g \circ f(x) = g(y)$, with $y = f(x)$, for any $x \in F$, i.e. $g \circ f(x) = g(f(x))$. The graph corresponding to $g \circ f$ has $F \cup H$ as set of vertices, and its edges correspond to paths of length 2 in the union of the graphs[7] associated with f and g.

Note that the composition of two bijections is a bijection, the composition of two permutations is a permutation, the composition of two involutions is a permutation (cf. Figs. 2.25), and the composition of an involution with itself

[7] The union of two graphs (V_1, E_1) and (V_2, E_2) is $(V_1 \cup V_2, E_1 \cup E_2)$.

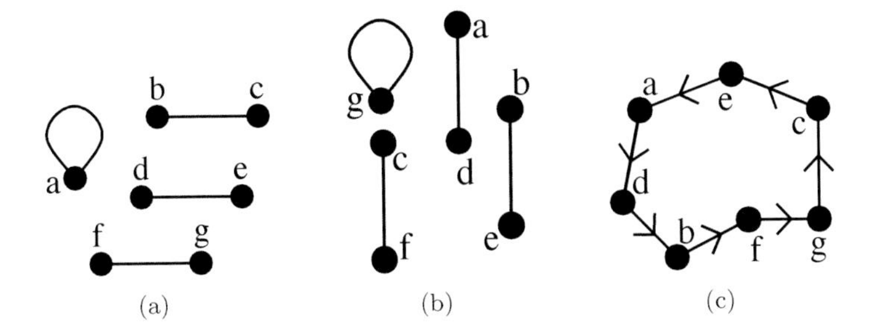

FIGURE 2.25
$F = \{a, b, c, d, e, f, g\}$.
(a) Involution $I = \{(a, a), (b, c), (c, b), (d, e), (e, d), (f, g), (g, f)\}$.
(b) Involution $I' = \{(a, d), (b, e), (c, f), (d, a), (e, b), (f, c), (g, g)\}$.
(c) $I' \circ I = \{(a, d), (b, f), (c, e), (d, b), (e, a), (f, g), (g, c)\}$.

is the identity (since it is equal to its inverse). Let P be a permutation. P^0 denotes Id, P^k denotes $P \circ P^{k-1} = P^{k-1} \circ P$ (thus P^1 denotes P, P^2 denotes $P \circ P$, etc.).

2.5.2 Hypermaps, Group of Permutations, Orbits

Let $S = \{P_1, \ldots, P_k\}$ be a set of permutations defined on a set D. $(D, P_1, \ldots, P_k)$ is a *k-dimensional hypermap*[8]; the associated graph is the union of the graphs (D, P_1), ..., (D, P_k): cf. Fig. 2.26.

Let $S' = \{P_{i_1}, \ldots, P_{i_j}\}$ be a subset of S. $\langle S' \rangle$ is the *group of permutations*, or *permutation group*, generated by S': it is the set of all permutations P such that P is either Id, a permutation of S' or its inverse, or any composition of permutations of S' and their inverses. For instance, let I and I' be the involutions depicted in Figs. 2.26(a) and (b). $\langle I, I' \rangle$ is $\{Id, I, I', I' \circ I, I \circ I', I \circ I' \circ I, I' \circ I \circ I', \ldots\}$.

The *orbit* of an element d of D relatively to S' is the set $\langle S' \rangle(d) = O(d) = \{P(d) | P \in \langle S' \rangle\}$. It is the set of all elements which can be reached, starting from d, by applying any composition of (inverses of) permutations of S' (note that d belongs to any of its orbits, since Id belongs to any permutation group). $\langle S' \rangle(d)$ denotes also, according to the context, the *sub-hypermap* $(O(d), P_{i_1|O(d)}, \ldots, P_{i_j|O(d)})$, where $P_{i_k|O(d)}$ is the restriction of P_{i_k} to $O(d)$

[8]As n-Gmaps and n-maps to which this book is mainly devoted, hypermaps are particular classes of *combinatorial maps*: the relations between these structures are explained in chapter 8.

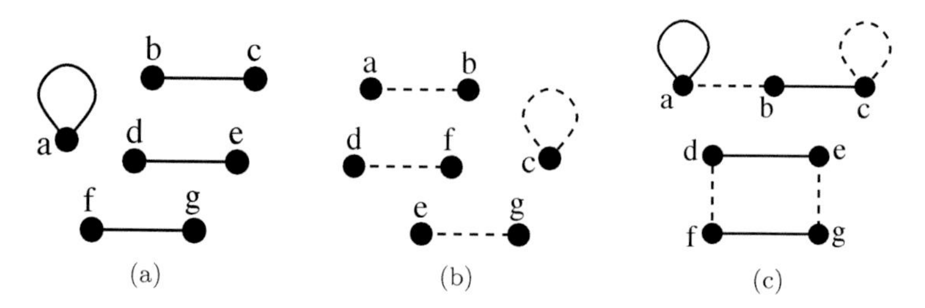

FIGURE 2.26
$D = \{a, b, c, d, e, f, g\}$.
(a) Involution $I = \{(a, a), (b, c), (c, b), (d, e), (e, d), (f, g), (g, f)\}$.
(b) Involution $I' = \{(a, b), (b, a), (c, c), (d, f), (e, g), (f, d), (g, e)\}$.
(c) 2-dimensional hypermap (D, I, I'), containing two connected components.

(note that these restrictions are well defined, due to the orbit definition). When $S' = S$, the corresponding orbit of d is the *connected component* incident to d. For instance, look at the 2-dimensional hypermap depicted in Fig. 2.26(c): the connected component incident to a (resp. f) contains elements $a = Id(a)$, $b = I'(a)$, $c = I \circ I'(a)$ (resp. $f = Id(f)$, $d = I'(f)$, $e = I \circ I'(f)$, $g = I(f)$). At last, note that several compositions can produce the same element, for instance $e = I \circ I'(f) = I' \circ I(f)$, $f = I \circ I' \circ I \circ I'(f)$, etc.

Let $H = (D, P_1, \ldots, P_k)$ and $H' = (D', P'_1, \ldots, P'_k)$ be two hypermaps. $\phi : D \to D'$ is an *isomorphism* between H and H' if ϕ is a bijection between D and D', such that: $\forall d \in D$, $\forall i \in \{1, \ldots, k\}$, $\phi(P_i(d)) = P'_i(\phi(d))$. H and H' are *isomorphic (by ϕ)*. The corresponding graphs are isomorphic, and for each i, edges corresponding to permutation P_i are associated with edges corresponding to permutation P'_i.

2.5.3 Partial Functions, Partial Permutations, Partial Involutions and Related Notions

Let F', F, G be three sets such that $F' \subset F$. If $f : F' \to G$ is a function, then it is a *partial function* from F to G; so, either $f(x) \in G$ if $x \in F' \subset F$, either $f(x)$ is undefined if $x \in F \setminus F'$: in this last case, denoted $f(x) = \varnothing$, no edge has x as origin in the associated oriented graph (cf. Fig. 2.27(a)). All previous notions related to functions can be extended for partial functions. For instance, an alternative definition of a *partial permutation* is the following (cf. Fig. 2.27(b)): a partial permutation is a function $P : F \cup \{\varnothing\} \to F \cup \{\varnothing\}$ such that:

1. $P(\varnothing) = \varnothing$;

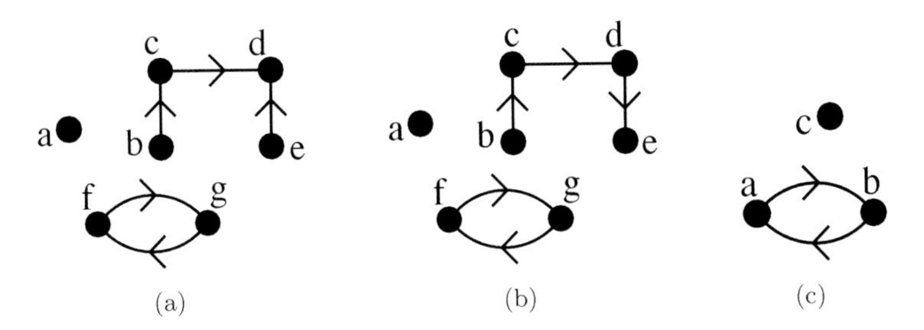

FIGURE 2.27
(a) $F = G = \{a, b, c, d, e, f, g\}$; $f = \{(b, c), (c, d), (e, d), (f, g), (g, f)\}$ is a partial function.
(b) $F = \{a, b, c, d, e, f, g\}$; $P = \{(a, \varnothing), (b, c), (c, d), (d, e), (e, \varnothing), (f, g), (g, f), (\varnothing, \varnothing)\}$ is a partial permutation.
(c) $F = \{a, b, c\}$; $I = \{(a, b), (b, a)(c, \varnothing), (\varnothing, \varnothing)\}$ is a partial involution.

2. $\forall x_1, x_2 \in F$, $P(x_1) = P(x_2) \neq \varnothing \implies x_1 = x_2$.

In other words, $\varnothing$ denotes the "undefinition". Let F' be the subset of F such that P is defined on the elements of F'. So, for any $x \in F'$, $P(x) \neq \varnothing$, and, for any $x \in F \setminus F'$, $P(x) = \varnothing$. P^{-1}, the inverse of P, is defined by:

1. $P^{-1}(\varnothing) = \varnothing$;

2. Let $y \in F$; if $x \in F$ exists such that $P(x) = y$, then $P^{-1}(y) = x$, else $P^{-1}(y) = \varnothing$.

Note that, for any $x \in F$, either $P(x) = \varnothing$, or $P^{-1}(P(x)) = x$.

A *partial involution* I is a partial permutation such that, $\forall x \in F$, either $P(x) = \varnothing$, or $P(x) \neq \varnothing \implies P(P(x)) = x$ (cf. Fig. 2.27(c)).

Hypermaps can be defined using partial permutations instead of permutations. Assuming that $f(\varnothing) = \varnothing$ for any partial function, it is easy to define the composition of partial permutations, and thus also the permutation group related to a set of partial permutations. The orbit of an element related to a set of partial permutations is defined as before, but $\varnothing$ does not belong to the orbit; let $S = \{P_1, \ldots, P_k\}$ be a set of partial permutations defined on F: $\langle S \rangle(x) = \{P(x) | P \in \langle S \rangle\} \setminus \{\varnothing\}$, for $x \in F$. As before, it denotes also this set of elements together with the partial permutations of S.

2.6 Incidence Graphs

Many discrete structures have been proposed in order to represent the topology of subdivided objects. An important subset of such data structures is that of *incidence graphs*, also referred to as *orders*, and several classes of incidence graphs have been defined, in order to represent different classes of subdivided objects [102, 195, 18, 84].

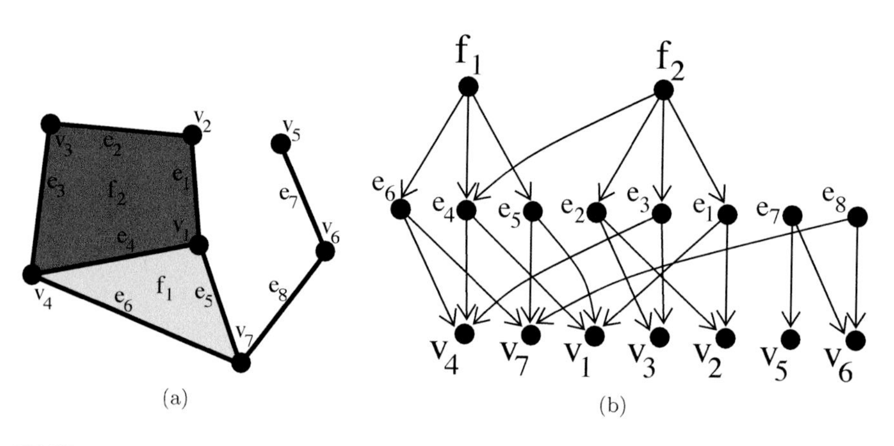

FIGURE 2.28
(a) A 2-dimensional complex. (b) The corresponding incidence graph.

For instance, look at the 2-dimensional complex depicted in Fig. 2.28(a); this complex satisfies an important topological property: there is *no multi-incidence* within this complex. More precisely, this property can be defined as follows for the 2-dimensional case:

1. any 2-dimensional (resp. 1-dimensional) cell is incident once to any 1-dimensional (resp. 0-dimensional) cell of its boundary. In other words, if you think at the construction of such a complex by identifying edges or vertices:

    - before any identification, each face has a boundary made by at least three edges and three vertices;
    - before any identification, the boundary of each edge is made by exactly two vertices;
    - two edges (free or not) can be identified if they are not incident to a same face;
    - no loop is produced by any identification (of edges, of vertices);

2. for a given face and a given vertex, at most one "corner" of the face is incident to the vertex (cf. Fig. 2.29(a) for a counter-example).

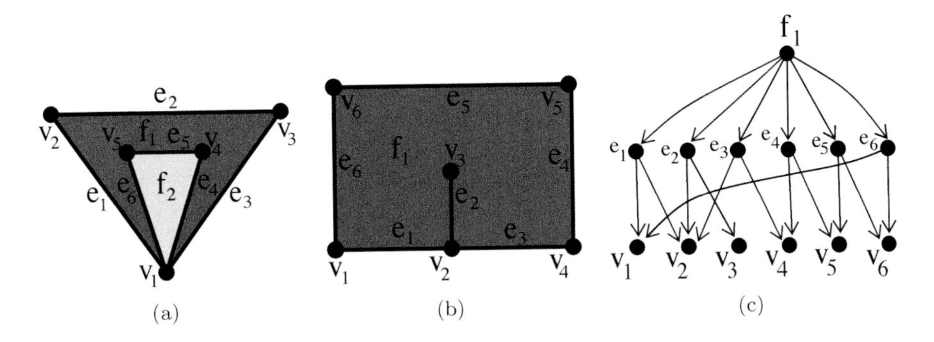

FIGURE 2.29
(a) Two corners of face f_1, namely (e_1, v_1, e_6) and (e_3, v_1, e_4) share vertex v_1: face f_1 is adjacent to itself through this vertex.
(b) A surface subdivision in which face f_1 is incident twice to edge e_2.
(c) The corresponding incidence graph.

Incidence graphs can describe the topology of n-dimensional complexes in which no multi-incidence occurs (see chapter 8 for a precise study of incidence graphs and their associated complexes). In order to construct the graph associated with such a complex (cf. Fig. 2.28(b)):

1. associate a vertex of the graph with any cell; moreover, associate also the dimension of the cell with the corresponding cell (this is not a requirement, several definitions of incidence graph omit dimensions, which can be retrieved from other informations represented by the graph): in Fig. 2.28(b), the dimension of cells is represented by the height of the graph vertices;

2. for each pair of incident i- and $(i-1)$-dimensional cells, associate an oriented edge between the corresponding vertices of the graph.

In fact, this graph is that of the incidence relation between i- and $(i-1)$-dimensional cells; in the graph, a vertex corresponding to an i-dimensional cell is the origin of edges the extremities of which are the vertices associated with the $(i-1)$-dimensional cells of its boundary. The idea at the basis of incidence graphs is thus simple and intuitive. Note that other variants have been proposed in order to represent classes of complexes or more general objects, for instance complexes from which subparts have been removed [195].

But, as far as we know, multi-incidence is never taken into account. Such structures are not *ordered* models, in the meaning of [39]; for instance, the order of the cells in the boundary of a higher-dimensional cell is not explicitly represented, it has to be computed from an incidence graph. Look at face f_1 of the complex depicted in Fig. 2.28(a): its boundary is made by vertices v_1, v_4 and v_7, edges e_4, e_5 and e_6. If you turn around the face, you encounter these cells in a sequence equivalent to $(v_1, e_4, v_4, e_6, v_7, e_5)$ (or its inverse sequence), which describes the "order" of these cells in the boundary of f_1. In order to extract this sequence from the incidence graph, you have to travel through the graph, starting from v_1, then going to its incident edge e_4, then going to the other incident vertex v_4, and so on.

Look now at the complex and its corresponding incidence graph depicted in Figs. 2.29(b) and (c). First, note that face f_1 is incident twice to e_2; thus there is multi-incidence, and the associated graph does not describe unambiguously the topology of this complex! Anyway, try to turn around face f_1, starting from vertex v_1 along edge e_1; then we have to go to the other incident vertex v_2, but now there are two incident edges different from e_1, i.e. e_2 and e_3: so, what information in the graph makes it possible to choose e_2 in order to follow the boundary of f_1?

Note also that it may be difficult to define subclasses of incidence graphs dedicated to particular complexes. For instance, it is possible to construct incidence graphs in which a 1-cell is incident to more than two 0-cells; so, what are the properties that an incidence graph has to satisfy in order to represent a quasi-manifold? Such properties have been defined (cf. [84], for instance), but they are not intrinsic to the definition of incidence graphs, contrary to "ordered" structures as n-Gmaps and n-maps (cf. chapter 4 and chapter 5).

Since only complexes in which no multi-incidence occurs can be represented with incidence graphs, only particular embeddings can be associated, usually linear embeddings, i.e. a 0-cell is associated with a point, a 1-cell is associated with a line segment, a 2-cell is associated with a part of a plane, etc., and these embeddings of cells satisfy the incidence relations between cells (e.g. the point associated with a 0-cell which is incident to a 1-cell is the extremity of the line segment associated with the 1-cell).

Many applications in geometric modeling, computational geometry, discrete geometry, computer graphics, image processing and analysis require to handle nonlinear objects, e.g. parametric curves and surfaces, etc., and objects in which multi-incidence can occur. Thus, many "ordered" models have been proposed (e.g. [17, 214, 203, 39, 165, 104]). Note also that it is often more easy to conceive construction operations for handling ordered models than for handling incidence graphs. n-Gmaps (resp. n-maps) are an archetype of ordered models dedicated to the representation of the topology of quasi-manifolds (resp. oriented quasi-manifolds), that is why this book is devoted to these structures.

3

Intuitive Presentation

In this intuitive introduction, we show how to represent the topology of subdivisions of quasi-manifolds, introduced in chapter 2. n-maps and n-Gmaps are presented in Section 3.1 and Section 3.2. Both models have their own advantages and drawbacks. n-maps allow to describe oriented subdivisions while n-Gmaps allow to describe oriented and nonoriented subdivisions. n-maps need twice less memory than n-Gmaps for representing oriented quasi-manifolds. However algorithms are more complex to write for n-maps than for n-Gmaps due to an inhomogeneous definition. For these reasons, both models are interesting and could be chosen depending on the specific needs of a particular application.

In this chapter, the different models are introduced, starting from the object to describe, by applying successive decompositions to cells. Note that this construction is only useful to understand the concept of darts, which is the basic element of combinatorial maps, and the different links between these darts. In practice, n-maps and n-Gmaps are handled through high level operations.

3.1 n-maps

n-maps can be introduced in the following way: given an oriented quasi-manifold, the intuitive construction consists in decomposing all cells by decreasing dimensions until obtaining isolated oriented edges, and in keeping at each step the relations between decomposed elements.

3.1.1 Objects without Boundary

First let us start by considering oriented quasi-manifolds without boundary.

3.1.1.1 1D

Look at the 1D oriented quasi-manifold without boundary given in Fig. 3.1(a), i.e. a closed polygonal curve with all the edges oriented in a coherent way. The quasi-manifold is composed by five edges (labeled from e_1 to e_5) and five

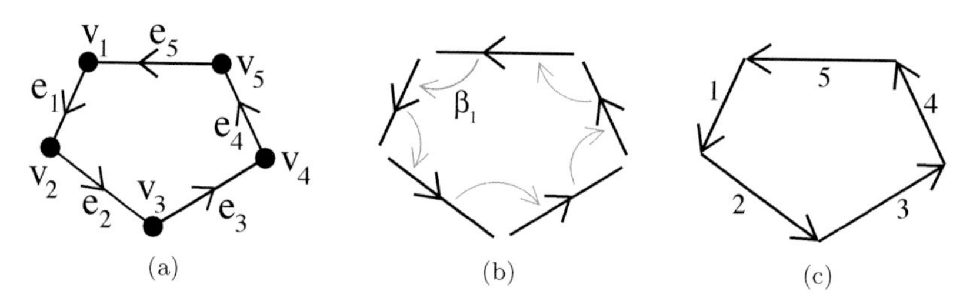

FIGURE 3.1
(a) A 1D oriented quasi-manifold without boundary.
(b) The decomposition of the polygonal curve produces a set of isolated oriented edges linked by β_1.
(c) A representation of the corresponding 1-map. The darts are drawn as oriented edges; β_1 is not explicitly represented: it is suggested by the edge adjacencies.

vertices (labeled from v_1 to v_5). In order to obtain the 1-map describing this polygonal curve, the curve is decomposed into edges by splitting each vertex in two, producing the set of isolated oriented edges drawn in Fig. 3.1(b). These oriented edges are the atomic elements, called *darts*, which are at the basis of the 1-map definition. A relation is defined on these darts, called β_1, in order to link an oriented edge and the next oriented edge of the polygonal curve that shared a common vertex before the split. Note that, since the quasi-manifold is without boundary, it is sure that for each edge, there is exactly one such edge. The set of darts plus the relation β_1 is the 1-map describing the initial 1D oriented quasi-manifold (see Fig. 3.1(c)). β_1 is a permutation, i.e. a bijection from the set of darts to the set of darts.

Each vertex and each edge of the 1D quasi-manifold is described by exactly one dart in the corresponding 1-map. In the example of Fig. 3.1, vertex v_1 *and* edge e_1 are described by dart 1: more generally each dart corresponds exactly to a pair of a vertex and an oriented edge (dart 1 corresponds to the pair (v_1, e_1)). Note that only the first extremity of the oriented edge is described by the dart; the second extremity will be described by the dart linked by β_1.

Multi-incidence can be taken into account, as for the example shown in Fig. 3.2(a) where vertex v_1 is incident twice to edge e_1 which is thus a loop. In this case, the corresponding 1-map shown in Fig. 3.2(c) contains only one dart which is linked with itself by β_1.

Several 1D oriented quasi-manifolds can be simultaneously represented. In such a case, the whole set of polygonal curves can be described by one 1-map, each polygonal curve being described as a distinct *connected component* of the 1-map. By definition, two darts d_1 and d_2 belong to the same connected component if there is a path of darts two by two linked by β_1 between d_1 and d_2. For any dart d, $\langle\beta_1\rangle(d)$ is the orbit of d related to β_1: it is the connected

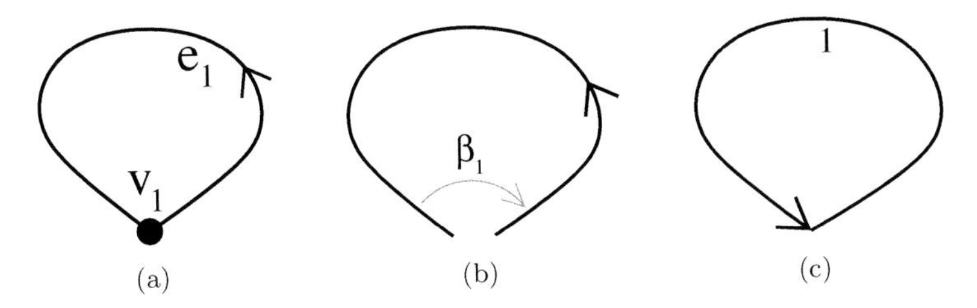

FIGURE 3.2
(a) A 1D oriented quasi-manifold without boundary which is loop (i.e. an edge having the same vertex for its two extremities).
(b) The decomposition produces one oriented edge linked with itself by β_1.
(c) Representation of the corresponding 1-map.

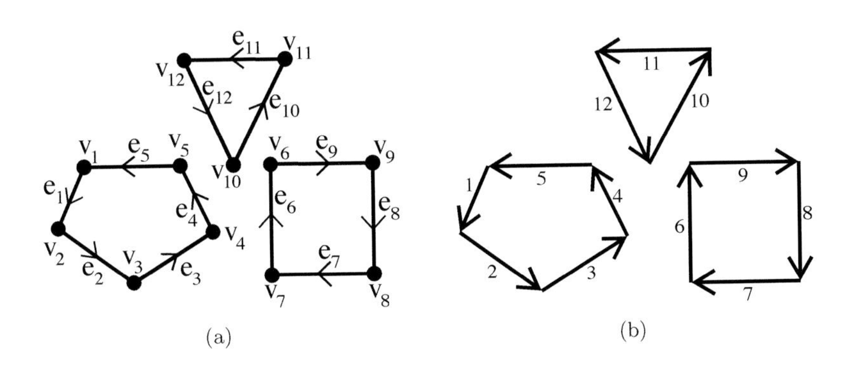

FIGURE 3.3
(a) Three disjoint 1D oriented quasi-manifolds without boundary.
(b) The corresponding 1-map contains three connected components, one for each polygonal curve.

component containing d, and it is easy, starting from d, to retrieve all the darts of its connected component, by applying β_1 as many times as possible.

An example of three disjoint 1D oriented quasi-manifolds is presented in Fig. 3.3(a), and the corresponding 1-map in Fig. 3.3(b). This 1-map contains three connected components, one for each 1D oriented quasi-manifold, which correspond to the sets of darts $\{1, 2, 3, 4, 5\}$, $\{6, 7, 8, 9\}$ and $\{10, 11, 12\}$.

3.1.1.2 2D

The same process can be applied for 2D oriented quasi-manifolds without boundary, i.e. closed oriented surfaces, as for the example given in Fig. 3.4(a). The quasi-manifold is composed by four faces, labeled from f_1 to f_4, nine edges and seven vertices. Note the presence of the unbounded face f_1 which

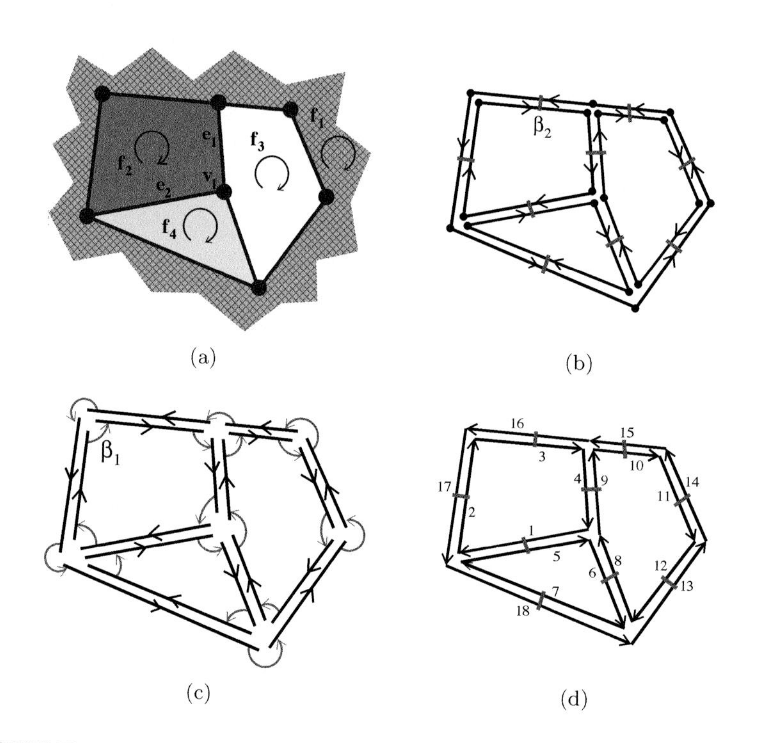

FIGURE 3.4
(a) A 2D oriented quasi-manifold without boundary.
(b) The decomposition of the surface produces a set of isolated oriented faces linked by β_2.
(c) The decomposition of the oriented polygonal curves describing the boundary of the isolated oriented faces produces a set of isolated oriented edges linked by β_1.
(d) Representation of the corresponding 2-map. Darts and β_1 are represented as above. Two darts associated by β_2 are linked by a gray line segment.

is required since the object is without boundary. First, the oriented surface is decomposed by splitting each edge in two, producing the set of isolated oriented faces drawn in Fig. 3.4(b). Since the initial quasi-manifold is coherently oriented (i.e. all faces have the same orientation), a correct orientation is retrieved for each split edge, i.e. two edges describing the same edge before the split are oriented in reverse directions. The information that two split edges described the same edge in the initial object is represented by adding a relation β_2 linking the two edges. Note that this relation is symmetric and thus β_2 in an involution, i.e. a bijection from the set of edges onto the set of edges, which is equal to its inverse.

The boundary of each isolated face is a 1D oriented quasi-manifold thus the construction presented above for the 1D case applies: each closed oriented polygonal curve is decomposed by splitting each vertex in two, producing a set of isolated oriented edges (cf. Fig. 3.4(c)) which is the set of darts of the 2-map describing the initial 2D oriented quasi-manifold. As in 1D, a β_1 relation is added in order to link an oriented edge and the next edge of the same face that shared a common vertex before the split.

Note that the two relations β_1 and β_2 are directly defined on the oriented edges (which are the darts) during the decomposition process. Thus the 2-map is the set of darts plus β_1, a permutation, and β_2, an involution (cf. Fig. 3.4(d)). Note also that each edge of the 2D quasi-manifold is described by exactly two darts in the corresponding 2-map. Darts are drawn by black arrows, sometimes numbered or labeled. β_2 relation is represented by gray segments across the two darts in relation (for example $\beta_2(1) = 5$, $\beta_2(2) = 17$), and β_1 relation is implicitly represented because two darts in relation are drawn consecutively (for example $\beta_1(1) = 2$, $\beta_1(2) = 3$).

Each dart of a 2-map describes a part of a vertex v, a part of an oriented edge e and a part of an oriented face f of the initial 2D oriented quasi-manifold, these three cells v, e and f being incident to each other. Moreover, v is the origin of the oriented edge e. For example dart 1 in Fig. 3.4(d) describes in Fig. 3.4(a) a part of vertex v_1, a part of edge e_2 and a part of face f_2 (and v_1, e_2 and f_2 are all together incident). The parts of vertex, edge and face corresponding to each dart of the 2-map of Fig. 3.5(a) are represented in Fig. 3.5(b).

The fact that each dart describes a part of a vertex, a part of an edge and a part of a face has an important consequence on the definition of cells in n-maps. Indeed, it is necessary to consider the set of all the darts which are part of a same cell in order to describe the whole cell. These sets of darts can be retrieved from a given dart d thanks to some specific orbits:

- $\langle\beta_1\rangle(d)$ is the face containing dart d;
- $\langle\beta_2\rangle(d)$ is the edge containing dart d;
- $\langle\beta_2 \circ \beta_1^{-1}\rangle(d)$ is the vertex containing dart d.

For example, in the 2-map given in Fig. 3.5(a), the vertex containing dart

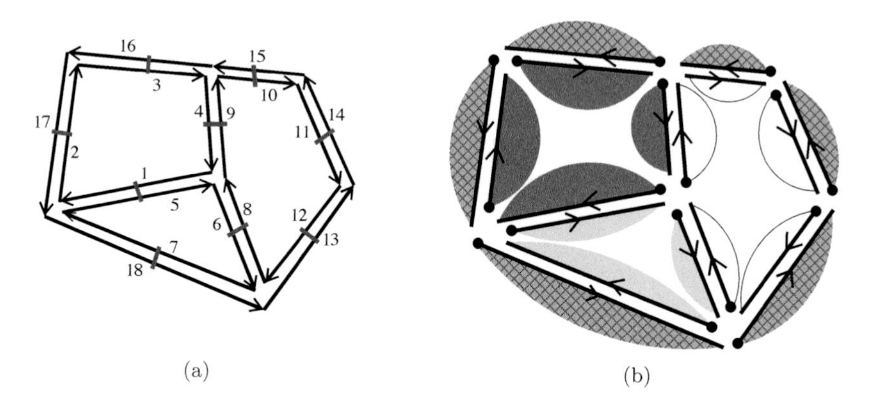

FIGURE 3.5
(a) 2-map of the example of Fig. 3.4.
(b) Representation showing for each dart, its corresponding parts of vertex, oriented edge and oriented face.

1 is $\langle\beta_2 \circ \beta_1^{-1}\rangle(1) = \{1,6,9\}$; the edge containing dart 1 is $\langle\beta_2\rangle(1) = \{1,5\}$ and the face containing dart 1 is $\langle\beta_1\rangle(1) = \{1,2,3,4\}$. More generally, each cell is composed by the set of all the darts that are part of the same cell; the structure of the cell is given by the β_i relations (see Fig. 3.6).

This decomposition process can be applied for any 2D oriented quasi-manifold without boundary. A subdivision of a plane is depicted in Fig. 3.4, a subdivided torus is depicted in Fig. 3.7.

As for the 1D case, several 2D oriented quasi-manifolds can be simultaneously represented. Each quasi-manifold is described by a distinct connected component, the whole 2-map describes all the considered objects. Two darts d_1 and d_2 belong to the same connected component if there is a path of darts two by two linked by β_1 or by β_2 between d_1 and d_2. For any dart d, $\langle\beta_1,\beta_2\rangle(d)$ denotes the connected component containing d: starting from d, the set of all the darts of its connected component is retrieved by applying β_1 and β_2 as many times as possible.

3.1.1.3 3D

Of course, a similar process can be used to decompose a 3D oriented quasi-manifold without boundary in order to obtain a 3-map. The 3D quasi-manifold depicted in Fig. 3.8(a) is composed by three volumes (a hexahedron in dark gray, a pyramid with square base in light gray plus an unbounded volume), ten faces, sixteen edges and nine vertices. First, the quasi-manifold is decomposed into isolated oriented volumes by splitting each face in two. This produces the set of isolated oriented volumes shown in Fig. 3.8(b); the information that two split faces described the same face before the split is represented by adding a

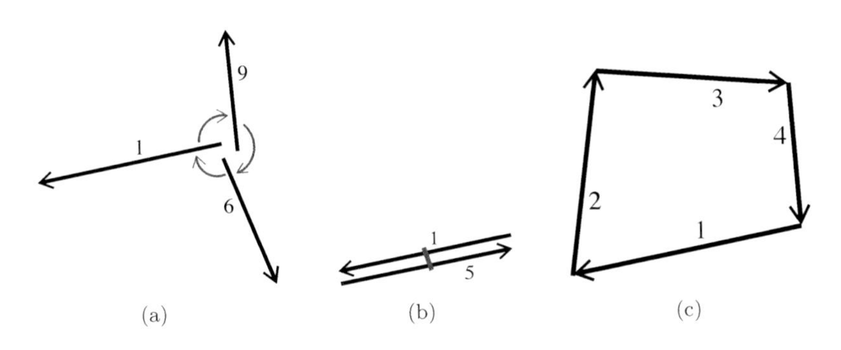

FIGURE 3.6
Examples of cells in the 2-map of Fig. 3.5(a).
(a) The 0-cell (vertex) containing dart 1.
(b) The 1-cell (edge) containing dart 1.
(c) The 2-cell (face) containing dart 1.

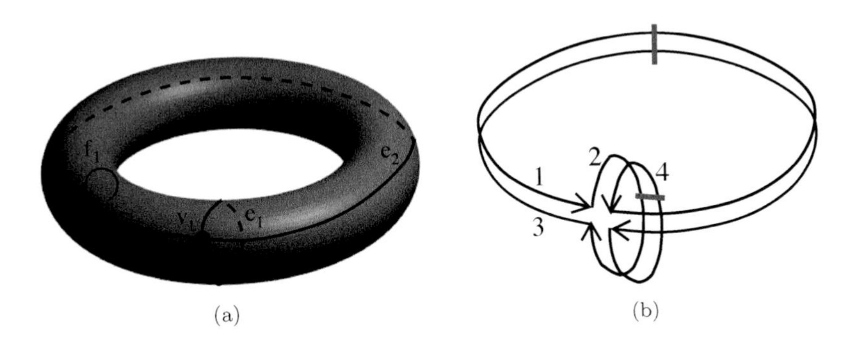

FIGURE 3.7
(a) A 2D oriented quasi-manifold without boundary, which is a subdivision of a torus surface in one face, two edges and one vertex.
(b) This 2-map containing four darts describes the previous subdivision.

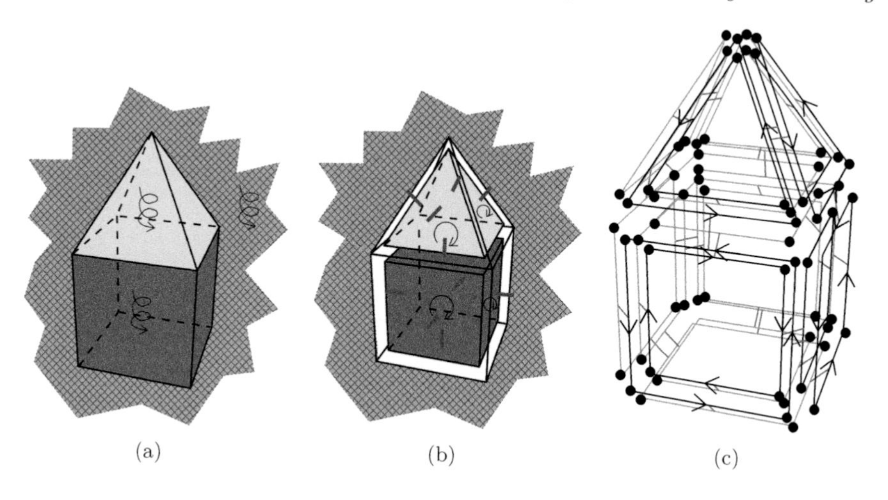

FIGURE 3.8
(a) A 3D oriented quasi-manifold without boundary.
(b) The decomposition of the 3D oriented quasi-manifold produces a set of isolated oriented volumes linked by β_3.
(c) The decomposition of the oriented surfaces describing the boundary of the isolated volumes produces a set of isolated oriented faces linked by β_2.

relation β_3 linking the two faces (which is symmetric). Each volume can be described by its boundary which is an oriented surface, i.e. a 2D oriented quasi-manifold without boundary. Thus the decomposition process presented above for 2D quasi-manifolds can be applied. Each oriented surface is decomposed into isolated faces by splitting each edge in two. This produces the set of isolated oriented faces given in Fig. 3.8(c). Each face can be described by its boundary which is an oriented polygonal curve.

Then each oriented polygonal curve is decomposed into oriented edges by splitting each vertex in two, producing the set of isolated oriented edges given in Fig. 3.9(a): this is the set of darts of the 3-map. In order to obtain the 3-map describing the original 3D oriented quasi-manifold, all β relations are transferred on the darts. This is straightforward for β_1 and β_2 since they are already defined on the oriented edges which correspond exactly to the darts. But this is not the case for β_3 which links pairs of oriented faces. Since these oriented faces are split in darts by the same decomposition process, two isomorphic sets of darts are obtained at the end. Thus the β_3 relation between two oriented faces can be directly reported on the darts that describe the two faces. Two darts are linked by β_3 if they belong to two faces linked by β_3, and if they belong to two split edges that were the same edge before the decomposition. Since the two split parts of a same face are coherently oriented (i.e. in reverse directions), and since all the darts of these two parts are linked

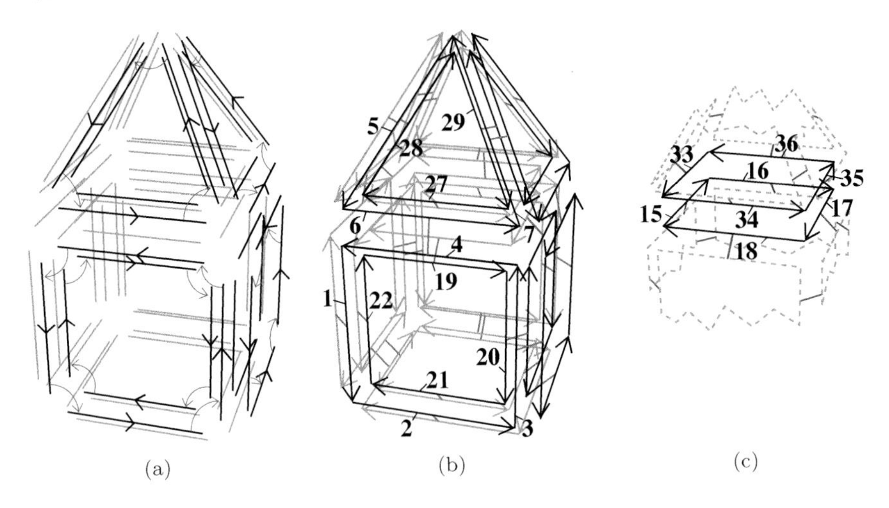

FIGURE 3.9
(a) The decomposition of the oriented polygonal curves describing the boundary of the isolated faces of Fig. 3.8(c) produces a set of isolated oriented edges linked by β_1.
(b) Corresponding 3-map.
(c) Zoom on the face separating the hexahedron and the pyramid.

two by two by β_3, we get the important property that $\beta_1 \circ \beta_3$ is an involution, i.e. $\beta_1 \circ \beta_3 = \beta_3 \circ \beta_1^{-1}$. This property can be checked in Fig. 3.9(c) which is a zoom on the face separating the hexahedron and pyramid: $\beta_3(15) = 33$, $\beta_3(16) = 36$, $\beta_3(17) = 35$ and $\beta_3(18) = 34$. Note that $\beta_1 \circ \beta_3$ is an involution; for example $\beta_1 \circ \beta_3(15) = 34$ and $\beta_1 \circ \beta_3(34) = 15$.

Figure 3.9(b) shows the representation of 3-maps, where each dart is drawn as an oriented segment; two darts linked by β_1 are drawn consecutively; two darts linked by β_2 are drawn in parallel with reverse orientations, and joined by a small gray segment; generally β_3 relations are not graphically represented, since they can be retrieved from the shapes of faces. However, some β relations are sometimes explicitly drawn in order to emphasize some specific relations.

Figure 3.10 shows the three volumes of the 3-map given in Fig. 3.9(b). Figure 3.10(a) represents the volume describing the unbounded volume surrounding the hexahedron and the pyramid.

In a 3-map, each dart corresponds to a part of a vertex v, a part of an edge e, a part of a face f and a part of a volume *vol* of the initial 3D oriented quasi-manifold, these four cells v, e, f and *vol* being incident to each other. As in the previous 2D case, the set of the darts describing a cell corresponds to a specific orbit:

- $\langle\beta_1, \beta_2\rangle(d)$ is the volume containing dart d;

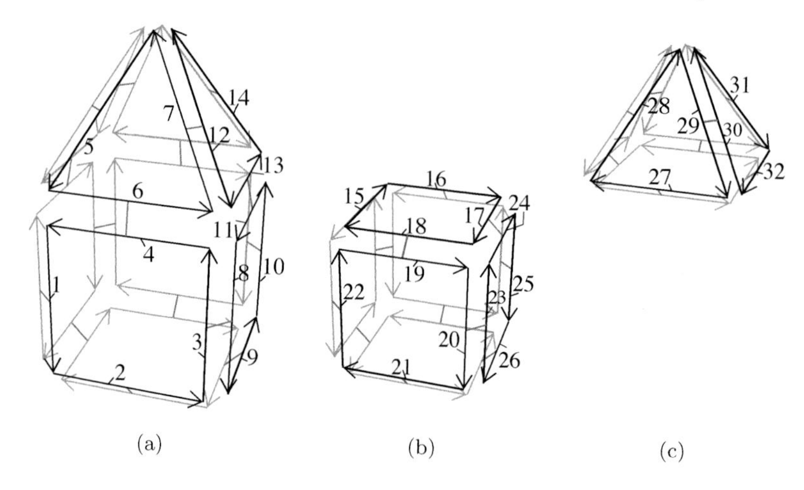

FIGURE 3.10
(a), (b) and (c): separated representations of the three volumes of the 3-map given in Fig. 3.9(b).

- $\langle\beta_1, \beta_3\rangle(d)$ is the face containing d;
- $\langle\beta_2, \beta_3\rangle(d)$ is the edge containing d;
- $\langle\beta_2 \circ \beta_1^{-1}, \beta_3 \circ \beta_1^{-1}\rangle(d)$ is the vertex containing d.

Some examples of cells of the 3-map depicted in Fig. 3.9(b) are represented in Figs. 3.10 and 3.11. Figure 3.10 gives the three volumes of the 3-map. In Fig. 3.11, darts belonging to the considered cell are drawn in bold black. (a) represents the face containing dart 18, described by the eight darts of $\langle\beta_1, \beta_3\rangle(18)$, (b) the edge containing dart 18, described by the six darts of $\langle\beta_2, \beta_3\rangle(18)$, and (c) the vertex containing dart 18, described by the ten darts of $\langle\beta_2 \circ \beta_1^{-1}, \beta_3 \circ \beta_1^{-1}\rangle(18)$.

As for the previous dimensions, several 3D oriented quasi-manifolds can be simultaneously represented. In this case, each quasi-manifold is described by a distinct connected component of the same 3-map that describes all the considered objects. Two darts d_1 and d_2 belong to the same connected component if there is a path of darts two by two linked by β_1, β_2 or by β_3 between d_1 and d_2. For any dart d, $\langle\beta_1, \beta_2, \beta_3\rangle(d)$ is the connected component containing dart d.

3.1.2 Objects with Boundary

The process of successive decompositions introduced in the previous section can be applied also in order to extract n-maps from oriented subdivisions of

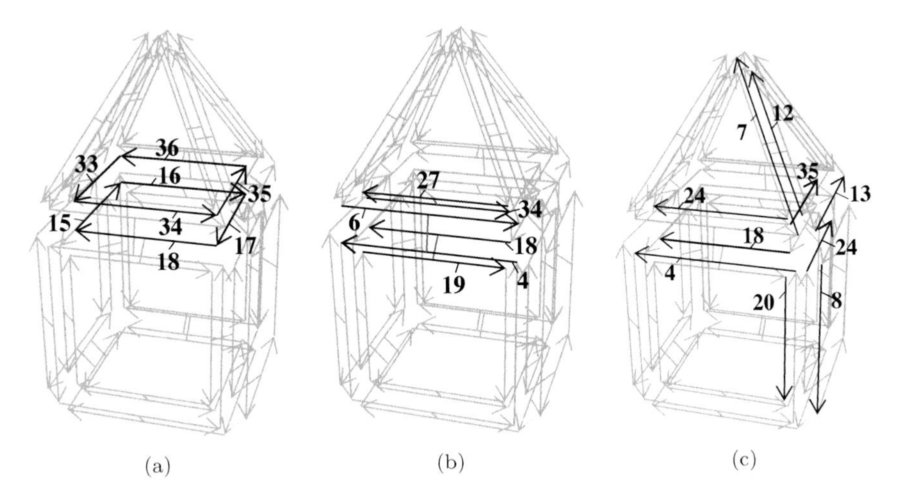

FIGURE 3.11
Examples of cells in a 3-map.
(a) Face containing dart 18.
(b) Edge containing dart 18.
(c) Vertex containing dart 18.

quasi-manifolds with boundaries. The principle is similar, i.e. cells are progressively decomposed by decreasing dimensions, the only difference being that i-cells that belong to a boundary are not split, because these cells belong to the boundary of only one $(i+1)$-cell.

3.1.2.1 1D

In 1D, an oriented quasi-manifold with boundary is an open oriented polygonal curve. In the example depicted in Fig. 3.12(a), the object is composed by four edges (labeled from e_1 to e_4) and five vertices (labeled from v_1 to v_5). The curve is decomposed into edges by splitting each vertex in two, except vertices that belong to a boundary (vertex v_1 and v_5 in this example). Indeed, such a vertex belongs to the boundary of only one edge and thus does not need to be split. As before, a dart represents an oriented edge and its origin vertex. Thus, four darts exist, corresponding to (v_1, e_1), (v_2, e_2), (v_3, e_3) and (v_4, e_4). v_5 is not explicitly represented in this case[1].

The dart describing a nonsplit vertex has no dart before it in the oriented

[1]This is generally not a problem because this case does not occur in practice. Moreover, if needed, this case can be considered specifically to retrieve the missing vertex. Indeed, a dart corresponds to an oriented edge; it is thus possible to explain all notions by associating an oriented edge and its two incident vertices with a dart. We chose to describe all notions related to n-maps by associating an oriented edge and its origin vertex with a dart in order to simplify several explanations.

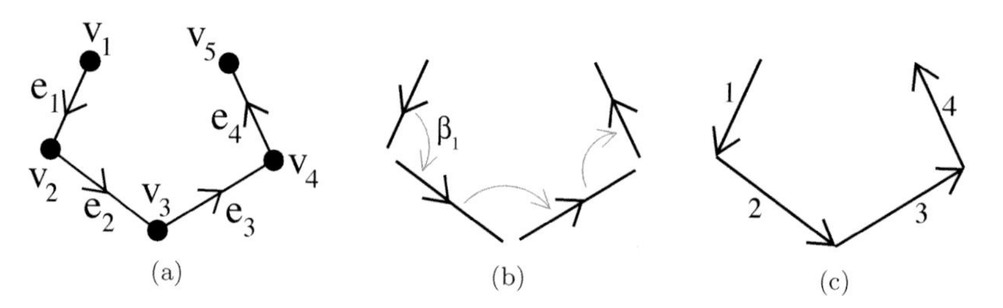

FIGURE 3.12
(a) A 1D oriented quasi-manifold with boundary.
(b) The decomposition of the polygonal curve produces a set of isolated oriented edges linked by β_1.
(c) Corresponding 1-map.

polygonal curve (dart 1 in the example). At the other extremity of the polygonal curve, there is one edge (e_4 in the example) which has only its starting vertex described. In the corresponding 1-map, the dart describing this edge has no successor for β_1 (dart 4 in the example): to encode this fact, the particular value $\varnothing$ is used, and $\beta_1(4) = \varnothing$. More generally dart d is 1-*free* iff $\beta_1(d) = \varnothing$.

As for 1D quasi-manifolds without boundary, each dart describes a vertex and an edge. Due to the presence of 1-free darts, β_1 relation is no more a permutation but a partial permutation (several darts can be linked to $\varnothing$ by β_1).

3.1.2.2 2D

In 2D, an oriented quasi-manifold with boundary (an oriented surface with boundary, see example in Fig. 3.13) is decomposed into isolated faces by splitting each edge in two, except edges that belong to a boundary (such as edge e_3 in this example). Indeed, such an edge belongs to the boundary of only one face and thus does not need to be split. By definition, the corresponding dart is free and it has $\varnothing$ as image by β_2. Then each polygonal curve, with or without boundary, is decomposed. This produces the 2-map shown in Fig. 3.13(d) that contains twelve darts. Darts $2, 3, 7, 10, 11, 12$ belong to a boundary, so they are 2-free as they have $\varnothing$ as image by β_2 (for example $\beta_2(2) = \varnothing$ and dart 2 is said 2-free).

By comparison with the case of 2D oriented quasi-manifold without boundaries, β_2 is now a *partial involution* and no more an involution. There is no modification for the definition of cells, however different types of cells can be distinguished. A vertex and a face can be either a cycle of darts (case without boundary) or a sequence of darts (case with boundary); and an edge can be composed by one dart (case with boundary) or two darts (case without boundary). There is also no difference for the definition of connected components.

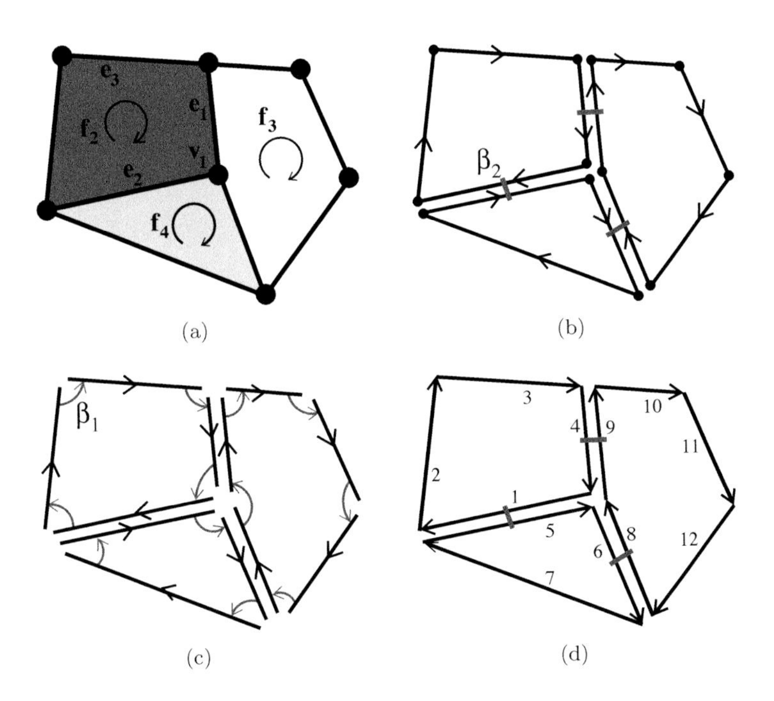

FIGURE 3.13
(a) A 2D oriented quasi-manifold with boundary.
(b) The decomposition of the surface produces a set of isolated oriented faces linked by β_2.
(c) Result of the decomposition of the oriented polygonal curves describing the boundary of the isolated oriented faces.
(d) Corresponding 2-map.

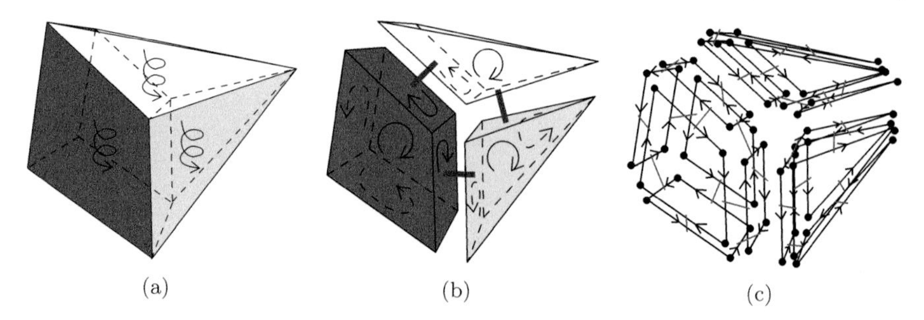

FIGURE 3.14
(a) A 3D oriented quasi-manifold with boundary.
(b) The decomposition of the 3D oriented quasi-manifold produces a set of isolated oriented volumes linked by β_3.
(c) The decomposition of the oriented surfaces describing the boundary of the isolated volumes produces a set of isolated oriented faces linked by β_2.

3.1.2.3 3D

The 3D oriented quasi-manifold with boundary given in Fig. 3.14 is composed by three volumes (a hexahedron in dark gray and two pyramids with square bases), thirteen faces, eighteen edges and nine vertices. It is decomposed into isolated volumes (see Fig. 3.14(b)) by splitting each face in two, except faces that belong to a boundary. In the example, only three faces do not belong to a boundary (two faces between the cube and the two pyramids, and one face between the two pyramids). Indeed, a face that belongs to a boundary belongs to only one volume and thus does not need to be split. For these faces, there is no other face in relation by β_3 and thus by definition their darts have $\varnothing$ as image by β_3: these darts are 3-*free*. Then each oriented surface, with or without boundary, is decomposed, producing a set of isolated oriented faces (see Fig. 3.14(c)).

Lastly each oriented curve, with or without boundary, is decomposed into a set of isolated oriented edges that correspond to darts (see Fig. 3.15(a)). The different β relations are reported on darts similarly than for the case without boundary, producing the 3-map represented in Fig. 3.15(b). Darts belonging to a boundary do not have another dart in relation by β_3. For such a dart d, $\beta_3(d) = \varnothing$ and d is 3-free (e.g. darts $1, 2, 3, 4$ in this example). Look at the face separating the hexahedron and the white pyramid (see Fig. 3.15(c)): $\beta_3(5) = 12$, $\beta_3(6) = 11$, $\beta_3(7) = 10$ and $\beta_3(8) = 9$. It can be checked that $\beta_1 \circ \beta_3$ is a partial involution (for example $\beta_1 \circ \beta_3(6) = 12$ and $\beta_1 \circ \beta_3(12) = 6$).

By comparison with the case of 3D oriented quasi-manifold without boundary, β_3 is no more an involution but a partial involution, and $\beta_1 \circ \beta_3$ is also a partial involution (as a partial permutation and a partial involution are com-

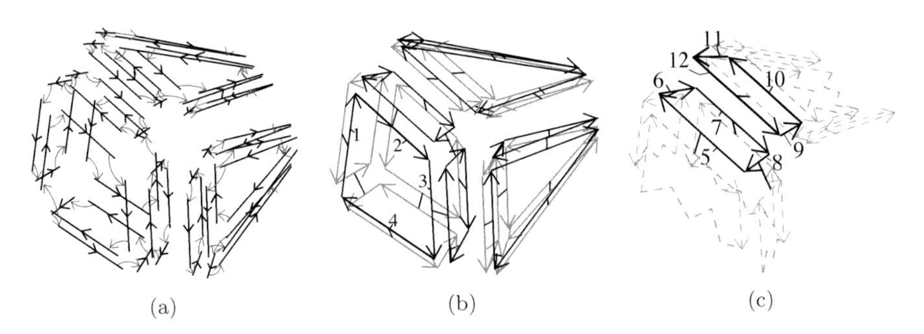

FIGURE 3.15
(a) The decomposition of the oriented polygonal curves describing the boundary of the isolated faces of Fig. 3.14(c) produces a set of isolated oriented edges linked by β_1.
(b) Corresponding 3-map.
(c) Zoom on the face separating the hexahedron and the white pyramid.

posed). Note that the 3D oriented quasi-manifold can contain one or several connected components.

There is no modification for the definitions of edges, faces and volumes, however the definition of vertices is different. Indeed, when the 3-map has no boundary, the set of darts that belong to the same vertex than d is $\langle \beta_2 \circ \beta_1^{-1}, \beta_3 \circ \beta_1^{-1} \rangle(d)$. By definition of orbits, this set of darts contains all the darts that can be obtained from d using any combination of $\beta_2 \circ \beta_1^{-1}$ and $\beta_3 \circ \beta_1^{-1}$ and their inverses. A possible combination is $\beta_2 \circ \beta_1^{-1} \circ (\beta_3 \circ \beta_1^{-1})^{-1}$ equals to $\beta_2 \circ \beta_3$. However when considering 3-maps with boundary, this equality is no more satisfied for 1-free darts. If dart d is such that $\beta_3(d)$ is 1-free, $\beta_1 \circ \beta_3^{-1}(d) = \varnothing$ and thus $\beta_2 \circ \beta_1^{-1} \circ \beta_1 \circ \beta_3^{-1}(d) = \varnothing$; and if $\beta_3(d)$ is not 2-free, $\beta_2 \circ \beta_3(d) \neq \varnothing$.

For this reason, if only the darts of the orbit $\langle \beta_2 \circ \beta_1^{-1}, \beta_3 \circ \beta_1^{-1} \rangle(d)$ are taken into account, some darts that describe the same vertex than d could be omitted. This problem is illustrated in Fig. 3.16 for a 3-map having two volumes sharing a face having only two darts $\{5, 9\}$ such that $\beta_3(5) = 9$. This face has a 1-boundary because $\beta_1(5) = \varnothing$ and $\beta_1(9) = \varnothing$. $\langle \beta_2 \circ \beta_1^{-1}, \beta_3 \circ \beta_1^{-1} \rangle(1)=\{1,2\}$ because $\beta_2 \circ \beta_1^{-1}(1) = 2$, and all other combinations give $\varnothing$ or dart 1: $\beta_2 \circ \beta_1^{-1}(2) = \varnothing$, $\beta_3 \circ \beta_1^{-1}(2) = \varnothing$, $\beta_3 \circ \beta_1^{-1}(1) = \varnothing$, $\beta_1 \circ \beta_2(1) = \varnothing$, $\beta_1 \circ \beta_3(1) = \varnothing$, $\beta_1 \circ \beta_2(2) = 1$, $\beta_1 \circ \beta_3(2) = \varnothing$: all the darts that describe the same vertex than dart 1 are not reached.

To solve this problem, $\beta_2 \circ \beta_3$ is added to the considered orbit. Thus, the set of darts describing the same vertex than d is given by the orbit $\langle \beta_2 \circ \beta_1^{-1}, \beta_3 \circ \beta_1^{-1}, \beta_2 \circ \beta_3 \rangle(d)$. In the example given in Fig. 3.16, adding $\beta_2 \circ \beta_3$ to the generators of the orbit solve the problem: $\langle \beta_2 \circ \beta_1^{-1}, \beta_3 \circ \beta_1^{-1}, \beta_2 \circ \beta_3 \rangle(1) = \{1, 2, 3, 4, 5\}$. Note that, as for the other cells, the structure of a

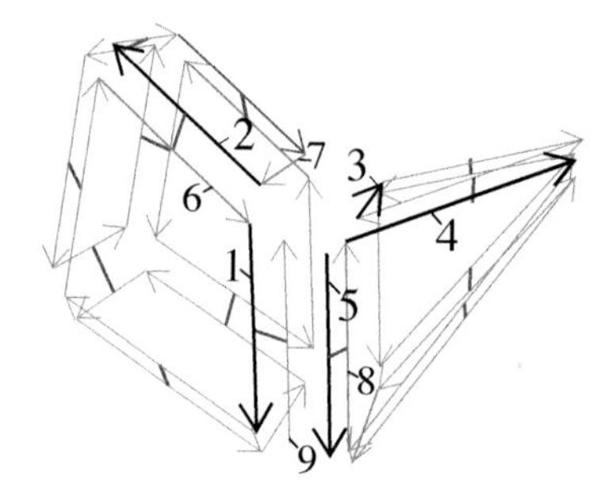

FIGURE 3.16
The definition of vertices for 3-maps without boundary cannot be directly applied for 3-maps with 1-boundaries.

vertex is given by the relations between the darts of the corresponding orbit. However, contrary to cells of dimension different to 0, these relations are here the composition of two permutations, and thus some darts are used, that do not belong to the vertex.

Several cells for the 3-map depicted in Fig. 3.15(b) are represented in Fig. 3.17. Darts belonging to the considered cell are drawn in bold black. (a) represents the volume containing dart 13, described by the twenty four darts of $\langle\beta_1, \beta_2\rangle(13)$, (b) the face containing dart 13, described by the eight darts of $\langle\beta_1, \beta_3\rangle(13)$, (c) the edge containing dart 13, described by the six darts of $\langle\beta_2, \beta_3\rangle(13)$, and (d) the vertex containing dart 13, described by the nine darts of $\langle\beta_1 \circ \beta_2, \beta_1 \circ \beta_3, \beta_2 \circ \beta_3\rangle(13)$.

3.1.3 Generalization in Any Dimension

The process of successive decompositions can be generalized in any dimension. Starting from an nD oriented quasi-manifold with or without boundary, all the cells are successively decomposed by decreasing dimensions, from n to 2. At the end of this process, a set of isolated oriented edges is obtained, which correspond to the darts of the n-map, plus n relations on these darts. β_1 is a partial permutation while $\beta_2, \ldots, \beta_n$ are partial involutions.

Each dart d corresponds to an $n+1$ tuple of cells $(c_0, \ldots, c_n)$ and all these cells are incident to each other. A dart describes a part of a vertex, a part of an oriented edge, a part of an oriented face ... a part of an oriented n-cell. Each i-cell containing dart d is thus a set of darts that can be retrieved thanks to some specific orbits.

For objects with boundaries, a particular value $\varnothing$ is used and $\beta_i(d) = \varnothing$ encodes the fact that dart d has no neighbor dart for β_i.

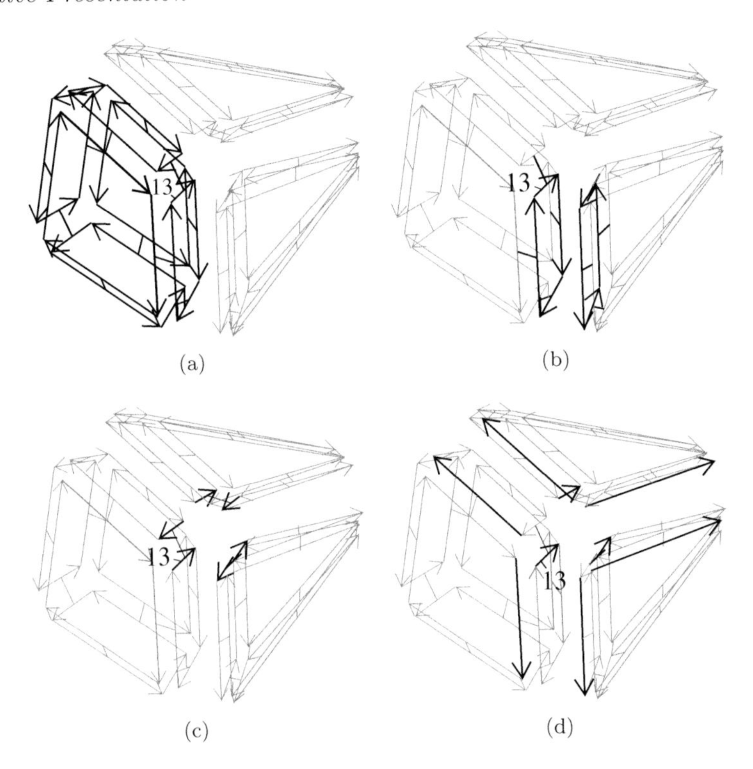

FIGURE 3.17
Examples of cells in a 3-map.
(a) Volume containing dart 13.
(b) Face containing dart 13.
(c) Edge containing dart 13.
(d) Vertex containing dart 13.

3.2 n-Gmaps

For n-Gmaps, the process of successive decompositions is similar to the one explained in the previous section for n-maps. There are only two differences:

1. the original object is not oriented (it can be orientable or not);

2. one more decomposition is applied: instead of ending the process with edges, the edges are decomposed into isolated vertices.

Due to these similarities, n-Gmaps are explained using the same examples presented in the previous section, and all the decomposition steps are not explained again, but only the last one.

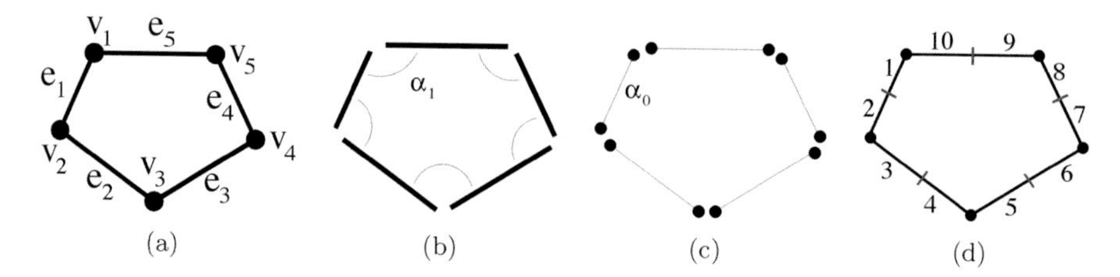

FIGURE 3.18
(a) A 1D quasi-manifold without boundary.
(b) The decomposition of the polygonal curve produces a set of isolated edges linked by α_1.
(c) The decomposition of the pair of vertices describing the boundary of the isolated edges produces a set of isolated vertices linked by α_0.
(d) Representation of the corresponding 1-Gmap. Darts are represented by black segments, each one being incident to a black disc symbolizing α_1, and to a small gray segment symbolizing α_0.

3.2.1 Objects without Boundary

First let us start with quasi-manifolds without boundary.

3.2.1.1 1D

Look at the 1D quasi-manifold without boundary represented in Fig. 3.18(a). The curve is decomposed into isolated edges by splitting each vertex in two: Fig. 3.18(b). A relation α_1 links two edges that shared a common vertex before the split. Note that since the quasi-manifold is without boundary, it is sure that all vertices are paired. Then the boundary of each edge is decomposed into vertices, producing the set of isolated vertices, called darts, drawn in Fig. 3.18(c): a relation α_0 links the two darts that belonged to the same edge before the split. The set of darts plus the two relations α_0 and α_1 is the 1-Gmap describing the initial 1D quasi-manifold (see Fig. 3.18(d)). α_0 and α_1 being symmetric, they are involutions.

Darts are drawn by segments, sometimes numbered or labeled. Two darts linked by α_1 are drawn consecutively, joined by a small disk (for example $\alpha_1(1) = 10$ and $\alpha_1(2) = 3$), and two darts linked by α_0 are drawn consecutively, separated by a small gray segment (for example $\alpha_0(1) = 2$ and $\alpha_0(10) = 9$).

Each vertex and each edge of the 1D quasi-manifold is described by exactly two darts in the corresponding 1-Gmap. In the example of Fig. 3.18, vertex v_1 is described by darts 1 and 10, and edge e_1 is described by darts 1 and 2. For this reason, each dart corresponds exactly to a pair of a vertex and an edge (dart 1 corresponds to the pair (v_1, e_1)).

The initial 1D quasi-manifold can have multi-incident cells, as the exam-

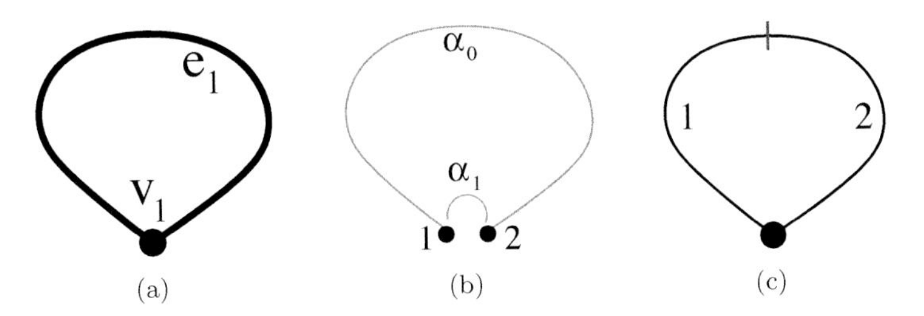

FIGURE 3.19
(a) A 1D quasi-manifold without boundary which is a loop.
(b) The decomposition produces two vertices linked together both by α_0 and α_1.
(c) Representation of the corresponding 1-Gmap.

ple shown in Fig. 3.19(a) where vertex v_1 is incident twice to edge e_1. The corresponding 1-Gmap shown in Fig. 3.19(c) contains only two darts which are linked together both by α_0 and α_1.

A 1-Gmap can represent several 1D quasi-manifolds and thus can contain several connected components: see for instance three disjoint 1D quasi-manifolds (Fig. 3.20(a)) and the corresponding 1-Gmap (Fig. 3.20(b)) which contains three connected components corresponding to sets of darts $\{1, \ldots, 10\}$, $\{11, \ldots, 18\}$ and $\{19, \ldots, 24\}$. Two darts d_1 and d_2 belong to the same connected component if there is a path of darts between d_1 and d_2, i.e. a sequence of darts linked by α_0 and α_1. For any dart d, $\langle \alpha_0, \alpha_1 \rangle(d)$ is the connected component containing d, and it is easy, starting from d, to retrieve all the darts of its connected component, by applying α_0 and α_1 as many times as possible.

3.2.1.2 2D

The same process can be applied for 2D quasi-manifolds without boundary. The quasi-manifold given in Fig. 3.21(a) is composed of four faces, labeled from f_1 to f_4, nine edges and seven vertices. The decomposition process applied to the example given in Fig. 3.4 page 56 for 2-maps produces here the set of isolated edges given in Fig. 3.21(b) plus two relations: α_2 links two split edges and α_1 links two split vertices. The only difference with 2-maps is the fact that the quasi-manifold is not oriented and thus no element is oriented, and each relation is symmetric. Thus α_1 and α_2 are involutions.

Then one more decomposition is added: the boundary of each edge is decomposed into vertices, producing a set of isolated vertices plus the relation α_0 that links two vertices that belonged to the same edge before the decomposition (cf. Fig. 3.21(c)). These isolated vertices are the darts of the 2-Gmap describing the initial 2D quasi-manifold.

In order to obtain the 2-Gmap describing the original 2D quasi-manifold, all the α relations are propagated onto darts. This is straightforward for α_0

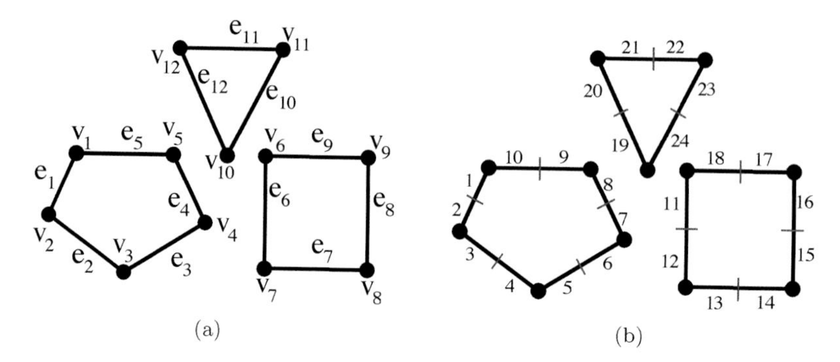

FIGURE 3.20
(a) Three disjoint 1D quasi-manifolds without boundary.
(b) The corresponding 1-Gmap which contains three connected components: one for each polygonal curve.

and α_1 since they are already defined on the isolated vertices which correspond exactly to the darts. But this is not the case for α_2 which links edges. Since two linked edges are both described by two darts, the α_2 relation between the two edges can be directly propagated on the corresponding four darts. As these four darts are linked two by two by α_2, $\alpha_0 \circ \alpha_2$ is an involution.

The corresponding 2-Gmap is represented in Fig. 3.21(d). α_2 is represented by gray segments across the extremities of the two darts in relation (for example $\alpha_2(1) = 19$ and $\alpha_2(2) = 20$). Note again that the composition of α_0 and α_2 is an involution, i.e. $\alpha_0 \circ \alpha_2 = (\alpha_0 \circ \alpha_2)^{-1} = \alpha_2 \circ \alpha_0$: for example, $\alpha_2 \circ \alpha_0(1) = \alpha_2(2) = 20$, and $\alpha_0 \circ \alpha_2(1) = \alpha_0(19) = 20$.

Each dart of a 2-Gmap describes a part of a vertex v, a part of an edge e and a part of a face f of the initial 2D quasi-manifold, these three cells v, e and f being incident to each other. For example dart 2 in Fig. 3.21(d) describes in Fig. 3.21(a) a part of vertex v_1, a part of edge e_2 and a part of face f_2. Fig. 3.22(b) represents the parts of vertex, edge and face of each dart of the 2-Gmap of Fig. 3.22(a).

The fact that each dart describes a part of a vertex, a part of an edge and a part of a face has an important consequence for the definition of cells in n-Gmaps. Indeed, the set of all the darts which are part of a same cell has to be taken into account in order to describe the whole cell. Such sets of darts can be retrieved from a given dart d thanks to some specific orbits:

- $\langle \alpha_0, \alpha_1 \rangle(d)$ is the face containing dart d;
- $\langle \alpha_0, \alpha_2 \rangle(d)$ is the edge containing dart d;
- $\langle \alpha_1, \alpha_2 \rangle(d)$ is the vertex containing dart d.

For example, in the 2-Gmap given in Fig. 3.22(a), the vertex containing dart 2 is $\langle \alpha_1, \alpha_2 \rangle(2) = \{2, 3, 10, 11, 20, 21\}$; the edge containing dart 2 is

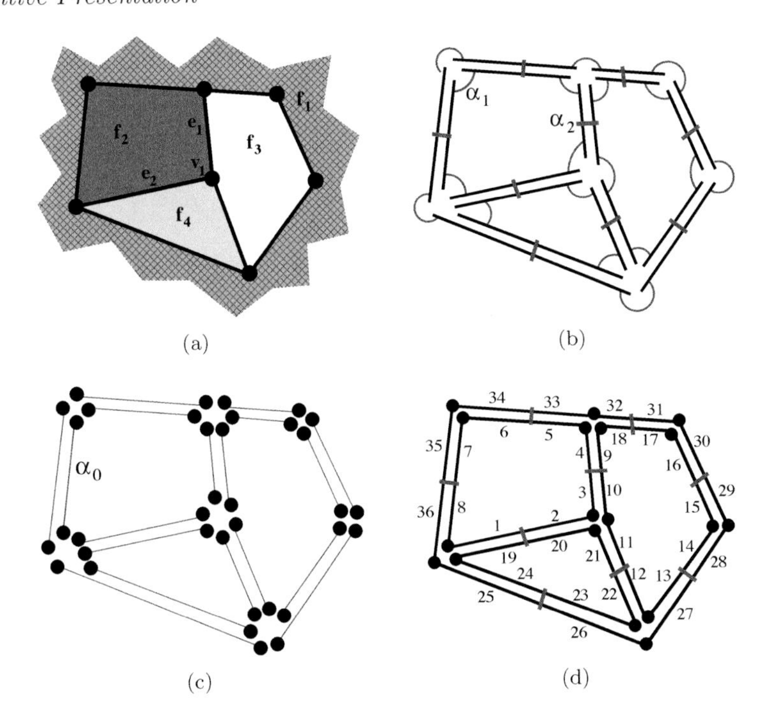

FIGURE 3.21
(a) A 2D quasi-manifold without boundary.
(b) The decomposition of the surface produces isolated faces, and the decomposition of the polygonal curves describing the boundary of the isolated faces produces isolated edges linked by relations α_1 and α_2.
(c) The decomposition of the pair of vertices describing the boundary of the isolated edges produces a set of isolated vertices linked by α_0.
(d) Representation of the corresponding 2-Gmap.

$\langle\alpha_0, \alpha_2\rangle(2) = \{1, 2, 19, 20\}$ and the face containing dart 2 is $\langle\alpha_0, \alpha_1\rangle(2) = \{1, 2, 3, 4, 5, 6, 7, 8\}$. More generally, each cell is composed by the set of all the darts that are part of the same cell; the structure of the cell is given by the α_i relations (see Fig. 3.23).

This decomposition process is valid for any 2D quasi-manifold without boundary: cf. Fig. 3.24 for an example with a torus.

A 2-Gmap can have several connected components. Two darts d_1 and d_2 belong to the same connected component if there is a path of darts two by two linked by α_0, α_1 or by α_2 between d_1 and d_2. For any dart d, $\langle\alpha_0, \alpha_1, \alpha_2\rangle(d)$ denotes the connected component containing d: starting from d, the set of all the darts of its connected component can be retrieved by applying α_0, α_1 and α_2.

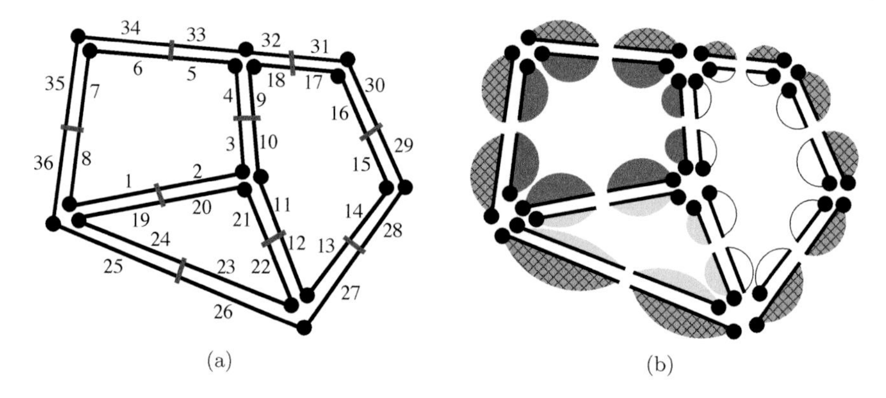

FIGURE 3.22
(a) 2-Gmap of the example of Fig. 3.21.
(b) Representation showing for each dart, its corresponding parts of vertex, edge and face.

3.2.1.3 3D

Of course, a similar process can be applied to a 3D quasi-manifold without boundary in order to obtain a 3-Gmap. The decomposition process applied to the example given in Figs. 3.8 and 3.9 page 60 for 3-maps produces the set of isolated edges given in Fig. 3.25(b) plus three relations: α_3 links two split faces, α_2 links two split edges and α_1 links two split vertices. The only difference with the same process for 3-maps is the fact that each relation is symmetric since the 3D quasi-manifold is not oriented and thus the three relations α_1, α_2 and α_3 are involutions.

Then, the boundary of each edge is decomposed into vertices, producing a set of isolated vertices plus the relation α_0 that links two vertices that belonged to the same edge before the decomposition (cf. Fig. 3.25(c)). These isolated vertices are the darts of the 3-Gmap.

In order to obtain the 3-Gmap describing the original 3D oriented quasi-manifold, all α relations are propagated onto the darts. For α_0, α_1 and α_2, this is done as above for the 2D case. The α_3 relation links split faces into pairs: since these faces are split into darts by the same decomposition process, each split face corresponds to an orbit for α_0 and α_1, and two corresponding split faces are isomorphic. Thus the α_3 relation between two split faces can be directly propagated on the corresponding darts. Two darts are linked by α_3 if they belong to two corresponding split faces, and if they belong to two split vertices (resp. edges) that were the same vertex (resp. edge) before the decomposition. Since all the darts of the two split parts of a same face are linked two by two by α_3, $\alpha_0 \circ \alpha_3$ and $\alpha_1 \circ \alpha_3$ are two involutions.

The corresponding 3-Gmap is given in Fig. 3.26(a). α_3 is generally not represented in figures since it can be retrieved from the shapes of faces. Look

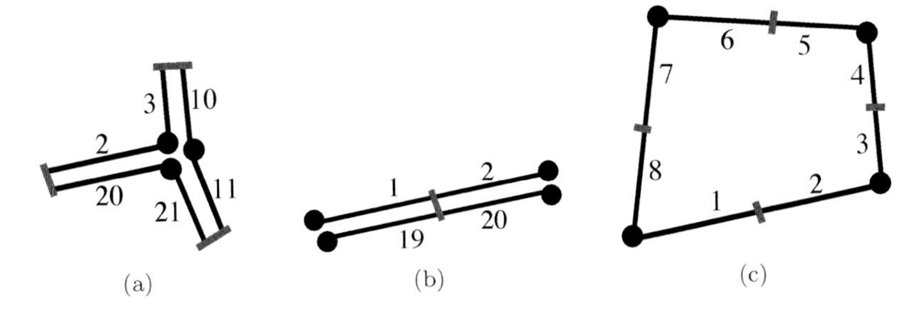

FIGURE 3.23
Examples of cells for the 2-Gmap of Fig. 3.22(a).
(a) The 0-cell (vertex) containing dart 2.
(b) The 1-cell (edge) containing dart 2.
(c) The 2-cell (face) containing dart 2.

at the face separating the hexahedron and pyramid given in Fig. 3.26(b): $\alpha_3(1) = 9$, $\alpha_3(2) = 10$, $\ldots$, $\alpha_3(8) = 16$ (and conversely). Thus, it can be checked that $\alpha_0 \circ \alpha_3$ and $\alpha_1 \circ \alpha_3$ are involutions: for example $\alpha_0 \circ \alpha_3(1) = \alpha_0(9) = 10$ and $\alpha_0 \circ \alpha_3(10) = \alpha_0(2) = 1$, and $\alpha_1 \circ \alpha_3(1) = \alpha_1(9) = 16$ and $\alpha_1 \circ \alpha_3(16) = \alpha_1(8) = 1$.

Figure 3.27 shows the three volumes of the 3-Gmap given in Fig. 3.26(a). Figure 3.27(a) is the volume describing the unbounded volume surrounding the hexahedron and the pyramid.

In a 3-Gmap, each dart corresponds to a part of a vertex v, a part of an edge e, a part of a face f and a part of a volume vol of the initial 3D quasi-manifold, these four cells v, e, f and vol being incident to each other. As in the previous 2D case, each cell corresponds to a specific orbit:

- $\langle \alpha_0, \alpha_1, \alpha_2 \rangle(d)$ is the volume containing dart d;
- $\langle \alpha_0, \alpha_1, \alpha_3 \rangle(d)$ is the face containing d;
- $\langle \alpha_0, \alpha_2, \alpha_3 \rangle(d)$ is the edge containing d;
- $\langle \alpha_1, \alpha_2, \alpha_3 \rangle(d)$ is the vertex containing d.

Several cells of the 3-Gmap given in Fig. 3.26(a) are represented in Figs. 3.27 and 3.28. Figure 3.27 gives the three volumes of the 3-Gmap. In Fig. 3.28, darts belonging to the considered cell are drawn in bold black. (a) represents the face containing dart 1, described by the sixteen darts of $\langle \alpha_0, \alpha_1, \alpha_3 \rangle(1)$, (b) the edge containing dart 1, described by the twelve darts of $\langle \alpha_0, \alpha_2, \alpha_3 \rangle(1)$, and (c) the vertex containing dart 1, described by the twenty darts of $\langle \alpha_1, \alpha_2, \alpha_3 \rangle(1)$.

As for the previous dimensions, several 3D quasi-manifolds can be described by one 3-Gmap containing several connected components. Two darts d_1 and d_2 belong to the same connected component if there is a path of darts

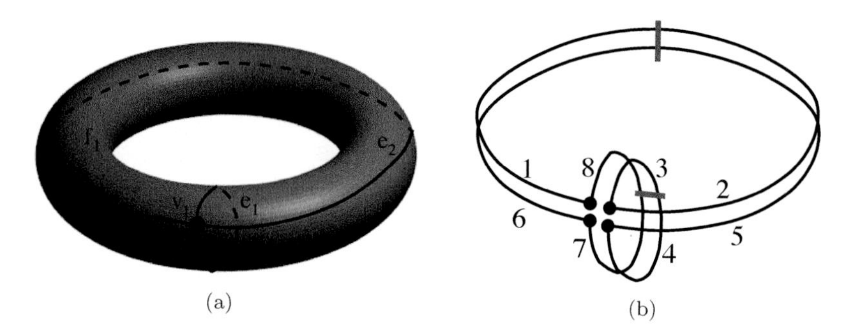

(a) (b)

FIGURE 3.24
(a) A 2D quasi-manifold without boundary which is a subdivision of a torus surface into one face, two edges and one vertex.
(b) The 2-Gmap containing eight darts which describes the previous subdivision.

two by two linked by α_0, α_1, α_2 or by α_3 between d_1 and d_2. For any dart d, $\langle\alpha_0, \alpha_1, \alpha_2, \alpha_3\rangle(d)$ is the connected component containing dart d.

3.2.2 Objects with Boundary

The principle is similar for subdivisions of quasi-manifolds with boundaries. Cells are progressively decomposed by decreasing dimensions, except i-cells that belong to a boundary, to obtain n-Gmaps.

3.2.2.1 1D

The 1D quasi-manifold with boundary given in Fig. 3.29 is decomposed into edges by splitting each vertex in two, except vertices that belong to a boundary (v_1 and v_5 in this example). Then each edge is decomposed into vertices producing the darts. Only one dart corresponds to a nonsplit vertex: to encode this fact, $\alpha_1(d) = d$ (in the example, $\alpha_1(1) = 1$ and $\alpha_1(8) = 8$): dart d is 1-*free*. As for 1D quasi-manifolds without boundary, each dart describes a part of a vertex and of an edge. Moreover, thanks to the convention used to describe free darts, α_0 and α_1 are still involutions.

3.2.2.2 2D

The 2D quasi-manifold with boundary given in Fig. 3.30 is decomposed, using the same decomposition process than for the example given in Fig. 3.13 page 65 for 2-maps, producing the set of isolated edges given in Fig. 3.30(b) plus two relations: α_2 links two split edges and α_1 links two split vertices. Edges that belong to a boundary are not split, and for such an edge α_2 is equal to the identity. Then each edge with or without boundary is decomposed, producing

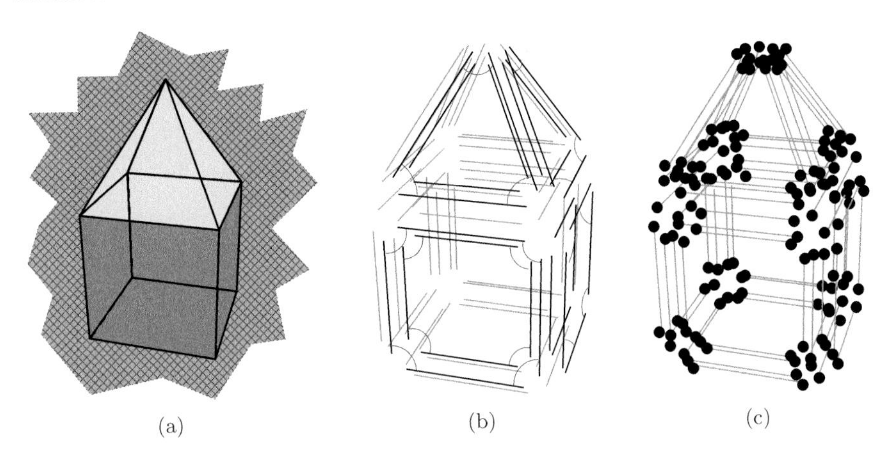

(a) (b) (c)

FIGURE 3.25
(a) A 3D quasi-manifold without boundary.
(b) The decomposition of the 3D quasi-manifold, then the decomposition of the surfaces describing the boundary of the isolated volumes and then the decomposition of the polygonal curves describing the boundary of the isolated faces produces a set of isolated edges linked by relations α_1, α_2 and α_3.
(c) The decomposition of the pair of vertices describing the boundary of the isolated edges produces a set of isolated vertices linked by α_0.

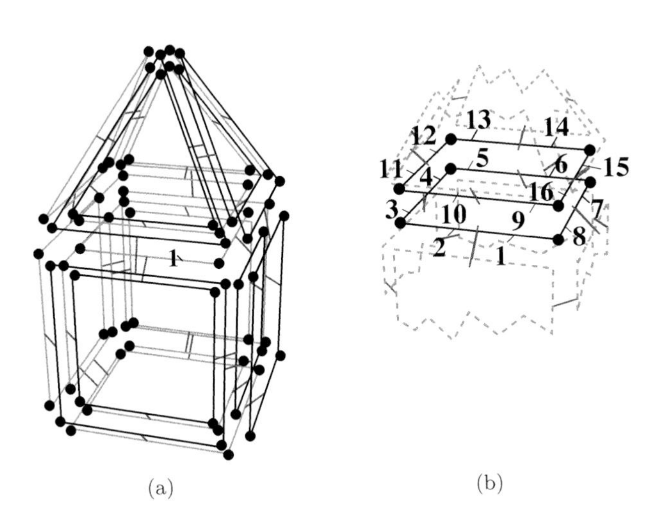

(a) (b)

FIGURE 3.26
(a) 3-Gmap describing the 3D quasi-manifold without boundary given in Fig. 3.25(a).
(b) Zoom on the face separating the hexahedron and pyramid.

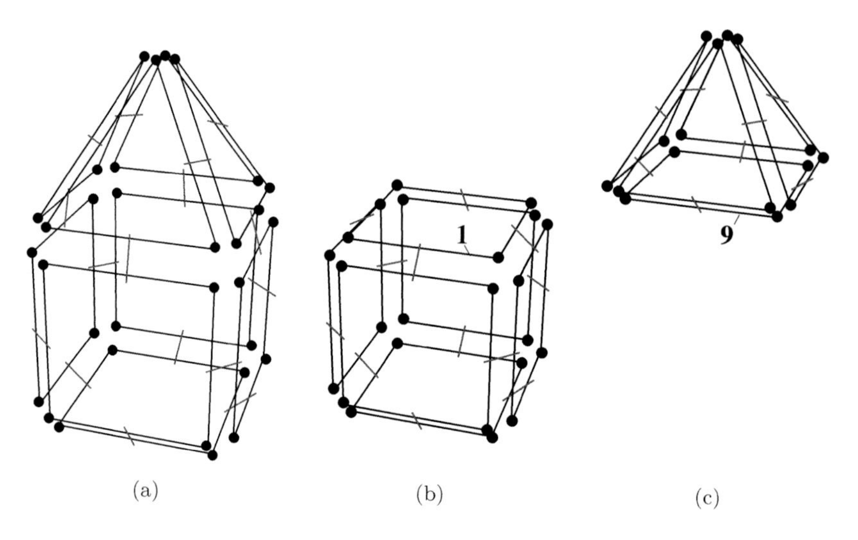

FIGURE 3.27
(a), (b) and (c): separated representations of the three volumes belonging to the 3-Gmap given in Fig. 3.26(a).

the darts (see Fig. 3.30(c)). The 2-Gmap given in Fig. 3.30(d) contains twenty-four darts. The twelve darts corresponding to the boundary of the surface are $5, 6, 7, 8, 13, 14, 15, 16, 17, 18, 23, 24$: for example $\alpha_2(5) = 5$ and dart 5 is said 2-*free*.

As for the case of 2D quasi-manifold without boundaries, α_0, α_1 and α_2 are all involutions. Cells are defined as before, however different types of 1-cells can be distinguished: an edge can be composed by one, two or four darts. Note also that a 2-Gmap can still describe several connected components of 2D quasi-manifolds with or without boundaries.

3.2.2.3 3D

The decomposition process applied to the 3D quasi-manifold with boundary given in Fig. 3.31 (using the same decomposition process than for the example given in Figs. 3.14 and 3.15 page 66 for 3-maps) produces the set of isolated edges given in Fig. 3.31(b) plus three relations: α_3 links two split faces, α_2 links two split edges and α_1 links two split vertices. Each relation is symmetric and thus α_1, α_2 and α_3 are involutions. In the example, only three faces do not belong to a boundary (two faces between the cube and the two pyramids, and one face between the two pyramids). All other faces are not split and for them, α_3 is equal to identity. Then each edge with or without boundary is decomposed, producing the darts (see Fig. 3.31(c)).

The final 3-Gmap describing the 3D quasi-manifold with boundary given in

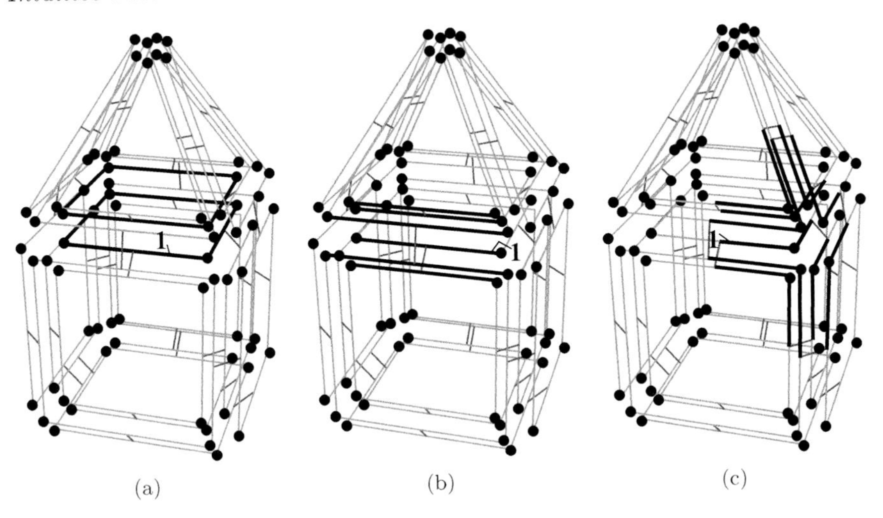

FIGURE 3.28
Examples of cells in a 3-Gmap.
(a) Face containing dart 1.
(b) Edge containing dart 1.
(c) Vertex containing dart 1.

Fig. 3.31(a) is represented in Fig. 3.32. Darts corresponding to the boundary of the 3D quasi-manifold are 3-free, i.e. they are fixed points for α_3. Look at the face separating the hexahedron and pyramid, given in Fig. 3.32(b): $\alpha_3(1) = 9, \alpha_3(2) = 10, \ldots, \alpha_3(8) = 16$ (and conversely). $\alpha_0 \circ \alpha_3$ and $\alpha_1 \circ \alpha_3$ are involutions on these darts, and this is also the case for the darts corresponding to the boundary.

A 3-Gmap composed by several connected components can represent several 3D quasi-manifolds with or without boundaries and the definition of cells is unchanged.

Several cells of the 3-Gmap given in Fig. 3.32(a) are represented in Fig. 3.33. Darts belonging to the considered cell are drawn in bold black. (a) represents the volume containing dart 17, described by the forty eight darts of $\langle\alpha_0, \alpha_1, \alpha_2\rangle(17)$, (b) the face containing dart 17, described by the sixteen darts of $\langle\alpha_0, \alpha_1, \alpha_3\rangle(17)$, (c) the edge containing dart 17, described by the twelve darts of $\langle\alpha_0, \alpha_2, \alpha_3\rangle(17)$, and (d) the vertex containing dart 17, described by the eighteen darts of $\langle\alpha_0, \alpha_1, \alpha_2\rangle(17)$.

3.2.3 Generalization in Any Dimension

As for n-maps, the process of successive decompositions can be generalized in any dimension. Starting from an nD quasi-manifold with or without boundary,

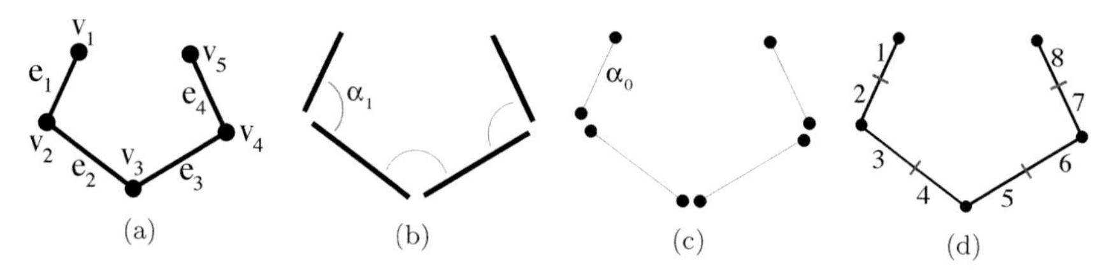

FIGURE 3.29
(a) A 1D quasi-manifold with boundary.
(b) The decomposition of the polygonal curve produces a set of isolated edges linked by α_1.
(c) The result of the decomposition of the pair of vertices describing the boundary of the isolated edges.
(d) Representation of the corresponding 1-Gmap.

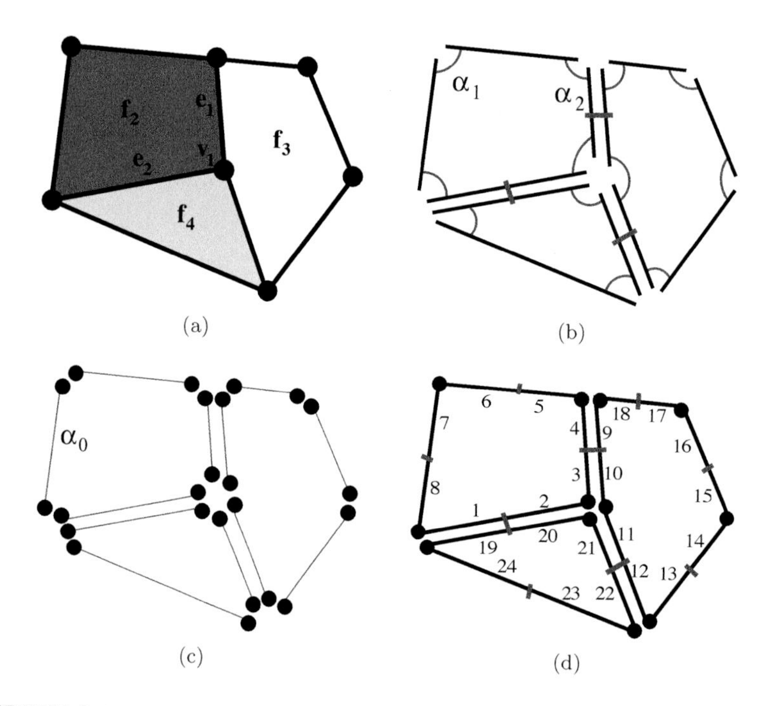

FIGURE 3.30
(a) A 2D quasi-manifold with boundary.
(b) The decomposition of the surface in isolated faces, then the decomposition of the polygonal curves describing the boundary of the isolated faces produces a set of isolated edges linked by relations α_1 and α_2.
(c) Result of the decomposition of the pair of vertices describing the boundary of the isolated edges.
(d) Representation of the corresponding 2-Gmap.

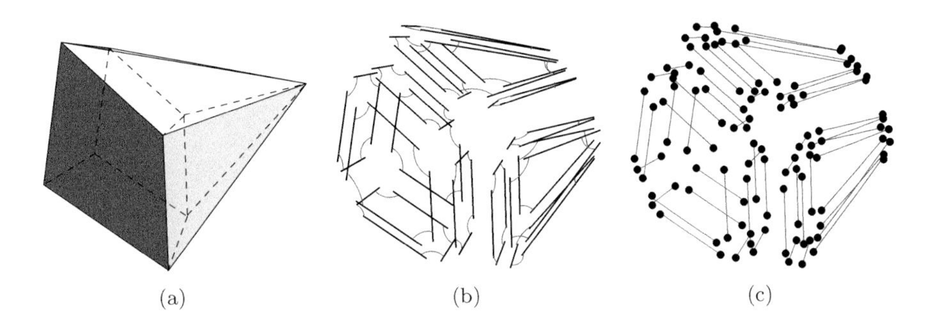

FIGURE 3.31
(a) Example of 3D quasi-manifold with boundary.
(b) Result of the decomposition of the 3D quasi-manifold into isolated volumes, the decomposition of the surfaces describing the boundary of the isolated volumes and the decomposition of the polygonal curves describing the boundary of the isolated faces.
(c) Result of the decomposition of the pair of vertices describing the boundary of the isolated edges.

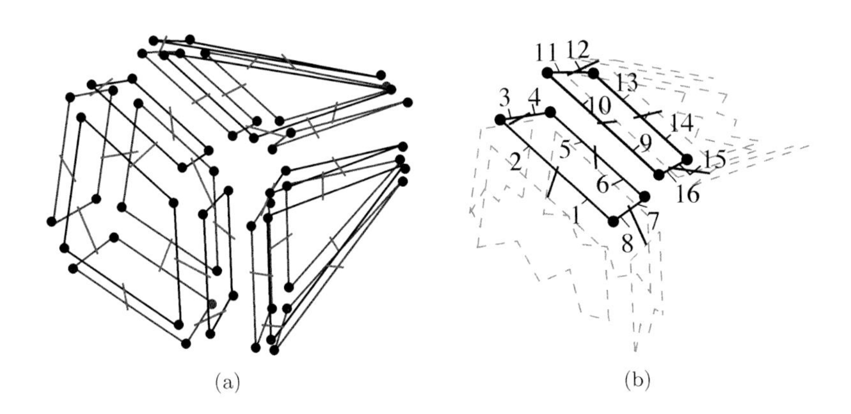

FIGURE 3.32
(a) 3-Gmap describing the 3D quasi-manifold with boundary given in Fig. 3.31(a).
(b) Zoom on the face separating the hexahedron and the white pyramid.

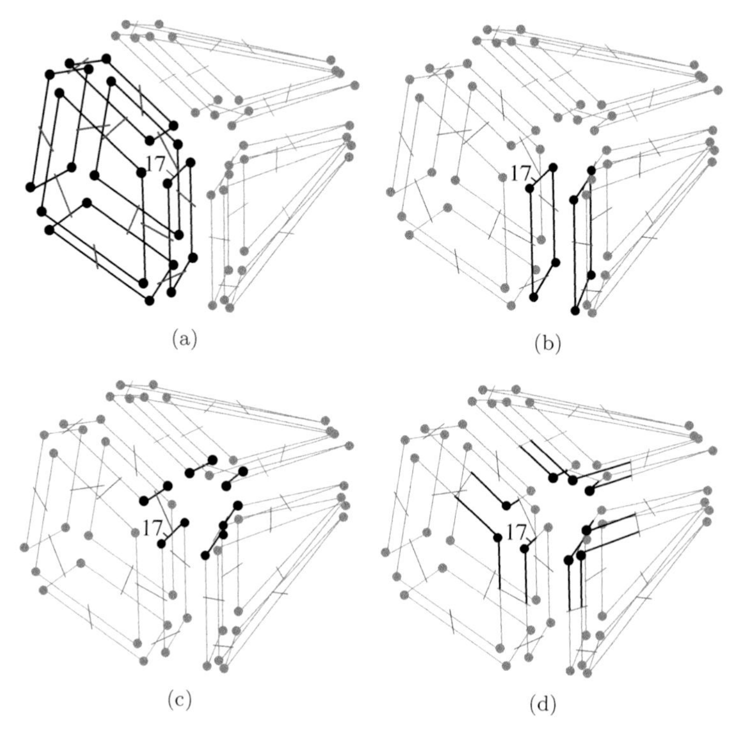

FIGURE 3.33
Examples of cells in a 3-Gmap.
(a) Volume containing dart 17.
(b) Face containing dart 17.
(c) Edge containing dart 17.
(d) Vertex containing dart 17.

all the cells are successively decomposed by decreasing dimensions, from n to 1. At the end of this process, a set of isolated vertices is obtained, which correspond to the darts of the n-Gmap, plus $n+1$ involutions on these darts $\alpha_0, \ldots, \alpha_n$.

Each dart d corresponds to an $n+1$ tuple of cells $(c_0, \ldots, c_n)$ and all these cells are incident to each other. A dart describes a part of a vertex, a part of an edge, a part of a face ... a part of an n-cell. Each i-cell containing dart d is thus a set of darts that can be retrieved thanks to some specific orbits. As before, the notion of incidence and of adjacency between cells can be transposed directly on the corresponding set of darts.

For objects with boundaries, $\alpha_i(d) = d$ encodes the fact that dart d has no neighbor dart for α_i. This is different for n-maps, for which the convention is: $\beta_i(d) = \varnothing$. Indeed, for n-maps, a dart d such that $\beta_1(d) = d$ is a loop (cf. chapter 8 for explanations based on simplicial interpretation).

4

n-Gmaps

An n-Gmap is a combinatorial data structure allowing to describe an n-dimensional orientable or nonorientable quasi-manifold with or without boundary. An intuitive presentation of n-maps and n-Gmaps is provided in the previous chapter.

In this chapter, n-Gmaps are defined in Section 4.1, as the basic notions of cells, incidence and adjacency relations between the cells. Some basic operations allowing to modify existing n-Gmaps are presented in Section 4.2. These operations allow to add/remove darts, increase/decrease the dimension of an n-Gmap, merge/split n-Gmaps; the sew/unsew operations allow to identify cells. It is shown in Section 4.3 that these operations make a small basis of operations allowing to build any n-Gmap. Moreover, multi-incidence between cells can be taken into account with n-Gmaps, and some specific configurations related to multi-incidence are illustrated, such as dangling cells and folded cells. In Section 4.4, a possible data structure for implementing n-Gmaps is presented, and also some algorithms allowing to develop a computer software handling n-Gmaps. Section 4.5 contains some additional notions related to n-Gmaps.

In this book, n-Gmaps are studied before n-maps, because their definition is simpler: it is homogeneous, i.e. all functions involved in their definition are involutions. This simplifies many definitions and algorithms.

4.1 Basic Definitions

As explained in the previous chapter, an n-Gmap is defined by a set of darts on which act $n+1$ involutions, satisfying some composition constraints. This leads to the following definition of n-Gmaps.

Definition 1 (n-Gmap) *An n-dimensional generalized map, or n-Gmap, with $0 \leq n$, is an $(n+2)$-tuple $G = (D, \alpha_0, \ldots, \alpha_n)$ where:*

1. *D is a finite set of darts;*
2. *$\forall i \in \{0, \ldots, n\}$: α_i is an involution on D;*
3. *$\forall i \in \{0, \ldots, n-2\}$, $\forall j \in \{i+2, \ldots, n\}$: $\alpha_i \circ \alpha_j$ is an involution.*

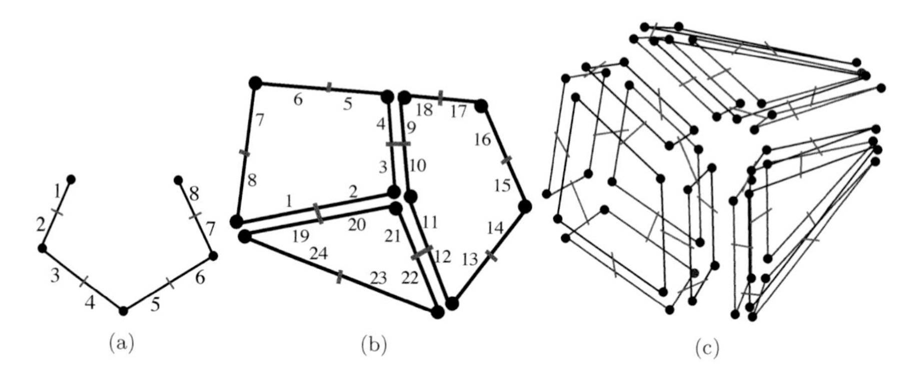

FIGURE 4.1
Examples of 1-Gmap, 2-Gmap and 3-Gmap.
(a) A 1-Gmap (D, α_0, α_1).
(b) A 2-Gmap $(D, \alpha_0, \alpha_1, \alpha_2)$.
(c) A 3-Gmap $(D, \alpha_0, \alpha_1, \alpha_2, \alpha_3)$.

A 0-Gmap (D, α_0) represents the structure (the topology) of a set of isolated vertices, or pairs of vertices (corresponding thus to 0-spheres); a 1-Gmap (D, α_0, α_1) represents the structure of a set of polygonal curves with or without boundaries (cf. Fig. 4.1(a)); a 2-Gmap $(D, \alpha_0, \alpha_1, \alpha_2)$ represents the structure of a set of surfaces (cf. Fig. 4.1(b)); a 3-Gmap $(D, \alpha_0, \alpha_1, \alpha_2, \alpha_3)$ represents the structure of assemblies of volumes (cf. Fig. 4.1(c))...

As seen before, darts can be linked with themselves for some involutions when describing objects with boundaries. Dart d is *i-free* if it is linked with itself by α_i, and it is *i-sewn* otherwise (cf. Def. 2 and example in Fig. 4.2).

Definition 2 (i-free, i-sewn) *Let $G = (D, \alpha_0, \dots, \alpha_n)$ be an n-Gmap, $d \in D$, and $i \in \{0, \dots, n\}$:*

- *d is i-free if $\alpha_i(d) = d$;*
- *d is i-sewn with dart $d_2 \in D$ if $\alpha_i(d) = d_2$ and $d \neq d_2$.*

When a dart is free for an involution, it is a *fixed point* for this involution.

For practical applications, it may be useful to handle objects with boundaries. Usually, a 3-dimensional subdivided object embedded into the usual 3-dimensional Euclidean space has 2-dimensional boundaries, which are surfaces without boundaries: it is composed by volumes, which boundaries are also surfaces without boundaries. So, these objects can be represented by a 3-Gmap, such that all darts are 0-sewn, 1-sewn, 2-sewn, and some darts can be 3-free. More generally, all involutions are often without fixed points, except α_n.

However, in all generality, boundaries can exist in all dimensions: for instance, assume an n-dimensional object is cut by another object, for instance

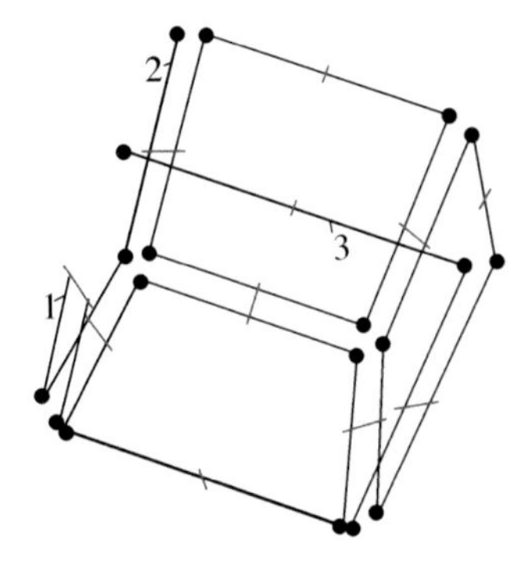

FIGURE 4.2
A 2-Gmap with 0-boundary, 1-boundary and 2-boundary.
For example, dart 1 is 0-free, i.e. $\alpha_0(1) = 1$. Thus dart 1 belongs to an edge incident to only one vertex.
Dart 2 is 1-free, i.e. $\alpha_1(2) = 2$, so it belongs to an edge, such that no other edge is incident to the same vertex and to the same face.
Dart 3 is 2-free, i.e. $\alpha_2(3) = 3$, thus dart 3 belongs to an edge which is incident to only one face.

by an $(n-1)$-dimensional hyperplane: this hyperplane can cut cells of any dimensions, creating boundaries of any dimensions. An i-boundary (or a boundary in dimension i) corresponds to the fact that darts can be i-free (for any i).

As we have seen in the intuitive presentation, an n-Gmap can have different connected components. The set of darts that belong to the same connected component than a given dart can be retrieved, thanks to Def. 3: a connected component corresponds to an orbit for all involutions.

Definition 3 (connected component) *Let $G = (D, \alpha_0, \ldots, \alpha_n)$ be an n-Gmap and $d \in D$. $\langle \alpha_0, \ldots, \alpha_n \rangle(d)$ is the connected component containing d.*

As said before, according to the context, an orbit denotes either a set of darts or this set of darts with the corresponding involutions. Thanks to this definition, we can directly test if a given n-Gmap is connected, i.e. made of only one connected component, simply by testing if the connected component of a dart is the set of all the darts of the n-Gmap (cf. Def. 4).

Definition 4 (connected n-Gmap) *Let $G = (D, \alpha_0, \ldots, \alpha_n)$ be an n-Gmap. G is connected iff $\langle \alpha_0, \ldots, \alpha_n \rangle(d)$ is equal to D, with $d \in D$.*

Remember that a graph can be associated with a discrete structure such as an n-Gmap: a node of the graph corresponds to a dart, and an edge numbered i links two nodes corresponding to two darts linked by α_i (when a dart is a fixed point for some involution, the corresponding edge in the graph is a loop). So, many notions related to graph theory can be directly applied to n-Gmaps. For instance, as for graph theory, the notion of connexity is related to the notion of path given in Def. 5.

Definition 5 (path) *Let $G = (D, \alpha_0, \ldots, \alpha_n)$ be an n-Gmap and $d, d' \in D$. A path between d and d' is a sequence of darts $(d_1, \ldots, d_k)$ such that $d_1 = d$, $d_k = d'$ and $\forall i \in \{2, \ldots, k\}$, $d_i = \alpha_{k_i}(d_{i-1})$, with $k_i \in \{0, \ldots, n\}$.*

Given any pair of darts d and d' that belong to the same connected component, there exists a path between d and d'. Reciprocally, if d and d' do not belong to the same connected component, there is no path between d and d'.

As seen in the previous chapter, a dart corresponds to an $(n+1)$-tuple of incident cells of different dimensions (vertex, edge, face, volume ...). In other words, a dart describes a "small part" of each cell, this explains why cells are not the basic elements of n-Gmaps, contrary to incidence graphs for instance. A consequence is the fact that very local notions and operations can be defined: for instance, if an edge is a loop, we can distinguish its two extremities, even if it is incident to only one vertex (cf. example in Fig. 3.19 page 71). More generally, n-Gmaps naturally take multi-incidence into account, and this can be useful for representing free-form objects.

As seen in the intuitive presentation, a cell containing a given dart corresponds to a specific orbit (cf. Def. 6).

Definition 6 (i-cell) *Let $G = (D, \alpha_0, \ldots, \alpha_n)$ be an n-Gmap, $d \in D$, and $i \in \{0, \ldots, n\}$. The i-dimensional cell (or i-cell) containing d is*

$$c_i(d) = \langle \alpha_0, \ldots, \alpha_{i-1}, \alpha_{i+1}, \ldots, \alpha_n \rangle (d).$$

The cells can equivalently be defined as connected components of the $(n-1)$-Gmap of cells defined in Def. 7.

Definition 7 ($(n-1)$-Gmap of i-cells) *Let $G = (D, \alpha_0, \ldots, \alpha_n)$ be an n-Gmap, $n > 0$, and $i \in \{0, \ldots, n\}$. The $(n-1)$-Gmap of i-cells is $G_{c_i} = (D, \alpha_0, \ldots, \alpha_{i-1}, \alpha_{i+1}, \ldots, \alpha_n)$. When $n = 0$, D is the set of 0-cells.*

Thanks to this definition of cells, incidence and adjacency relations between cells can be easily expressed (cf. Def. 8).

Definition 8 (incidence and adjacency) *Let $G = (D, \alpha_0, \ldots, \alpha_n)$ be an n-Gmap, $d, d' \in D$, and $i, j \in \{0, \ldots, n\}$, $i \neq j$.*

- *cells $c_i(d)$ and $c_j(d')$ are incident iff $c_i(d) \cap c_j(d') \neq \emptyset$;*
- *cells $c_i(d)$ and $c_i(d')$ are adjacent iff $\exists d_1 \in c_i(d)$ and $\exists d_2 \in c_i(d')$, $d_1 \neq d_2$, such that $d_1 = \alpha_i(d_2)$.*

Note that, according to this definition, a cell is never incident to itself since i must be different from j, but it can be adjacent to itself: for instance a loop is adjacent to itself, and more generally cells can be multi-incident to other cells: cf. Section 4.3.

It is possible to test if two n-Gmaps describe the topology of the same subdivision, i.e. the same set of cells and the same structure of incidence relations. This corresponds to the *isomorphism* notion. Two n-Gmaps are isomorphic if there is a one-to-one mapping between their darts which preserves all involutions (cf. Def. 9).

Definition 9 (isomorphism) *Two n-Gmaps $G = (D, \alpha_0, \ldots, \alpha_n)$ and $G' = (D', \alpha'_0, \ldots, \alpha'_n)$ are isomorphic if there exists an* isomorphism *mapping D onto D', i.e. a one-to-one mapping $f : D \to D'$, such that $\forall d \in D$, $\forall i \in \{0, \ldots, n\}$: $f(\alpha_i(d)) = \alpha'_i(f(d))$.*

It is also possible to test if an n-Gmap G' is included in a second n-Gmap G, or in other words if there exists a part of G being isomorphic to G'. An n-Gmap is subisomorphic to a second one if an injective mapping exists between the darts of the first n-Gmap onto the darts of the second n-Gmap that preserves all the involutions inside the concerned part of the second n-Gmap (cf. Def. 10 and example in Fig. 4.3).

Definition 10 (subisomorphism) *An n-Gmap $G' = (D', \alpha'_0, \ldots, \alpha'_n)$ is* isomorphic to a submap *of $G = (D, \alpha_0, \ldots, \alpha_n)$ if there exists a* subisomorphism *between D' and D, i.e. an injective mapping $f : D' \to D$, such that $\forall d \in D'$, $\forall i \in \{0, \ldots, n\}$:*

- *if d is not i-free, then $f(\alpha'_i(d)) = \alpha_i(f(d))$;*
- *otherwise, either $f(d)$ is i-free, or $\forall d_k \in D', f(d_k) \neq \alpha_i(f(d)))$.*

4.2 Basic Operations

In the previous section, several basic definitions concerning n-Gmaps have been introduced. Now some basic operations are defined, allowing to modify existing n-Gmaps. Note that many high level operations have been defined, sometimes based on these basic ones, allowing to directly construct and modify n-Gmaps. Even if the basic operations introduced here can be directly used to interactively construct an object, their interest is mainly theoretical by providing a small basis of operations allowing to build any n-Gmap. The high level operations usually handle orbits, as cells for instance, although basic operations handle directly the darts and the involutions. All these operations take as input an n-Gmap and produce another n-Gmap which is the result of the operation.

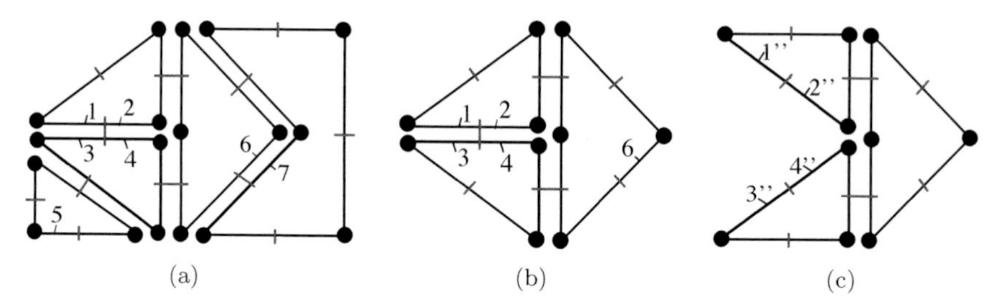

FIGURE 4.3
Submap isomorphism example.
(a) A 2-Gmap G.
(b) 2-Gmap G' which is a submap of G as it is obtained from G by deleting the six darts in $c_2(5)$ and the ten darts in $c_2(7)$.
(c) 2-Gmap G'' which is not isomorphic to a submap of G as the injection f that respectively matches darts $1''$ to $4''$ to darts 1 to 4 does not verify the conditions of Def. 10: dart $1''$ is 2-free and it is matched with dart 1 which is 2-sewn with dart 3 which is itself matched with 3" (i.e. $f^{-1}(3) = 3''$).

4.2.1 Basic Tools

The first basic tool consists in adding a new dart in an n-Gmap. To guaranty the validity of the n-Gmap, the new dart is isolated, i.e. it is free for all dimensions ($\forall i \in \{0, \ldots, n\}$, $\alpha_i(d) = d$). This operation can be used several times to add any number of isolated darts in the n-Gmap.

Definition 11 (add isolated dart) *Let $G = (D, \alpha_0, \ldots, \alpha_n)$ be an n-Gmap. The n-Gmap obtained from G by adding an isolated dart d, $d \notin D$, is $G_{+d} = (D \cup \{d\}, \alpha'_0, \ldots, \alpha'_n)$, where $\forall i \in \{0, \ldots, n\}$, α'_i is defined by:*

- $\forall e \in D$, $\alpha'_i(e) = \alpha_i(e)$;
- $\alpha'_i(d) = d$.

The inverse operation given in Def. 12 consists in removing an isolated dart.

Definition 12 (remove isolated dart) *Let $G = (D, \alpha_0, \ldots, \alpha_n)$ be an n-Gmap and $d \in D$ be an isolated dart, i.e. $\forall i \in \{0, \ldots, n\}$, $\alpha_i(d) = d$. The n-Gmap obtained from G by removing d is $G_{-d} = (D \setminus \{d\}, \alpha'_0, \ldots, \alpha'_n)$, where $\forall i \in \{0, \ldots, n\}$, α'_i is the restriction of α_i to the set of remaining darts, i.e. $\alpha'_i = \alpha_{i|D\setminus\{d\}}$.*

Since only isolated darts are removed, the result of the operation is an n-Gmap. Note that any dart can be removed, but it is necessary to isolate it by applying (maybe several times) the unsew operation (cf. Section 4.2.3).

The operation which increases the dimension of an n-Gmap is straightforward, and its definition is given in Def. 13. Increasing by one the dimension of an $(n-1)$-Gmap G leads to create one n-cell for each connected component of G. All i-cells, for $0 \leq i \leq n-1$, are not modified by this operation.

Definition 13 (increase dimension) *Let $G = (D, \alpha_0, \ldots, \alpha_{n-1})$ be an $(n-1)$-Gmap. The n-Gmap obtained from G by increasing its dimension is $G^+ = (D, \alpha_0, \ldots, \alpha_{n-1}, \alpha_n)$, where α_n is the identity on D, i.e. $\forall d \in D$, $\alpha_n(d) = d$.*

The operation which decreases the dimension of an n-Gmap G is defined in Def. 14. This operation is the inverse of the increase dimension operation. It leads to the disappearance of all the n-cells that were in G.

Definition 14 (decrease dimension) *Let $G = (D, \alpha_0, \ldots, \alpha_n)$ be an n-Gmap such that $\forall d \in D$, $\alpha_n(d) = d$. The n-Gmap obtained from G by decreasing its dimension is $G^- = (D, \alpha_0, \ldots, \alpha_{n-1})$.*

Note that this operation is restricted to an n-Gmap which has all its darts n-free, in order to get the inverse of the operation which increases the dimension. However, it is possible to first n-unsew all the non n-free darts, and then to decrease the dimension of the resulting n-Gmap.

Two distinct n-Gmaps can be merged into a same n-Gmap, which connected components are thus the connected components of the initial n-Gmaps (cf. Def. 15). The number of cells (resp. of connected components) of the resulting n-Gmap is the sum of the numbers of cells (resp. of connected components) of the two initial n-Gmaps.

Definition 15 (merge) *Let $G = (D, \alpha_0, \ldots, \alpha_n)$ and $G' = (D', \alpha'_0, \ldots, \alpha'_n)$ be two n-Gmaps such that $D \cap D' = \emptyset$. The n-Gmap obtained by merging G and G' is $G \cup G' = (D'', \alpha''_0, \ldots, \alpha''_n)$, defined by:*

- $D'' = D \cup D'$;
- $\forall d \in D''$, $\forall i \in \{0, \ldots, n\}$, $\alpha''_i(d) = \begin{cases} \alpha_i(d) & \text{if } d \in D; \\ \alpha'_i(d) & \text{if } d \in D'. \end{cases}$

The reverse operation consists in splitting a given n-Gmap in two. It is based on the restrict operation defined in Def. 16: it takes as input a given n-Gmap and a subset of darts which correspond to a set of connected components, and build the n-Gmap restricted to this set of darts.

Definition 16 (restrict) *Let $G = (D, \alpha_0, \ldots, \alpha_n)$ be an n-Gmap, and $D' \subseteq D$, such that $\forall d \in D'$, $\forall i \in \{0, \ldots, n\}$, $\alpha_i(d) \in D'$. The n-Gmap obtained by restricting G to D' is $G_{|D'} = (D', \alpha'_0, \ldots, \alpha'_n)$, where $\forall i \in \{0, \ldots, n\}$, $\alpha'_i = \alpha_{i|D'}$.*

Note that the restrict operation is only possible if the given set of darts D' corresponds to a set of connected components. Indeed, otherwise, we would obtain some darts linked with some elements that do not belong to D', contradicting the involution definition and thus the definition of n-Gmaps. However, we can use the restrict operation for any set of darts simply by using first the required unsew operations to isolate this set of darts.

The split operation, given in Def. 17 takes as input an n-Gmap G and a subset of darts D', and produces two n-Gmaps by using twice the restrict operation: a first time for D' and the second time for $D \setminus D'$.

Definition 17 (split) *Let $G = (D, \alpha_0, \ldots, \alpha_n)$ be an n-Gmap, and $D' \subseteq D$ such that $\forall d \in D'$, $\forall i \in \{0, \ldots, n\}$, $\alpha_i(d) \in D'$. The two n-Gmaps obtained by splitting G along D' are $G_{s1}(D') = G_{|D'}$ and $G_{s2}(D') = G_{|D \setminus D'}$.*

It is straightforward to prove that each operation is the inverse of its corresponding operation. For that, we have to show that starting from G, using an operation then the corresponding inverse operation, we obtain an n-Gmap G' which is isomorphic to G: for each operation, it is easy to choose the parameters of the operations which lead to this conclusion.

4.2.2 Sew Operations

Intuitively, the i-sew operation allows to "link" some i-free darts by α_i, for a given i. In order to satisfy the "composition constraint" of the n-Gmap definition (i.e. $\alpha_i \circ \alpha_j$ is an involution for $|i - j| \geq 2$), we often need to put in relation not only two darts but several darts: for instance, if we link two darts by α_2 in a 2-Gmap, it is necessary to link also the darts which are associated by α_0. For the i-sew operation, we choose to put in relation the minimal number of darts such that the result of the operation is an n-Gmap.

In fact, the i-sew operation implies some identifications of cells: for instance, when two darts are sewn by α_2 in a 2-Gmap, the corresponding edges are identified into a single edge: this identification can involve the identification of the vertices incident to the initial edges.

For this reason, it is not always possible to apply the i-sew operation: the cells to identify must have the same structure. More precisely, the two cells must be isomorphic to each other. Indeed, when two cells have different structures, we do not know how to do their identification. For example it is not possible to sew a cube and a tetrahedron since we do not know how to identify a square face with a triangle face. So, before defining the sew operation, we need to be able to test if a sew is possible. The property of "being i-sewable" is detailed in Def. 18.

Definition 18 (i-sewable) *Let $G = (D, \alpha_0, \ldots, \alpha_n)$ be an n-Gmap, $i \in \{0, \ldots, n\}$, $d, d' \in D$. Let o_d denotes $\langle \alpha_0, \ldots, \alpha_{i-2}, \alpha_{i+2}, \ldots, \alpha_n \rangle(d)$, and $o_{d'}$ denotes $\langle \alpha_0, \ldots, \alpha_{i-2}, \alpha_{i+2}, \ldots, \alpha_n \rangle(d')$. d is i-sewable with d' if $d \neq d'$, $\alpha_i(d) =$*

d, $\alpha_i(d') = d'$ and if a isomorphism[1] *f exists between o_d and $o_{d'}$, such that $f(d) = d'$. Moreover, if $o_d = o_{d'}$ then $f = f^{-1}$.*

Considering the whole orbits $\langle \alpha_0, \ldots, \alpha_{i-2}, \alpha_{i+2}, \ldots, \alpha_n \rangle(d)$ and $\langle \alpha_0, \ldots, \alpha_{i-2}, \alpha_{i+2}, \ldots, \alpha_n \rangle(d')$ is required in order to satisfy after the sew the condition that $\alpha_i \circ \alpha_j$ is an involution when $|i - j| \geq 2$: in fact, the i-sew operation will "link" together all the darts of o_d with the darts of $o_{d'}$. d and d' must be i-free, so all the darts in o_d and $o_{d'}$ are also i-free since $\alpha_i \circ \alpha_j$ are involutions when $|i - j| \geq 2$. d and d' must be different (otherwise nothing will be changed by the i-sew). At last, note that it is possible that $o_d = o_{d'}$: the corresponding sew will create *folded* cells (see Section 4.3).

Note that any pair of free darts is always sewable in any 0-Gmap and in any 1-Gmap. Indeed in these cases, orbits o_d and $o_{d'}$ contain only one dart, and the set $\{0, \ldots, i-2, i+2, \ldots, n\}$ is empty. This means that we can sew without precondition any pair of free darts in any 0-Gmap and in any 1-Gmap.

The notion of being i-sewable is used to define the i-sew operation in Def. 19. After applying the i-sew operation, all the darts in o_d and $o_{d'}$ are linked by α_i. This implies the identification of the two $(i-1)$-cells containing darts d and d'. Moreover, several other cells can be identified i.e. the j-cells incident to the two identified $(i-1)$-cells, with $j \in \{0, \ldots, i-2, i+1, \ldots, n\}$ (cf. below for some examples).

Definition 19 (i-sew) *Let $G = (D, \alpha_0, \ldots, \alpha_n)$ be an n-Gmap, $i \in \{0, \ldots, n\}$, $d, d' \in D$ such that d and d' are i-sewable by the isomorphism f of Def. 18. The n-Gmap obtained from G by the i-sewing of d and d' is $G_i\text{-}sew(d, d') = (D, \alpha_0, \ldots, \alpha_{i-1}, \alpha'_i, \alpha_{i+1}, \ldots, \alpha_n)$, where α'_i is defined by: $\forall e \in D$,*

- *if $e \in o_d$, $\alpha'_i(e) = f(e)$;*
- *if $o_d \neq o_{d'}$ and $e \in o_{d'}$, $\alpha'_i(e) = f^{-1}(e)$;*
- *otherwise, $\alpha'_i(e) = \alpha_i(e)$.*

Given an n-Gmap and two i-sewable darts, the i-sewing operation consists simply in linking all the darts in o_d with their "equivalent" darts in $o_{d'}$. More precisely, α'_i is directly given by α_i and isomorphism f. So, the i-sew of d and d' produces the same n-Gmap than the i-sew of any dart d'' of o_d and $f(d'')$ of $o_{d'}$.

An example of 0-sew operation in a 0-Gmap is shown in Fig. 4.4. Two isolated vertices are grouped to form a 0-sphere (i.e. a pair or vertices) by linking the two corresponding darts by α_0. As said above, the 0-sew operation in 0D concerns only the two given darts.

An example of 1-sew operation in a 1-Gmap is shown in Fig. 4.5. Two

[1] Remember that f is an isomorphism between o_d and $o_{d'}$ iff:

$$\forall e \in o_d, \forall j \in \{0, \ldots, i-2, i+2, \ldots, n\}, f(\alpha_j(e)) = \alpha_j(f(e))$$

FIGURE 4.4
0-sew operation in a 0-Gmap.
(a) 0-Gmap $G = (D, \alpha_0 = \mathrm{Id}_D)$ contains two isolated darts.
(b) The corresponding 0D quasi-manifold having two isolated vertices.
(c) 0-Gmap G0-sew$(1, 2)$: $\alpha_0(1) = 2$ (and conversely).
(d) The corresponding 0D quasi-manifold.

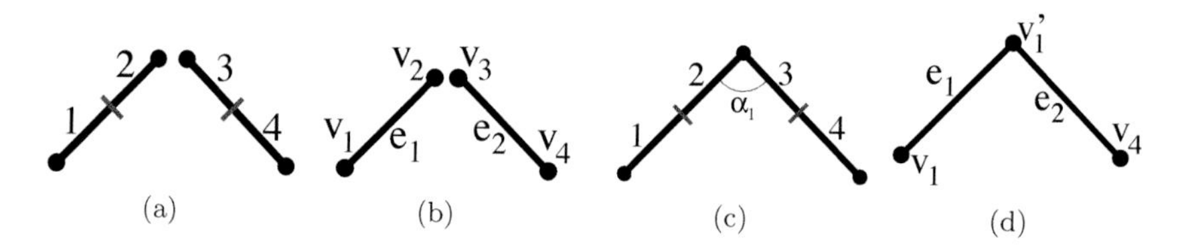

FIGURE 4.5
1-sew operation in a 1-Gmap.
(a) 1-Gmap $G = (D, \alpha_0, \alpha_1 = \mathrm{Id}_D)$ contains four darts.
(b) The corresponding 1D quasi-manifold having two edges and four vertices.
(c) 1-Gmap G1-sew$(2, 3)$: $\alpha_1(2) = 3$ (and conversely).
(d) The corresponding 1D quasi-manifold. Two vertices v_2 and v_3 in (b) are identified into vertex v_1'.

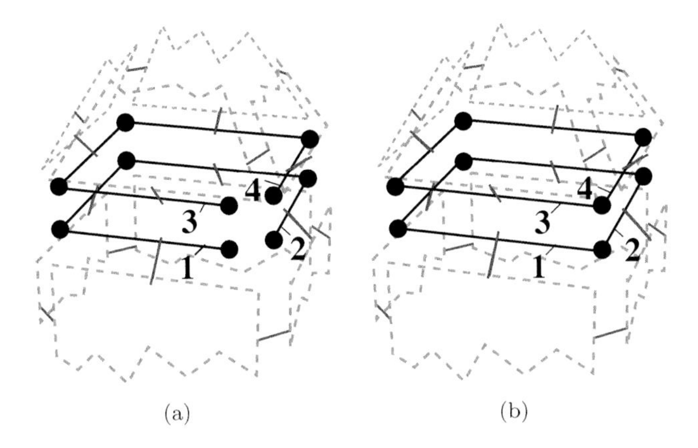

FIGURE 4.6
1-sew operation in a 3-Gmap.
(a) Partial representation of 3-Gmap G. Zoom on a face separating two volumes which has a 1-boundary.
(b) 3-Gmap $G_{1\text{-sew}}(1,2)$: $\alpha_1(1) = 2$ and $\alpha_1(3) = 4$ (and conversely).

distinct vertices are identified by linking the two corresponding darts by α_1. As said above, the 1-sew operation in 1D concerns only the two given darts.

A second example of 1-sew operation is presented in Fig. 4.6, but now in a 3-Gmap. In the inital 3-Gmap shown in Fig. 4.6(a), the face between the two volumes contains sixteen darts and has a 1-boundary. Indeed, darts 1, 2, 3 and 4 are 1-free. The 1-sew operation is applied to darts 1 and 2, producing the 3-Gmap shown in Fig. 4.6(b) where dart 1 is 1-sewn with dart 2, and dart 3 is 1-sewn with dart 4. Indeed, $o_1 = \langle\alpha_3\rangle(1) = \{1,3\}$, $o_2 = \langle\alpha_3\rangle(2) = \{2,4\}$ and the isomorphism f between o_1 and o_2 is such that $f(1) = 2$ and $f(3) = 4$. Note that the same result is obtained if the 1-sew operation is applied to darts 3 and 4.

An example of 2-sew operation in a 2-Gmap is presented in Fig. 4.7. The two 1-cells containing darts 1 and 7 are identified, involving the identification of their two extremities. The 2-sew operation is possible, since the two edges to identify have the same structure. Note that in a 2-Gmap, there are only two possible configurations for an edge having a 2-free dart: either it is made of a dart 0-free, or it is made of two darts linked by α_0. In the first case, there is only one way to 2-sew the darts of two edges. In the second case, there are two possibilities for sewing two edges e_1 and e_2, since one dart of e_1 can be 2-sewn with any of the two darts of e_2. These two possibilities explain why nonorientable objects can be represented: two darts can be sewn in order to make a torsion along an edge (see example in Fig. 4.8).

For 3-Gmaps, the 2-sew operation is similar and there is no additional constraint. Indeed, orbits o_d and $o_{d'}$ are related to involution α_0 as for 2-

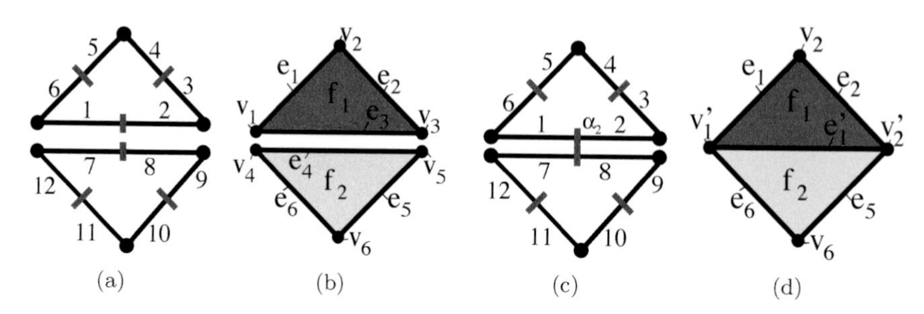

FIGURE 4.7
2-sew operation in a 2-Gmap.
(a) 2-Gmap $G = (D, \alpha_0, \alpha_1, \alpha_2 = \mathrm{Id}_D)$ containing twelve darts.
(b) The corresponding 2D quasi-manifold having two faces, six edges and six vertices.
(c) 2-Gmap $G_{\text{2-sew}}(1, 7)$: $\alpha_2(1) = 7$ and $\alpha_2(2) = 8$ (and conversely).
(d) The corresponding 2D quasi-manifold. Edges e_3 and e_4 are identified into edge e'_1; vertices v_1 and v_4 (resp. v_3 and v_5) are identified into vertex v'_1 (resp. v'_2).

Gmaps, and thus the two only possible configurations are either a dart 0-free, or two darts linked by α_0. The only difference for the 3D case is that the 2-sew operation can lead to the identification of two volumes if the two 2-sewn darts did not belong to the same volume before the operation.

An example of 3-sew operation in a 3-Gmap is presented in Fig. 4.9. Two 2-cells are identified by linking two by two by α_3 all the darts belonging to the two initial faces. This identification of faces involves the identification of the boundaries of the two faces (each pair of edges and each pair of vertices of the two faces in relation by the isomorphism f are identified two by two). Such 3-sew operation is not always possible: the two initial faces must be 3-

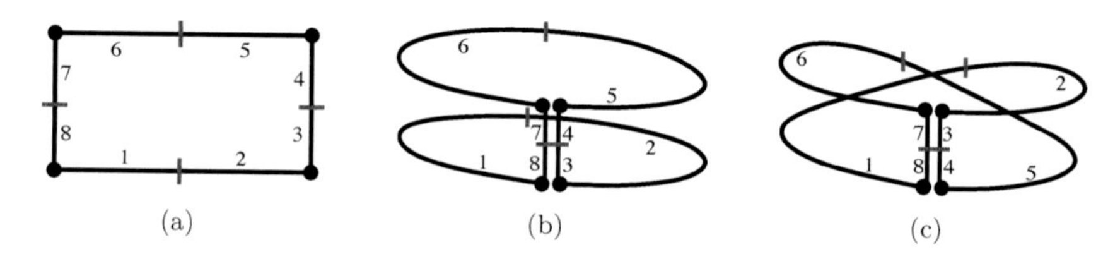

FIGURE 4.8
Illustration of the two possibilities to identify two edges.
(a) 2-Gmap G representing a square.
(b) 2-Gmap $G_{\text{2-sew}}(3, 8)$ which is orientable: it describes an annulus.
(c) 2-Gmap $G_{\text{2-sew}}(3, 7)$ which is nonorientable: it describes a Möbius strip.

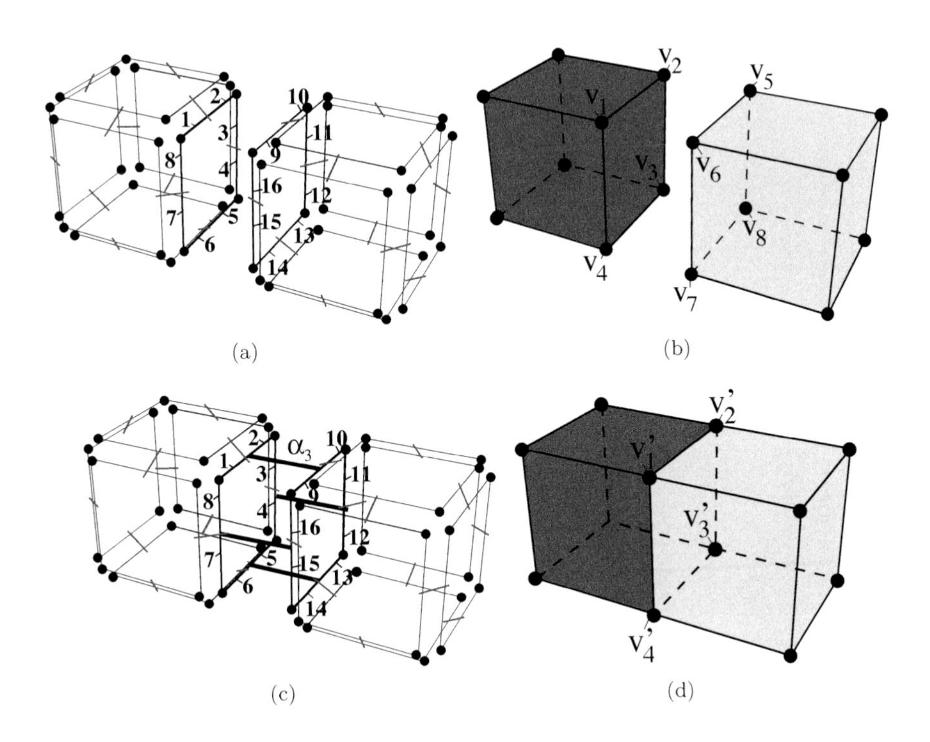

FIGURE 4.9
3-sew operation in a 3-Gmap.
(a) 3-Gmap $G = (D, \alpha_0, \alpha_1, \alpha_2, \alpha_3 = \mathrm{Id}_D)$ containing ninety-six darts.
(b) The corresponding 3D quasi-manifold having two volumes, twelve faces, twenty-four edges and sixteen vertices.
(c) 3-Gmap $G_{3\text{-sew}}(1, 9)$: $\alpha_3(1) = 9$, $\alpha_3(2) = 10$, $\alpha_3(3) = 11$, $\alpha_3(4) = 12$, $\alpha_3(5) = 13$, $\alpha_3(6) = 14$, $\alpha_3(7) = 15$ and $\alpha_3(8) = 16$ (and conversely).
(d) The corresponding 3D quasi-manifold. Let denote a cell by a sequence of its incident vertices. Faces (v_1,v_2,v_3,v_4) and (v_5,v_6,v_7,v_8) are identified into one face (v'_1,v'_2,v'_3,v'_4). Edges (v_1,v_2) and (v_6,v_5) (resp. (v_2,v_3) and (v_5,v_8), (v_3,v_4) and (v_8,v_7), (v_4,v_1) and (v_7,v_6)) are identified into edge (v'_1,v'_2) (resp. (v'_2,v'_3), (v'_3,v'_4), (v'_4,v'_1)). Vertices v_1 and v_6 (resp. v_2 and v_5, v_3 and v_8, v_4 and v_7) are identified into vertex v'_1 (resp. v'_2, v'_3, v'_4).

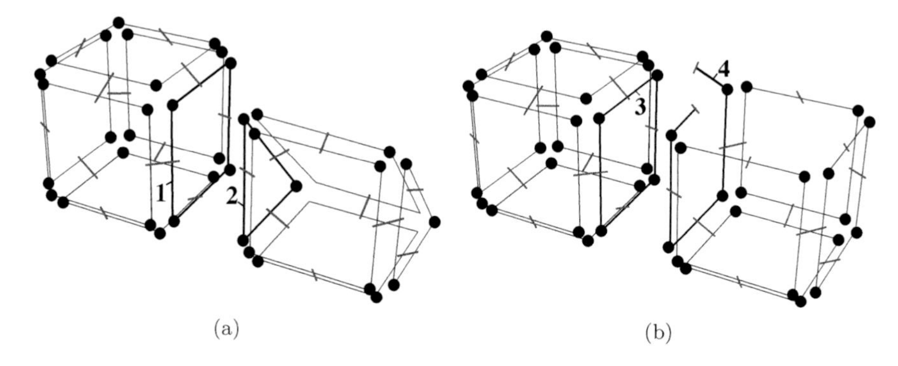

FIGURE 4.10
Two examples of non 3-sewable darts in a 3-Gmap.
(a) Darts 1 and 2 are not 3-sewable. We cannot define a one to one mapping since the two 2-cells do not have the same number of darts.
(b) Darts 3 and 4 are not 3-sewable. If we consider the one to one mapping f defined starting from $f(3) = 4$, we have $\alpha_0(f(3)) = \alpha_0(4) = 4$ while $f(\alpha_0(3)) \neq 4$. This is not an isomorphism.

sewable, i.e. they must have the same structure. In this example, it is easy to verify that darts 1 and 5 are 3-sewable. $o_1 = \{1, 2, 3, 4, 5, 6, 7, 8\}$ and $o_2 = \{9, 10, 11, 12, 13, 14, 15, 16\}$ are isomorphic, isomorphism f is defined by $f(i) = i + 8$, for $1 \leq i \leq 8$, and for each dart $e \in o_1$ we have $\alpha_0(f(e)) = f(\alpha_0(e))$ and $\alpha_1(f(e)) = f(\alpha_1(e))$.

Fig. 4.10 illustrates two cases where the 3-sew operation cannot be applied. In the first example, darts 1 and 2 are not 3-sewable since faces $c_2(1)$ and $c_2(2)$ do not have the same number of darts. Thus no one to one mapping exists between o_1 and o_2. In the second example, the two faces $c_2(3)$ and $c_2(4)$ do not have the same structure: the first one has no 0-boundary while the second one has one 0-boundary. Thus no isomorphism exists between o_3 and o_4.

4.2.3 Unsew Operations

The i-unsew operation, defined in Def. 20, is the inverse of the i-sew operation. It is always possible to i-unsew non i-free darts.

Definition 20 (i-unsew) *Let $G = (D, \alpha_0, \ldots, \alpha_n)$ be an n-Gmap, $i \in \{0, \ldots, n\}$ and $d \in D$ be a non i-free dart. The n-Gmap obtained from G by i-unsewing d and $\alpha_i(d)$ is G_i-unsew$(d) = (D, \alpha_0, \ldots, \alpha_{i-1}, \alpha'_i, \alpha_{i+1}, \ldots, \alpha_n)$, where α'_i is defined by: $\forall e \in D$,*

- *if $e \in \langle \alpha_0, \ldots, \alpha_{i-2}, \alpha_i, \alpha_{i+2}, \ldots, \alpha_n \rangle(d)$, $\alpha'_i(e) = e$;*
- *otherwise, $\alpha'_i(e) = \alpha_i(e)$.*

The i-unsew operation consists in setting α_i to the identity for each dart e in the orbit $\langle \alpha_0, \ldots, \alpha_{i-2}, \alpha_i, \alpha_{i+2}, \ldots, \alpha_n \rangle(d)$. So, the i-unsew applied to any dart of this orbit produces the same result.

It is easy to prove that i-sew and i-unsew are inverse from each other: applying i-sew to two i-free darts d and d', and then applying i-unsew to d, produces the original n-Gmap; applying i-unsew to dart d, which is non i-free, and then applying i-sew to d and d', such that d and d' were originally linked by α_i, produces the original n-Gmap. As a consequence, we can see examples of unsew operations by looking at the examples of sew operations and regarding figures in reverse direction. For example, considering the 3-Gmap given in Fig. 4.6(b), applying the 1-unsew operation to dart 1 produces the 3-Gmap given in Fig. 4.6(a). As another example, starting from the 3-Gmap given in Fig. 4.9(c), applying the 3-unsew operation to dart 1 produces the 3-Gmap given in Fig. 4.9(a).

4.3 Completeness, Multi-Incidence

In this section, we show that any n-Gmap can be built thanks to the basic operations. For that, two basic construction methods allowing to build an n-Gmap are proposed: the first one consists in applying an iterative creation process, which increases progressively the dimension of the n-Gmap starting from 0 until the dimension of the object, and at each step sews some darts for the new dimension; the second one consists in starting directly in the dimension of the object, in adding some isolated darts, and in sewing some darts for any dimension. Note that as for the basic operations, the main interest of these construction methods is mainly theoretical, and generally high level operations are used for practical applications.

4.3.1 Construction by Increasing Dimensions

The first construction method corresponds to the inverse of the decomposition process described for the intuitive presentation of n-Gmaps (cf. Section 3.2). It starts in 0D with a set of darts and α_0 equal to the identity. This set of darts describes a set of isolated vertices. Then some darts are 0-sewn, in order to group the corresponding vertices into pairs. The dimension is increased by adding α_1 initialized to the identity. This produces a set of isolated edges whose boundaries are the previous pairs of vertices. Then, some darts are 1-sewn in order to glue the corresponding edges together, by identifying vertices. This produces a set of subdivided curves. The dimension is increased by adding α_2 initialized to the identity, producing a set of isolated faces whose boundaries are the previous subdivided curves. Now some darts are 2-sewn in order to

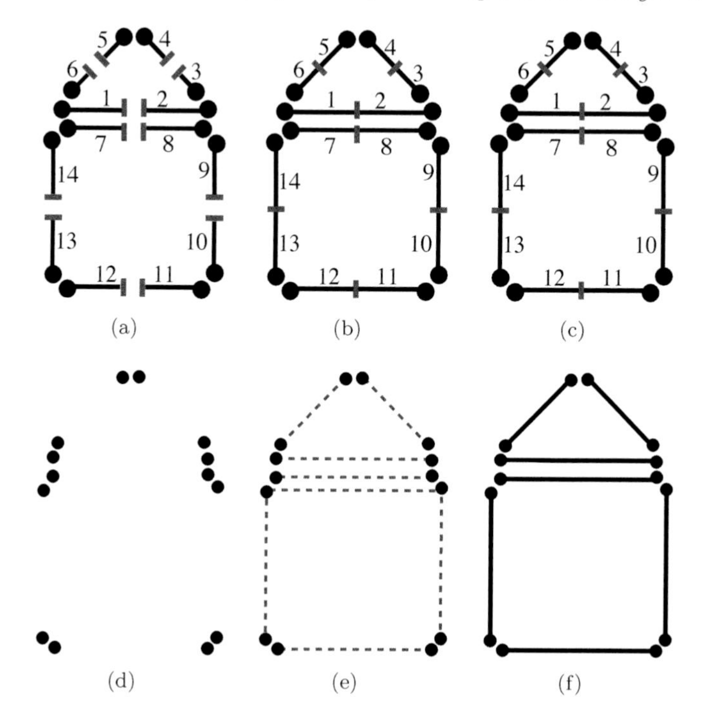

FIGURE 4.11
Construction of a 2D object by applying the first method. The first line shows the different n-Gmaps, and the second line the corresponding objects.
(a) 0-Gmap (D, Id_D) corresponds to (d), a set of isolated vertices.
(b) 0-Gmap (D, α_0) obtained after seven 0-sews. This 0-Gmap corresponds to the pairs of vertices shown in (e).
(c) 1-Gmap $(D, \alpha_0, \mathrm{Id}_D)$ corresponds to the set of isolated edges shown in (f).

identify some edges and maybe their boundaries. This glues the corresponding faces along their edges, producing a set of subdivided surfaces, and so on.

This process can be generalized in any dimension. For that, the two following operations are iteratively applied:

- increase the dimension by adding α_n initialized to the identity;
- n-sew some darts to identify some $(n-1)$-cells (and their boundaries). Remember that the identification of two $(n-1)$-cells is possible only if they have the same structure.

An example showing the whole process of a 2D object construction is presented in Figs. 4.11 and 4.12. It starts with a 0-Gmap containing only isolated darts (see Fig. 4.11(a)), corresponding to isolated vertices (see (d)). Seven 0-sew operations are applied in order to pair the corresponding vertices

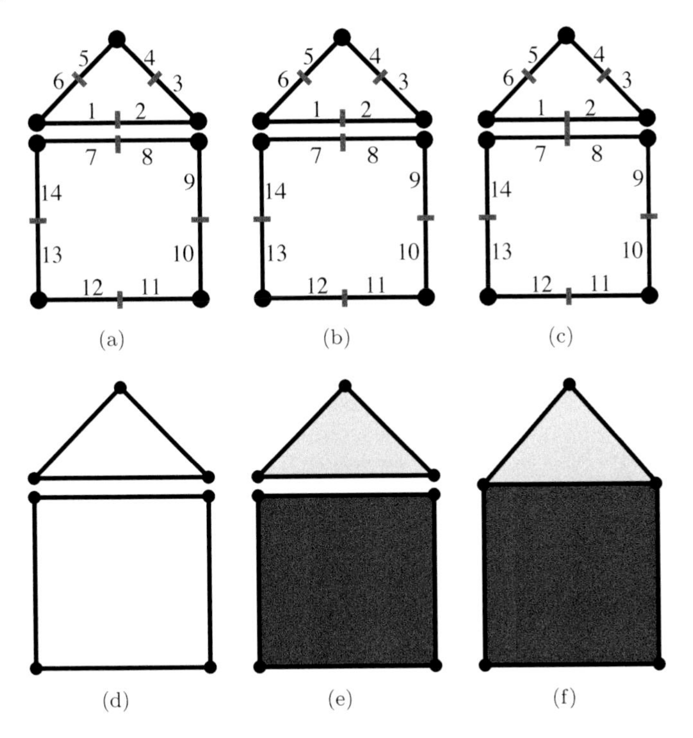

FIGURE 4.12
Following Fig. 4.11.
(a) 1-Gmap (D, α_0, α_1) obtained from Fig. 4.11(c) after seven 1-sew operations. This 1-Gmap corresponds to the two polygonal lines shown in (d).
(b) 2-Gmap $(D, \alpha_0, \alpha_1, \mathrm{Id}_D)$ which corresponds to the two faces shown in (e).
(c) 2-Gmap $(D, \alpha_0, \alpha_1, \alpha_2)$ obtained after one 2-sew operation. In (f) the corresponding 2D subdivided object.

together (for example between darts 1 and 2, or between darts 3 and 4). This produces the 0-Gmap shown in (b), which corresponds to the subdivision given in (e). Now the dimension of the 0-Gmap is increased in order to obtain the 1-Gmap shown in (c) where α_1 is equal to the identity. This operation transforms all the paired vertices into edges (see (f)).

The construction continues in Fig. 4.12. Seven 1-sew operations are applied in order to construct polygonal lines by grouping isolated edges (for example between darts 1 and 6, or between darts 2 and 3). This produces the 1-Gmap shown in Fig. 4.12(a) (corresponding to the subdivision given in (d)). Now the dimension of the 1-Gmap is increased in order to obtain the 2-Gmap shown in (b), where α_2 is equal to the identify. This operation transforms all the polygonal lines into faces (see (e)). Lastly the 2-sew operation is applied to darts 1 and 7. This produces the 2-Gmap shown in (c), which corresponds to the subdivided surface shown in (f).

A second example is presented in Fig. 4.13, showing the construction of a 3D object. The first steps are similar to the previous 2D example. Starting from a 0-Gmap made of isolated darts (not shown), 0-sew and increase dimension operations are applied, producing the 1-Gmap shown in (a). Then 1-sew operations are applied, and the dimension of the 1-Gmap is increased (see (b)). Then 2-sew operations are applied, and the dimension of the 2-Gmap is increased (see (c)). The last step is the application of 3-sew operation. This produces the 3-Gmap shown in (d), which corresponds to the 3D subdivided object shown in (h).

The inverse operations can be applied in order to destroy a given n-Gmap: this corresponds exactly to the decomposition process used for the intuitive presentation of n-Gmaps (cf. Section 3.2). First, n-unsew operations are applied until having all the darts n-free, then the dimension of the n-Gmap is decreased. This process is iterated until obtaining a 0-Gmap in which α_0 is equal to the identity. As examples for this "destruction process", we can consider the examples given for the construction process in reverse direction (given in Figs. 4.11, 4.12 and 4.13), by applying i-unsew (resp. decrease dimension) operations instead of i-sew (resp. increase dimension) operations.

4.3.2 Construction Directly in a Given Dimension

The second construction method consists in working directly in dimension n. Starting from an empty n-Gmap, the operation which adds an isolated dart is applied as many times as necessary. Next, i-sew operations can be applied in this n-Gmap, for $i \in \{0, \ldots, n\}$ (contrary to the previous construction where only k-sew operations can be applied to a k-Gmap).

An example of construction is presented in Fig. 4.14: it corresponds to the same object than in Figs. 4.11 and 4.12, but now the second construction method is applied. The main differences with the first one are:

- the initial n-Gmap has its final dimension, and contains isolated darts (here a 2-Gmap, see Fig. 4.14(a));
- sew operations can be applied in any order; for example 0-sew operations can be first applied (see Fig. 4.14(b)), then 2-sew operations (see Fig. 4.14(c)) and finally 1-sew operations (see Fig. 4.14(d)).

A second construction of the object depicted in Fig. 4.13 is presented in Fig. 4.15. The initial 3-Gmap contains a set of isolated darts (not shown). Then the sew operations can be applied in any order. For example, they can be applied in the same order than in Fig. 4.13: start by 0-sew operations, then 1-sew operations, then 2-sew operations and finish by 3-sew operation. The 3-Gmaps obtained after each step are similar to the ones in Fig. 4.13(a), (b), (c) and (d), the only difference being the dimension of the Gmaps, which is always

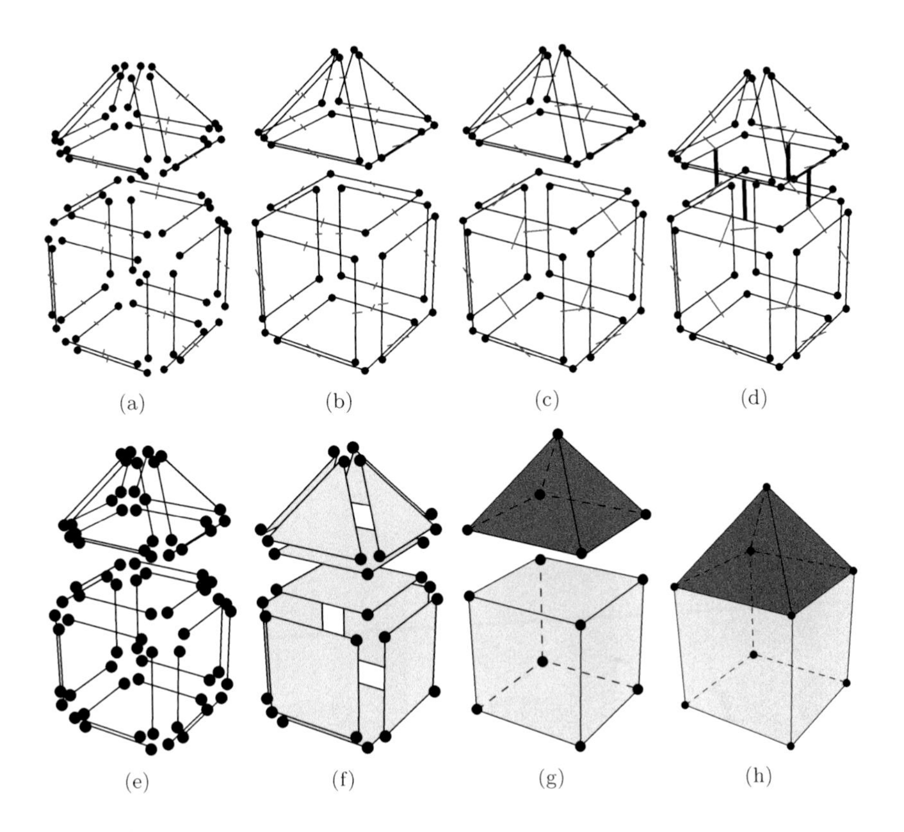

FIGURE 4.13
Construction of a 3D object by applying the first method (partial representation). The first line shows the different n-Gmaps, and the second line the corresponding objects.
(a) 1-Gmap $(D, \alpha_0, \mathrm{Id}_D)$ corresponds to (e), a set of isolated edges.
(b) 2-Gmap $(D, \alpha_0, \alpha_1, \mathrm{Id}_D)$ obtained after forty 1-sew operations, and one increase dimension operation. This 2-Gmap corresponds to eleven faces, shown in (f).
(c) 3-Gmap $(D, \alpha_0, \alpha_1, \alpha_2, \mathrm{Id}_D)$ obtained after twenty 2-sew operations and one increase dimension operation. This 3-Gmap corresponds to the 3D subdivided object shown in (g) containing two volumes.
(d) 3-Gmap $(D, \alpha_0, \alpha_1, \alpha_2, \alpha_3)$ obtained after one 3-sew operation, and (h) the corresponding 3D subdivided object.

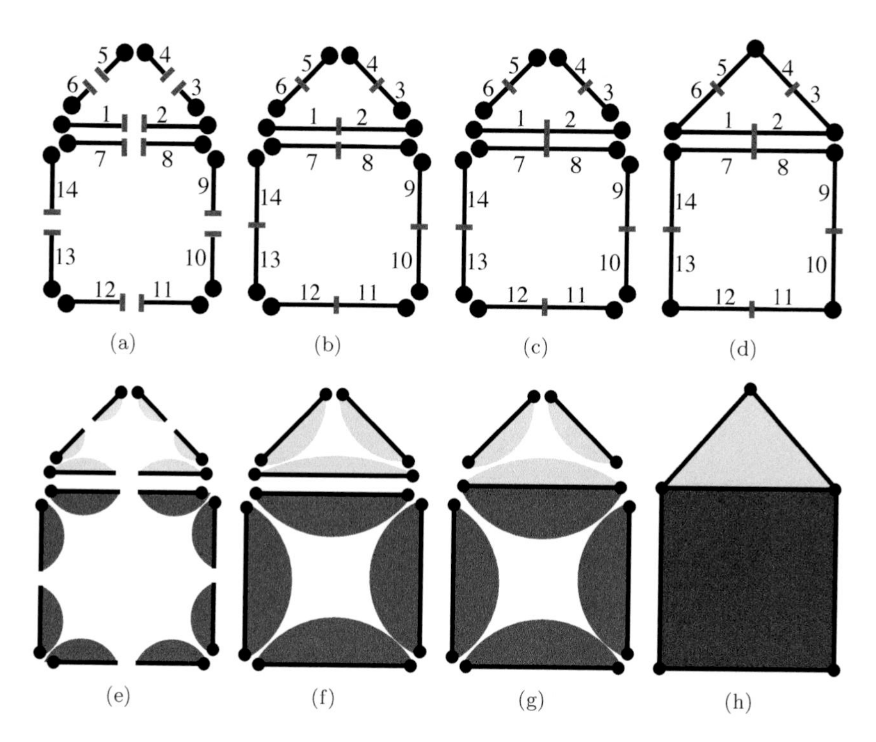

FIGURE 4.14
Construction of a 2D object by applying the second method. The first line shows the different 2-Gmaps, and the second line the corresponding objects.
(a) 2-Gmap $(D, \mathrm{Id}_D, \mathrm{Id}_D, \mathrm{Id}_D)$ corresponds to (e), a set of isolated faces with boundaries.
(b) 2-Gmap $(D, \alpha_0, \mathrm{Id}_D, \mathrm{Id}_D)$, obtained after seven 0-sew operations, corresponds to the object shown in (f).
(c) 2-Gmap $(D, \alpha_0, \mathrm{Id}_D, \alpha_2)$, obtained after one 2-sew operation, corresponds to the object shown in (g).
(d) 2-Gmap $(D, \alpha_0, \alpha_1, \alpha_2)$, obtained after seven 1-sew operations, corresponds to the object shown in (h).

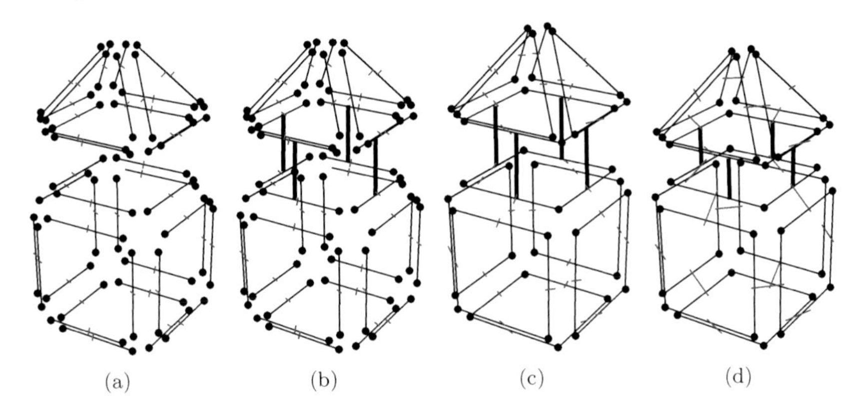

FIGURE 4.15
Construction of a 3D object by applying the second method (partial representation). In this example, the different 3-Gmaps are depicted, but not the corresponding objects.
(a) 3-Gmap $(D, \alpha_0, \mathrm{Id}_D, \mathrm{Id}_D, \mathrm{Id}_D)$.
(b) 3-Gmap obtained after four 3-sew operations.
(c) 3-Gmap obtained after thirty-six 1-sew operations.
(d) 3-Gmap obtained after twenty 2-sew operations.

3 along the whole construction process. For this reason, the corresponding cellular objects are different since each dart describes always a part of a vertex, an edge, a face and a volume (contrary to the construction shown in Fig. 4.13).

Another order is chosen in Fig. 4.15. First, 0-sew operations are applied, producing the 1-Gmap shown in (a). Then 3-sew operations are applied (see (b)). As these operations concern non 0-free darts, they modify α_3 for pairs of darts linked by α_0. This is required in order to satisfy the constraint that $\alpha_0 \circ \alpha_3$ is an involution. Then 1-sew operations are applied (see (c)). The four 1-sew operations between darts that are not 3-free imply the modification of α_1 for the darts linked by α_3 (in order to satisfy the constraint that $\alpha_1 \circ \alpha_3$ is an involution). Last, 2-sew operations are applied, producing the 3-Gmap shown in (d).

The inverse operations can be applied in order to destroy a given n-Gmap. i-unsew operations can be applied in any order, until having all the darts isolated, then all these isolated darts are removed, producing an empty n-Gmap. As examples of these destruction operations, we can consider the examples given for the construction operations in reverse direction (given in Figs. 4.14 and 4.15), by applying i-unsew instead of i-sew operations.

At last, note that for practical applications, working directly in the required dimension is more flexible, since sew (and/or unsew) operations can be applied in any order.

4.3.3 Completeness

Now we show that any n-Gmap can be constructed thanks to the two construction methods. First, we show that any n-Gmap can be destroyed by following the inverse processes; then, since any basic operation has an inverse operation, any n-Gmap can be constructed by applying the inverse sequence of inverse operations.

Let start with the first method.

Theorem 1 (Completeness of the first method for n-Gmap construction or destruction) *Any n-Gmap can be transformed into a set of isolated darts in dimension 0, by a sequence of unsew and decrease dimension operations, unsew being always applied for the highest possible dimension. Conversely, any n-Gmap can be constructed starting from a set of isolated darts in dimension 0, by using only the increase dimension and sew operations, sew being always applied for the highest possible dimension.*

The proof of this theorem is easy. For the first part of the theorem, let us start from an n-Gmap G. For each non n-free dart d, the n-unsew operation is applied. This operation can always be applied for any non n-free dart, and it increases at least by two the number of n-free darts. Thus, at the end, an n-Gmap having all its darts n-free is obtained. Now the decrease dimension operation can be applied, producing an $(n-1)$-Gmap. The process can be iterated until obtaining a 0-Gmap which is composed by a set of isolated darts of dimension 0. In order to prove the second part of the theorem, let us consider the previous operation sequence in reverse order, by replacing each operation by its inverse operation (decrease dimension by increase dimension and i-unsew by i-sew). This sequence transforms the set of isolated darts in dimension 0 into G.

A similar theorem and proof can be stated for the second construction method.

Theorem 2 (Completeness of the second method for n-Gmap construction or destruction) *Any n-Gmap can be transformed into an empty n-Gmap by a sequence of i-unsew and remove isolated darts operations. Conversely, any n-Gmap can be constructed starting from an empty n-Gmap by using only the add isolated darts and i-sew operations.*

The proof of the first part of this theorem is based upon the fact that an i-unsew operation can be applied to any non i-free dart, increasing at least by two the number of non i-free darts. And when a dart is isolated, it can be removed. This produces a sequence of operations which at last destructs the n-Gmap: in order to get the initial n-Gmap, the inverse operations can be applied in reverse order, proving thus the theorem.

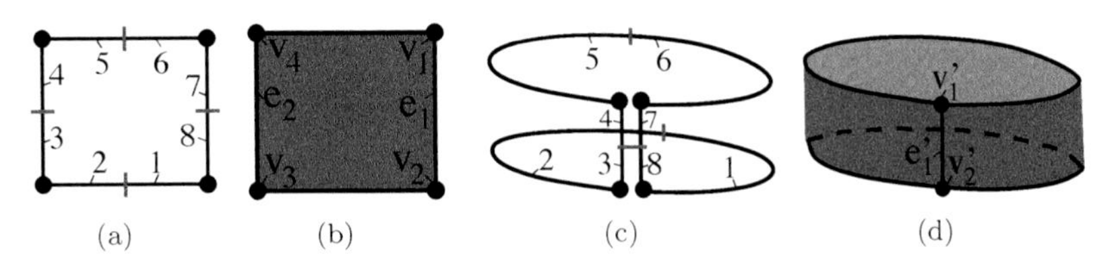

FIGURE 4.16
A 2-sew operation in a 2-Gmap that creates a multi-incidence.
(a) 2-Gmap $G = (D, \alpha_0, \alpha_1, \alpha_2 = \mathrm{Id}_D)$ contains eight darts.
(b) The corresponding 2D quasi-manifold contains one face, four edges and four vertices.
(c) 2-Gmap $G_{2\text{-sew}}(3, 8)$ describes a cylinder.
(d) The corresponding 2D quasi-manifold. Edges e_1 and e_2 are identified into edge e'_1; vertices v_1 and v_4 (resp. v_2 and v_3) are identified into vertex v'_1 (resp. v'_2). Edge e'_1 is incident twice to the face.

4.3.4 Multi-Incidence

Any n-Gmap can be constructed by the basic construction operations. Let us now study possible configurations that can be made. Indeed, there are two cases for sew operations: the first one is the identification of two distinct cells; the second one is the identification of one cell with itself. For practical applications, identification is (almost) always applied to two distinct cells.

4.3.4.1 Identification of Two Distinct Cells

This standard case has already been illustrated in all the examples given in Section 4.2.2. But note that an important property of n-Gmaps is the fact that it is possible to represent subdivisions containing multi-incident cells. It is thus possible to describe complex subdivisions using few cells: for instance, it is possible to construct a torus starting with a single four-sided face, leading to the minimal subdivision of the torus. The first part of the construction consists in applying the 2-sew operation to two darts that belong to the same square face, but not to two consecutive edges. In the example of Fig. 4.16, for instance, darts 3 and 8 are 2-sewn. This folds the face on itself, producing a cylinder (if darts 3 and 7 were 2-sewn, a Möbius strip would be obtained, cf. Fig. 4.8(c) page 94). In this cylinder, the face is incident twice to the edge containing dart 3, since 3 and $8 = \alpha_2(3)$ both belong to the edge and to the face. In order to get a torus, darts 1 and 6 are 2-sewn: this has for result to identify the two edges which make the boundary of the cylinder into one edge, which is also incident twice to the face.

A dangling cell corresponds to a particular case of multi-incidence. A 2D example is depicted in Fig. 4.17: two darts are 2-sewn, which belong to two consecutive edges of a same face. This creates a dangling edge, i.e. an edge

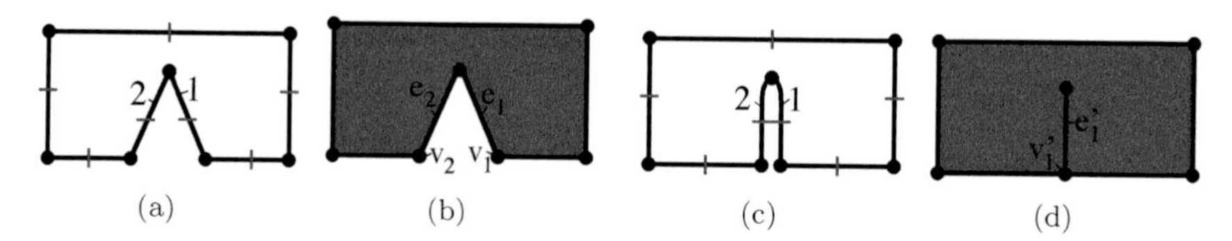

FIGURE 4.17
A 2-sew operation creating a dangling edge in a 2-Gmap.
(a) 2-Gmap $(D, \alpha_0, \alpha_1, \mathrm{Id}_D)$ containing fourteen darts.
(b) The corresponding 2D quasi-manifold has one face, seven edges and seven vertices.
(c) 2-Gmap $G_{2\text{-sew}}(1, 2)$.
(d) The corresponding 2D quasi-manifold. Edges e_1 and e_2 are identified into edge e_1'; vertices v_1 and v_2 are identified into vertex v_1'. Edge e_1' is dangling.

which is incident twice to a face (since darts 1 and $\alpha_2(1) = 2$ belong to the same face and to the same edge), and which has one of its two vertices incident only to this edge (since $\alpha_2(1) = 2 = \alpha_1(1)$). Dangling edges can be created in any dimension. In 3D, a similar construction can create dangling faces by using the 3-sew operation. More generally, dangling cells can be created in any dimension.

4.3.4.2 Identification of a Cell with Itself

The second possible case of the sew operation consists in identifying a cell with itself, if at least an automorphism different from the identity exists, i.e. an isomorphism mapping the cell onto itself. The identification of the cell with itself leads to a folded cell. Such identifications are illustrated in Figs. 4.18 and 4.19. In the first example, the 2-sew operation is applied in a 2-Gmap between a dart and its image by α_0: so, initially, both darts belong to the same edge. Indeed, this is allowed by the definition of sewable and sewing operations, and as we can see in Fig. 4.18(c), this results into a 2-Gmap. However, the corresponding 2D quasi-manifold shown in Fig. 4.18(d) is quite unusual: indeed, edge e_1', which results from the identification of edge e_1 with itself, is incident to one face and to one vertex: this is not a dangling edge as in the example of Fig. 4.17; note also that it is not a loop, since a loop which does not belong to a boundary is incident to two different faces.

A similar example is presented in Fig. 4.19 for 3-sew operation in a 3-Gmap. Starting from a 3-Gmap representing a cube, a dart of a face (dart 1) is 3-sewn with the dart belonging to the same face but to the opposite edge (dart 6). This is possible because the face has seven automorphisms different from the identity, and this association between darts 1 and 6 corresponds to one of these automorphisms. After the 3-sew, $\alpha_3(1) = 6$, $\alpha_3(2) = 5$, $\alpha_3(3) = 4$, $\alpha_3(8) = 7$ (and conversely). This produces a 3-Gmap which has a folded face

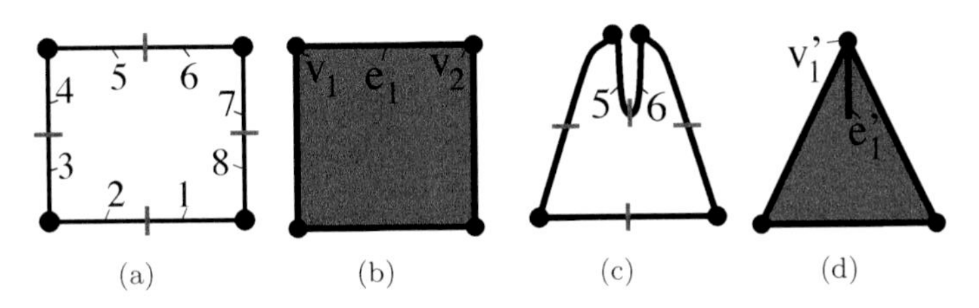

FIGURE 4.18
A 2-sew operation creating a folded cell in a 2-Gmap.
(a) 2-Gmap $(D, \alpha_0, \alpha_1, \mathrm{Id}_D)$ contains eight darts.
(b) The corresponding 2D quasi-manifold has one face, four edges and four vertices.
(c) 2-Gmap $G_{2\text{-sew}}(5,6)$: $\alpha_0(5) = \alpha_2(5) = 6$.
(d) The corresponding 2D quasi-manifold. Edge e_1 is folded on itself. It is incident to only one vertex v_1', which is the result of the identification of v_1 and v_2, although it is not a loop.

and two folded edges. As for the previous case in 2D, the face is not a dangling face, since its boundary is made of three edges, one edge being incident to two vertices and the other edges being folded ones. The original face (v_1, v_2, v_3, v_4), is transformed into a face incident to two vertices v_1' and v_2'. Edges (v_1, v_2) and (v_4, v_3) are identified into edge (v_1', v_2'); edge (v_1, v_4) is folded into an edge incident to vertex v_1' and edge (v_2, v_3) is folded into an edge incident to vertex v_2'. Note that, as seven automorphisms different from the identity map the face onto itself, seven different identifications of the cell with itself can be done, each one leading to a different 3-Gmap. But in each case, the resulting face is folded.

Although n-Gmaps with folded cells seem strange, their topological interpretation is well defined (cf. chapter 8). Anyway, this specific identification is generally not applied for usual applications. It can be easily avoided, either in a constructive way, by adding a constraint to the definition of the sew operation, or directly by adding a constraint to the definition of n-Gmaps. In the first case, the i-sew operation is restricted in order to take as input two darts belonging to two distinct orbits $\langle \alpha_0, \ldots, \alpha_{i-2}, \alpha_{i+2}, \ldots, \alpha_n \rangle$. In the second case, a constraint is added to the definition of n-Gmaps, which requires that any non i-free dart $d \notin \langle \alpha_0, \ldots, \alpha_{i-2}, \alpha_{i+2}, \ldots, \alpha_n \rangle(\alpha_i(d))$. Note that it is easy to test if the constraint is satisfied, by searching such darts during a traversal of the n-Gmap.

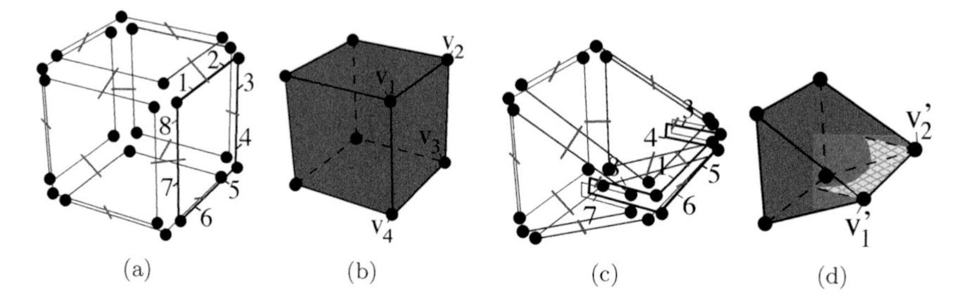

FIGURE 4.19
A 3-sew operation creating folded cells in a 3-Gmap.
(a) 3-Gmap $(D, \alpha_0, \alpha_1, \alpha_2, \mathrm{Id}_D)$ contains forty-eight darts.
(b) The corresponding 3D quasi-manifold has one volume, six faces, twelve edges and eight vertices.
(c) 3-Gmap $G_{3\text{-sew}}(1, 6)$.
(d) The corresponding 3D quasi-manifold. Face (v_1, v_2, v_3, v_4) is folded on itself. It is incident twice to the edge (v_1', v_2'), which is the result of the identification of the two edges (v_1, v_2) and (v_4, v_3). Moreover, the two edges containing darts 3 and 7 are folded as they are incident twice to the same vertex.

4.4 Data Structures, Iterators and Algorithms

In this section, data structures and algorithms are defined in order to handle n-Gmaps in computer softwares. We choose to use an algorithmic language to facilitate the coding in different programming languages. To simplify notations, we do not make the distinction in algorithms between a pointer to an element and the element itself.

The objective is here to describe one way to encode darts and involutions. Here we choose the solution based on structures with pointers, but it is easy to modify the algorithms given here to deal with other possibilities. Note that we describe here only the combinatorial part of the data structure. You can see chapter 7 for two examples showing how is it possible to represent the geometrical part. The solution presented here is the one used for example in `Moka`, a 3D geometrical modeler based on 3-Gmaps [211].

4.4.1 Data Structures

As we have seen in the definition of n-Gmaps, the basic element is the dart. Thus the main data structure is the one which implements the darts. In dimension n, a dart structure contains an array `Alphas` of $n+1$ pointers to represent the different α_i links.

In addition to the $n+1$ pointers, a dart structure contains an array `Marks` of Boolean which serves to mark darts. These marks are used in several algorithms in order to test in constant time if a dart is already processed. They also can be used to mark some part of an n-Gmap having a specific property. The number of marks is defined by a constant number, `NB_MARKS`, which can be fixed by the users depending on their requirements. Generally, 8 is enough for usual applications. In our algorithms, marks satisfy an important invariant: all darts are always unmarked for unused marks. This invariant allows us to avoid initialization of marks in algorithms, and it is often more efficient to unmark only the marked darts at the end of each algorithm instead of unmarking all the darts.

Listing 4.1

`Dart` data structure for n-Gmaps

```
structure Dart
{
  pointer to Dart Alphas[n+1] ;
  Boolean Marks[NB_MARKS] ;
};
```

The `nGMap` structure is mainly composed by a set of darts `Darts`, and a set of integer `FreeMarks` allowing to manage Boolean marks. Note that the dimension n of the `nGMap` could be a global constant, a template argument or a parameter of the function that creates an n-Gmap.

Listing 4.2

`nGMap` data structure

```
structure nGMap
{
  set of Dart Darts ;
  set of integer FreeMarks ;
};
```

An `nGMap` has a set of free marks, i.e. marks that are not in use. When creating a new `nGMap`, we add in the set of free marks all the integers between 0 and `NB_MARKS` − 1 which correspond to all the available marks (see Algorithm 1).

When an algorithm needs a mark, we just get an index from this set and remove it from this set (cf. Algorithm 2). As said above, marks satisfy the invariant: all darts of `gm` are unmarked for this new mark. We can use this mark during the algorithm, and at the end, unmark all the marked darts and free the mark by adding it in the set of free marks (cf. Algorithm 3). Before to free a mark, the user must ensure that all the darts of `gm` are unmarked for this mark. This test is not done in Algorithm 3 because it is costly, but it can be added for debugging purpose.

We can test if a mark is free or reserved by testing if it belongs to `FreeMarks` or not. Moreover, given a reserved mark `i`, we can test if a dart `d` is marked or

Algorithm 1: `createNGMap(n)`: create a new `n`-Gmap

Input: `n`: the dimension of the `n`-Gmap.
Output: An empty new `n`-Gmap.

```
1 gm ← a new nGMap;
2 gm.Darts ← ∅;
3 gm.FreeMarks ← ∅;
4 for i ← 0 to NB_MARKS − 1 do
5     add i in gm.FreeMarks;
6 return gm;
```

Algorithm 2: `reserveMarkNGMap(gm)`: reserve a new mark for n-Gmaps

Input: `gm`: an n-Gmap.
Output: The index of a new reserved mark;
-1 if there is no more available mark.

```
1 if gm.FreeMarks is empty then return -1;
2 i ← an element of gm.FreeMarks;
3 remove i from gm.FreeMarks;
4 return i;
```

Algorithm 3: `freeMarkNGMap(gm,i)`: free a reserved mark for n-Gmaps

Input: `gm`: an n-Gmap having all its darts unmarked for mark `i`;
`i`: the index of a reserved mark.
Result: Free the reserved mark `i`.

```
1 add i into gm.FreeMarks;
```

not by looking directly at d.Marks[i] (see Algorithm 4), and we can mark/unmark this dart by setting d.Marks[i] to true/false (see Algorithms 5 and 6).

Algorithm 4: `isMarkedNGMap(d,i)`: test if a dart is marked for n-Gmaps

Input: d: a dart;
i: the index of a reserved mark.
Output: True iff d is marked for mark i.
1 **return** d.Marks[i];

Algorithm 5: `markNGMap(d,i)`: mark a dart for n-Gmaps

Input: d: a dart;
i: the index of a reserved mark.
Result: Mark dart d for mark i.
1 d.Marks[i] ← true;

Algorithm 6: `unmarkNGMap(d,i)`: unmark a dart for n-Gmaps

Input: d: a dart;
i: the index of a reserved mark.
Result: Unmark dart d for mark i.
1 d.Marks[i] ← false;

The complexity of all these algorithms is in constant time because we use only atomic operations and no loop, except in Algorithm 1 but in this case NB_MARKS is a constant value. We use a stack to store the set of free marks, allowing constant time access to a new free mark and constant time insertion and removal of elements.

Note that it is possible to add in the nGMap data structure an array of Boolean masks which define, for each mark, the values corresponding to marked/unmarked. Thanks to these masks we are able to reverse in constant time the value of a mark for all the darts of the n-Gmap. This is very efficient to unmark all the darts of the n-Gmap when we know that they are all marked.

4.4.2 Iterators

As we have seen before, cells are defined thanks to the orbit notion. Thus it is important to be able to iterate through the darts belonging to a given orbit of an n-Gmap. Hopefully, this can be achieved easily thanks to Algorithm 7. This algorithm is generic since it allows to run through any orbit $\langle\alpha_{i_1},\ldots,\alpha_{i_k}\rangle(d)$

for any valid sequence $(\mathtt{i}_1, \ldots, \mathtt{i}_k)$ and any dart d. We say that the sequence $(\mathtt{i}_1, \ldots, \mathtt{i}_k)$ is valid if:

- $\forall j \in \{1, \ldots, k\}$: $\mathtt{i}_j \in \{0, \ldots, n\}$;
- $\forall j, l \in \{1, \ldots, k\}$: $j \neq l \Rightarrow \mathtt{i}_j \neq \mathtt{i}_l$.

A *sequence* of integers is used here, because the order of the integers can be taken into account in algorithms (cf. for example Algorithm 20 page 122).

Note that, by definition of orbits, we need to consider functions $\alpha_{\mathtt{i}_j}$ and their inverses $\alpha_{\mathtt{i}_j}^{-1}$. However as we only deal with involutions, $\alpha_{\mathtt{i}_j}^{-1} = \alpha_{\mathtt{i}_j}$.

Algorithm 7: Generic iterator for n-Gmaps

Input: gm: an n-Gmap;
d $\in$ gm.Darts: a dart;
$(\mathtt{i}_1, \ldots, \mathtt{i}_k)$: a *valid* sequence of integers between 0 and n.
Result: Run through all the darts of $\langle \alpha_{\mathtt{i}_1}, \ldots, \alpha_{\mathtt{i}_k} \rangle$(d).

```
1  ma ← reserveMarkNGMap(gm);
2  P ← an empty stack of pointer to Dart;
3  push(P,d);
4  markNGMap(d,ma);
5  while P is not empty do
6      cur ← top(P);
7      // process dart cur
8      pop(P);
9      for j ← 1 to k do
10         if not isMarkedNGMap(cur.Alphas[i_j],ma) then
11             markNGMap(cur.Alphas[i_j],ma);
12             push(P,cur.Alphas[i_j]);
13 unmark all marked darts for ma;
14 freeMarkNGMap(gm,ma);
```

Algorithm 7 makes a breadth first traversal. For that we use a stack of darts to process. We take a dart, cur, which belongs to $\langle \alpha_{\mathtt{i}_1}, \ldots, \alpha_{\mathtt{i}_k} \rangle$(d), and we process it. Then for each $\mathtt{i}_j$ in the sequence, we know that $\alpha_{\mathtt{i}_j}$(cur) belongs also to $\langle \alpha_{\mathtt{i}_1}, \ldots, \alpha_{\mathtt{i}_k} \rangle$(d). So, we test if this dart is marked. Indeed, if it is the case this dart was already processed. Otherwise, we need to process it and for that it is enough to add it in the stack of dart to process. So, we mark it in order to not reconsider it later and push it in the stack.

At the end of the algorithm, we must unmark all the marked darts in order to satisfy the property that all the darts are always unmarked for nonused marks. This can be achieved easily by reusing the same algorithm to iterate onto these darts and reversing the role of marked and unmarked darts.

The complexity of this algorithm is linear in the number of darts of $\langle \alpha_{\mathtt{i}_1}, \ldots, \alpha_{\mathtt{i}_k} \rangle$(d), because each dart is considered exactly twice (a first time during the main loop, a second time during the unmarking step), and all operations are atomic ones.

Algorithm 7 can for example be used with the sequence $(0, \ldots, i-1, i+1, \ldots, n)$ in order to iterate through all darts of the i-cell $c_i(d)$ containing a given dart, for any $i \in \{0, \ldots, n\}$. For instance, we can use the sequence $(0, 1, 2)$ to iterate through all the darts of the volume containing a given dart in a 3-Gmap.

Note that this generic algorithm can be optimized in several specific cases by using the properties of n-Gmaps. For example we can avoid the use of stack `P` to iterate through the darts of a face in a 3-Gmap because it exists a linear order allowing to iterate through these darts: for instance, given dart d, we apply successively α_0 and α_1 in order to traverse orbit $\langle \alpha_0, \alpha_1 \rangle(d)$, then apply α_3 to d, and apply again successively α_0 and α_1. This optimization is based upon the fact that $\alpha_0 \circ \alpha_3$ and $\alpha_1 \circ \alpha_3$ are involutions. The traversal of other orbits can be optimized in this way for similar reasons.

This algorithm is a basis for several other algorithms allowing to iterate through some specific cells based on incidence and adjacency relations. We illustrate this by showing three useful examples. In the first one, detailed in Algorithm 8, we run through one dart per each `i`-cell of an n-Gmap. This algorithm allows us to process each `i`-cell exactly once.

Algorithm 8: Iterator over one dart per each `i`-cell for n-Gmaps

Input: `gm`: an n-Gmap;
`i` $\in \{0, \ldots, n\}$.
Result: Run through one dart per each `i`-cell of `gm`.

```
1 ma ← reserveMarkNGMap(gm);
2 foreach dart d ∈ gm.Darts do
3     if not isMarkedNGMap(d,ma) then
4         // process dart d
5         mark all the darts in c_i(d) for ma;
6 unmark all darts for ma;
7 freeMarkNGMap(gm,ma);
```

Algorithm 8 runs through all darts of the n-Gmap, and for each nonmarked dart `d`, it marks the darts of the `i`-cell $c_{\mathtt{i}}(\mathtt{d})$ containing `d`. Note that these darts are marked by using Algorithm 7 with the sequence $(0, \ldots, \mathtt{i}-1, \mathtt{i}+1, \ldots, n)$. To optimize the algorithm, we use in Algorithm 7 the same mark `ma` reserved in Algorithm 8 (indeed, as the goal of line 5 is to mark all the darts in $c_{\mathtt{i}}(\mathtt{d})$, it is useless and inefficient to use a local mark in Algorithm 7 for iterating through these darts, and to unmark these darts at the end of the iterator). At the end of Algorithm 8, we need to unmark all the marked darts. Here we are in the case where all the darts of the n-Gmap are marked, thus we can simply negate the mask of mark `ma`.

The complexity of Algorithm 8 is linear in the number of darts of the n-Gmap because each dart is considered in the first loop (line 2 of the algorithm), and thus each dart is marked exactly once in line 5.

A second example given in Algorithm 9 consists in iterating through one dart per each i-cell incident to a given j-cell for two given dimensions i and j, with $\mathtt{i} \neq \mathtt{j}$.

Algorithm 9: Iterator over one dart per each i-cell incident to a j-cell for n-Gmaps

---

Input: gm: an n-Gmap;
d $\in$ gm.Darts: a dart;
i $\in \{0, \ldots, n\}$;
j $\in \{0, \ldots, n\}$: with $\mathtt{i} \neq \mathtt{j}$.

Result: Run through one dart per each i-cell incident to $c_{\mathtt{j}}(\mathtt{d})$.

```
1 ma ← reserveMarkNGMap(gm);
2 foreach dart e ∈ c_j(d) do
3     if not isMarkedNGMap(e,ma) then
4         // process dart e
5         mark all the darts in c_i(e) for ma;
6 unmark all marked darts for ma;
7 freeMarkNGMap(gm,ma);
```

This algorithm is similar to the previous one, by replacing the loop over all the darts of the n-Gmap by a loop over all the darts of $c_{\mathtt{j}}(\mathtt{d})$ (loop achieved by using Algorithm 7). Thanks to the definition of incidence in n-Gmaps, we are sure that during this loop we discover at least one dart of each incident i-cell, and by using a mark, we ensure to process each i-cell only once. To unmark all the marked darts, we reuse the same algorithm by reversing the role of marked and unmarked darts.

The complexity of this algorithm is linear in the number of darts discovered during the traversal, i.e. in the number of darts of $c_{\mathtt{j}}(\mathtt{d})$ plus the number of darts of the incident cells.

The last example given in Algorithm 10 allows to run through one dart per each i-cell adjacent to a given i-cell.

Once again we use the same principle than for the two previous examples, but here, we have to test dart $\alpha_{\mathtt{i}}(\mathtt{e})$ for each $\mathtt{e} \in c_{\mathtt{i}}(\mathtt{d})$. Indeed, due to the definition of adjacency relations in n-Gmaps, we know that all adjacent i-cells to $c_{\mathtt{i}}(\mathtt{d})$ can be reached by such a dart.

The complexity of this algorithm is linear in the number of darts discovered during the traversal, i.e. in the number of darts of $c_{\mathtt{i}}(\mathtt{d})$ plus the number of darts of the adjacent cells.

Note that we generally implement the iterators as data structures (like iterators in STL [12]) allowing to start a traversal, move to the next position, get the current element pointed by the iterator and test if the iterator has

Algorithm 10: Iterator over one dart per each `i`-cell adjacent to an `i`-cell for n-Gmaps

```
  Input: gm: an n-Gmap;
         d ∈ gm.Darts: a dart;
         i ∈ {0,...,n}.
  Result: Run through one dart per each i-cell adjacent to c_i(d).
1 ma ← reserveMarkNGMap(gm);
2 foreach dart e ∈ c_i(d) do
3 |  if not isMarkedNGMap(e.Alphas[i],ma) then
4 |  |  // process dart e.Alphas[i]
5 |  |  mark all the darts in c_i(e.Alphas[i]) for ma;
6 unmark all marked darts for ma;
7 freeMarkNGMap(gm,ma);
```

reached its end position (i.e. the traversal is finished). Some algorithms given in next sections use this type of implementations (for example Algorithms 19 and 20 page 121).

4.4.3 Basic Tools

We propose now algorithms implementing the basic tools defined in Section 4.2.

Algorithm 11 implements a method to test if a given dart is i-free or not. It only consists in looking if $\alpha_i(d)$ is equal to d or not.

Algorithm 11: `isFreeNGMap(d,i)`: test if a dart is `i`-free for n-Gmaps

```
  Input: d: a dart;
         i ∈ {0,...,n}.
  Output: True iff d is i-free.
1 return d.Alphas[i] = d;
```

The algorithms which add isolated darts and remove isolated darts are straightforward since they only add or remove elements in the set of darts. In Algorithm 12, when we create a dart, we initialize all the α links to the new dart itself, so that it is i-free in all dimensions. Moreover we initialize all its marks to false so that it is unmarked for all the possible marks.

Remember that the remove dart operation can be applied only for isolated darts. Thus Algorithm 13 takes an isolated dart as input and removes it from the set of darts. Testing if `d` is isolated is easy: we test if `d` is i-free for all $i \in \{0, \ldots, n\}$.

The complexity of these three algorithms is in constant time as n and

Algorithm 12: `createDartNGMap(gm)`: create a new isolated dart in an n-Gmap

```
   Input: gm: an n-Gmap.
   Output: A new isolated dart added in gm.
 1 Dart d ← a new Dart;
 2 add d in gm.Darts;
 3 for i ← 0 to n do
 4   d.Alphas[i] ← d;
 5 for i ← 0 to NB_MARKS − 1 do
 6   d.Marks[i] ← false;
 7 return d;
```

Algorithm 13: `removeIsolatedDartNGMap(gm,d)`: remove an isolated dart in an n-Gmap

```
   Input: gm: an n-Gmap;
          d ∈ gm.Darts: an isolated dart.
   Result: d is removed from gm.
 1 remove d from gm.Darts;
```

`NB_MARKS` are constants, and we can use a container of darts allowing constant time insertion and removal of elements.

Now we present the two algorithms that respectively increase and decrease the dimension of a given n-Gmap. In practice, these operations create a new n'-Gmap which is a copy of the initial n-Gmap, but in the case of increase dimension we add a new involution, and in the case of decrease dimension we remove one. Thus we start first by defining Algorithm 14, which copies a given n-Gmap into a second n'-Gmap, by copying only the α_i links for i between 0 and the smaller dimension between n and n'.

During the copy, we need to keep an association between each dart in the original n-Gmap and its corresponding dart in the copy. This is done through an associative array (which can be for example a hash table or a binary search tree). Note that if $n < n'$, we do not need to initialize α_i for the new darts when $n < i \leq n'$ as this is done by the algorithm that creates a new dart. Note also that we copy the free marks, and the values of all the marks for all the darts. Depending on specific needs, we can remove these copies of marks so that `gm'` have no reserved mark.

Algorithm 15 implements the operation that increase the dimension of a given n-Gmap: it only creates a new $(n+1)$-Gmap `gm'` and then copies `gm` into `gm'`.

Algorithm 16 implements the operation that decrease the dimension of a given n-Gmap: it only creates a new $(n-1)$-Gmap `gm'` and then copies `gm` into `gm'`. Note that we require here that the given n-Gmap has all its darts n-free.

Algorithm 14: `copyNGMap(gm,gm′)`: copy an n-Gmap into a second n'-Gmap

```
Input: gm: an n-Gmap;
       gm′: an n′-Gmap.
Result: Copy gm into gm′ until minimum(n,n').
1  gm′.Darts ← ∅;
2  gm′.FreeMarks ← gm.FreeMarks;
3  assoc ← an empty associative array between darts;
4  foreach dart d ∈ gm.Darts do
5  |  assoc[d] ← createDartNGMap(gm′);
6  k ← min(n, n′);
7  foreach dart d ∈ gm.Darts do
8  |  d′ ← assoc[d];
9  |  for i ← 0 to k do
10 |  |  d′.Alphas[i] ← assoc[d.Alphas[i]];
11 |  for i ← 0 to NB_MARKS − 1 do
12 |  |  d′.Marks[i] ← d.Marks[i];
```

Algorithm 15: `increaseDimNGMap(gm)`: increase the dimension of an n-Gmap

```
Input: gm: an n-Gmap.
Output: The (n + 1)-Gmap obtained from gm by increasing its
        dimension.
1  gm′ ← createNGMap(n + 1);
2  copyNGMap(gm,gm′);
3  return gm′;
```

We can add a test in the algorithm to check if this precondition is satisfied. Moreover we can also remove this requirement as it is possible to decrease the dimension of an n-Gmap even if it has some darts n-sew. In this case, we obtain the same result than if we first n-unsew all these darts, then decrease the dimension.

Algorithm 16: `decreaseDimNGMap(gm)`: decrease the dimension of an n-Gmap

Input: gm: an n-Gmap having all its darts n-free.
Output: The $(n-1)$-Gmap obtained from gm by decreasing its dimension.

1 gm′ ← `createNGMap`$(n-1)$;
2 `copyNGMap(gm,gm′)`;
3 **return** gm′;

The complexity of these three algorithms is linear in the number of darts $\#d$ of the given n-Gmap times the complexity for accessing an element in the associative array. This access can be done in $O(\log \#d)$ for hash table or binary search tree, thus the complexity is in $O(\#d. \log \#d)$. Note that hash tables allow constant time access in average by using a correct hash function.

Now we present the two algorithms allowing to merge two given n-Gmaps, and to restrict a given n-Gmap to a subset of its darts. As for the two previous algorithms, these operations build as result a new n-Gmap which in the case of merge is a copy of the two initial n-Gmaps, and in the case of restrict is a partial copy of the initial n-Gmap. For this reason, the two algorithms use once again an associative array to link the original darts and the copy ones.

Algorithm 17 implements the merge operation of two given n-Gmaps. This operation is always possible, as in our data structure, two different n-Gmaps have necessarily two different sets of darts (while this could be not the case for another implementation of n-Gmaps).

In the new n-Gmap, we consider that a mark is free if it is free in both n-Gmaps gm and gm′: this explains line 2 which initializes gm″.`FreeMarks` as the intersection of both sets of free marks. In the associative array, we associate with each original dart its corresponding copy, both for gm and for gm′. After having created all the darts, we only need to make the same links between the new darts than the original links in the initial n-Gmaps. This is directly done thanks to the associative array.

Algorithm 18 implements the operation that restricts an n-Gmap to a given subset of darts. As seen in Def. 16 page 89, this operation is only possible if the subset of darts corresponds to a set of connected components. Thus we start to test if this property is satisfied, and construct the restricted n-Gmap only when it is the case. Note that we could propose different variants, for example that unsew all the nonfree darts linked with a dart that does not

---

Algorithm 17: `mergeNGMaps(gm,gm')`: merge two given n-Gmaps

```
   Input: gm, gm': two n-Gmaps.
   Output: The n-Gmap obtained by merging gm and gm'.
 1 gm'' ← createNGMap(n);
 2 gm''.FreeMarks ← gm.FreeMarks ∩ gm'.FreeMarks;
 3 assoc ← an empty associative array between darts;
 4 foreach dart d ∈ gm.Darts ∪ gm'.Darts do
 5 |  assoc[d] ← createDartNGMap(gm'');
 6 foreach dart d ∈ gm.Darts ∪ gm'.Darts do
 7 |  d' ← assoc[d];
 8 |  for i ← 0 to n do
 9 |  |  d'.Alphas[i] ← assoc[d.Alphas[i]];
10 |  for i ← 0 to NB_MARKS − 1 do
11 |  |  d'.Marks[i] ← d.Marks[i];
12 return gm'';
```

---

belong to $\mathtt{D}'$, or a nonsafe version that does not make the test and assume the user gives always a set of connected components...

The complexity of the merging operation is linear in the number of darts of the two given n-Gmaps ($\#d_1$ and $\#d_2$) times the complexity for accessing an element in the associative array, i.e. in $O((\#d_1 + \#d_2).\log(\#d_1 + \#d_2))$. The complexity of the restrict operation is linear in the number of darts in $\mathtt{D}'$ times $\log |\mathtt{D}'|$ (this corresponds to the cost of the copy), plus the cost of the initial test (checking whether the subset of darts corresponds to a set of connected components: lines $1-3$). The test complexity is $O(|\mathtt{D}'|.\log|\mathtt{D}'|)$ if $\mathtt{D}'$ is a container where searching an element is achieved in log. However this test can be done in $O(|\mathtt{D}'|)$ by first marking all the darts in $\mathtt{D}'$, then reiterate through all these darts and just verify if for each dart d, $\alpha_i(d)$ is marked. The global complexity of Algorithm 18 is thus in $O(|\mathtt{D}'|.\log|\mathtt{D}'|)$.

4.4.4 Sew/Unsew Operations

Now we present the i-sew operation and its inverse which is the i-unsew operation. As seen in Section 4.2.2 when introducing the sew operation, it is not always possible to i-sew two darts: they must be i-sewable. Algorithm 19 allows to test if two given darts are i-sewable or not. This algorithm follows directly the i-sewable definition given in Def. 18 page 90. Its main principle is to run simultaneously through the two orbits o_d and $o_{d'}$, to progressively build the isomorphism while testing it is indeed an isomorphism.

Firstly in line 1, we verify that $\mathtt{d} \neq \mathtt{d}'$ and that $\mathtt{d}$ and $\mathtt{d}'$ are `i`-free. If it is not the case, the two darts are not `i`-sewable. Secondly we create two generic iterators that correspond to the two orbits $o_d = \langle\alpha_0, \ldots, \alpha_{\mathtt{i}-2}, \alpha_{\mathtt{i}+2}, \ldots, \alpha_n\rangle(\mathtt{d})$

Algorithm 18: `restrictNGMap(gm,D′)`: restrict an n-Gmap to a subset of darts

```
   Input: gm: an n-Gmap;
          D′ ⊆ gm.Darts: a subset of darts.
   Output: The n-Gmap obtained by restricting gm to D′.
 1 foreach dart d ∈ D′ do
 2     for i ← 0 to n do
 3         if d.Alphas[i] ∉ D′ then error;
 4 gm′ ← createNGMap(n);
 5 gm′.FreeMarks ← gm.FreeMarks;
 6 assoc ← an empty associative array between darts;
 7 foreach dart d ∈ D′ do
 8     assoc[d] ← createDartNGMap(gm′);
 9 foreach dart d ∈ D′ do
10     d′ ← assoc[d];
11     for i ← 0 to n do
12         d′.Alphas[i] ← assoc[d.Alphas[i]];
13     for i ← 0 to NB_MARKS − 1 do
14         d′.Marks[i] ← d.Marks[i];
15 return gm′;
```

Algorithm 19: `sewableNGMap(gm,d,d',i)`: test if two darts are i-sewable for n-Gmaps

Input: gm: an n-Gmap;
d, d′ ∈ gm.Darts: two darts;
i ∈ $\{0, \ldots, n\}$.

Output: True iff d is i-sewable with d′.

```
1  if d = d' or not isFreeNGMap(d,i) or not isFreeNGMap(d',i) then
2      return false;
3  it ← generic iterator(gm,d,(0,...,i − 2,i + 2,...,n));
4  it' ← generic iterator(gm,d',(0,...,i − 2,i + 2,...,n));
5  f ← an empty associative array between darts;
6  while it is not to its end and it' is not to its end do
7      f[it] ← it';
8      foreach j ∈ {0,...,i − 2,i + 2,...,n} do
9          if it.Alphas[j] is defined in f and
           f[it.Alphas[j]] ≠ it'.Alphas[j] then
10             return false;
11     advance it to its next position;
12     advance it' to its next position;
13 if it is not to its end or it' is not to its end then
14     return false;
15 return true;
```

and $o_{d'} = \langle\alpha_0, \ldots, \alpha_{i-2}, \alpha_{i+2}, \ldots, \alpha_n\rangle(\mathtt{d'})$. During the main loop, for each pair of darts pointed by `it` and `it'`, we consider all the darts in the neighborhood of `it` that are already considered. Indeed, we already have built the isomorphism between these darts through the associative array `f`, and we need to test if the local property of the `i`-sewable operation is satisfied (i.e. $\mathtt{f}(\alpha_{\mathtt{j}}(\mathtt{it})) = \alpha_{\mathtt{j}}(\mathtt{it'})$, with $\mathtt{f}(\mathtt{it}) = \mathtt{it'}$). When it is not the case, we directly return false as we know that the two darts `d` and `d'` are not i-sewable. Otherwise we continue the main loop until finishing at least one iterator. If only one iterator has finished, the two darts are not `i`-sewable (the two orbits do not have the same number of darts); otherwise, they are `i`-sewable, and we return true.

The complexity of the `i`-sewable algorithm is linear in the number of darts $\#d$ in the smaller orbit among o_d and $o_{d'}$ times $\log \#d$. Indeed, we stop the main loop as soon as one iterator has reached its end position, and all operations are atomic except the access to an element in the associative array which is in log, and the inner loop (line 8) is bounded by n which is a constant.

The `i`-sew operation is given in Algorithm 20, for $\mathtt{i} \in \{0, \ldots, n\}$. This algorithm follows directly the definition of the `i`-sew operation (see Def. 19 page 91). The main loop consists to run through the two orbits $o_d = \langle\alpha_0, \ldots, \alpha_{i-2}, \alpha_{i+2}, \ldots, \alpha_n\rangle(\mathtt{d})$ and $o_{d'} = \langle\alpha_0, \ldots, \alpha_{i-2}, \alpha_{i+2}, \ldots, \alpha_n\rangle(\mathtt{d'})$, and put in relation all pairs of darts by $\alpha_{\mathtt{i}}$.

Algorithm 20: `sewNGMap(gm,d,d',i)`: `i`-sew two darts for n-Gmaps

Input: `gm`: an n-Gmap;
  `d`, $\mathtt{d'} \in$ `gm.Darts`: two `i`-sewable darts;
  $\mathtt{i} \in \{0, \ldots, n\}$.
Result: `i`-sew darts `d` and `d'`.

1 `it` ← generic iterator(gm,d,$(0, \ldots, \mathtt{i}-2, \mathtt{i}+2, \ldots, n)$);
2 `it'` ← generic iterator(gm,d',$(0, \ldots, \mathtt{i}-2, \mathtt{i}+2, \ldots, n)$);
3 **while** `it` *is not to its end* **do**
4   `it.Alphas[i]` ← `it'`;
5   `it'.Alphas[i]` ← `it`;
6   advance `it` to its next position;
7   advance `it'` to its next position;

The key point of this algorithm is the use once again of the two generic iterators (gm,d,$(0, \ldots, \mathtt{i}-2, \mathtt{i}+2, \ldots, n)$) and (gm,d',$(0, \ldots, \mathtt{i}-2, \mathtt{i}+2, \ldots, n)$) to iterate through the two orbits by using the same α links. Since we know that `d` and `d'` are `i`-sewable, and since the generic iterator uses the indexes given in the sequence in the same order, at each step of the while loop, the two iterators point to two darts in relation by isomorphism f of Def. 19. For this reason, we only need to put the two pointed darts in relation by $\alpha_{\mathtt{i}}$, then move the two iterators to their next positions. Note that the relation is done in both directions as $\alpha_{\mathtt{i}}$ is an involution. The algorithm ends when the two iterators reach their end positions.

The complexity of this algorithm is linear in the number of darts in the orbit $\langle\alpha_0,\ldots,\alpha_{i-2},\alpha_{i+2},\ldots,\alpha_n\rangle$(d).

The i-unsew operation is given in Algorithm 21, for i $\in \{0,\ldots,n\}$. As for the i-sew algorithm, it follows directly the i-unsew definition (see Def. 20 page 96). It runs through all the darts of the orbit $\langle\alpha_0,\ldots,\alpha_{i-2},\alpha_{i+2},\ldots,\alpha_n\rangle$(d), and we i-link dart α_i(e) to itself, and dart e to itself.

Algorithm 21: unsewNGMap(gm,d,i): i-unsew two darts for n-Gmaps

Input: gm: an n-Gmap;
    d $\in$ gm.Darts: a non i-free dart;
    i $\in \{0,\ldots,n\}$.
Result: i-unsew darts d and α_i(d).

```
1 foreach dart e ∈ ⟨α0,...,αi−2,αi+2,...,αn⟩(d) do
2     (e.Alphas[i]).Alphas[i] ← e.Alphas[i];
3     e.Alphas[i] ← e;
```

The complexity of the unsew algorithm is linear in the number of darts of the traversed orbit $\langle\alpha_0,\ldots,\alpha_{i-2},\alpha_{i+2},\ldots,\alpha_n\rangle$(d).

4.5 Complements

In this section, some additional definitions are provided: boundary map, duality, orientability, and we conclude by providing a classification of 2-Gmaps, corresponding to the classification of surfaces.

Note that the algorithms corresponding to these new notions are not provided, but they are not complicated to retrieve. For the boundary map and the dual map, the principle is similar to the copy of an n-Gmap, while modifying the α links according to the definitions.

4.5.1 Boundary Map

For many applications, n-Gmaps are such that all involutions, except α_n, are without fixed points, i.e. no dart d exists, such that $\alpha_i(d) = d$ with $i < n$: so, the boundary of each n-cell corresponds to an $(n-1)$-quasi-manifold without boundary, but the n-dimensional quasi-manifold corresponding to the n-Gmap can have boundaries, which correspond to the connected components of the *$(n-1)$-Gmap of the boundaries*.

Let us consider an n-Gmap G without boundary except for dimension n. G_∂ is the $(n-1)$-Gmap of the boundaries of G: it corresponds thus to an $(n-1)$-quasi-manifold without boundary. It is defined by keeping all the n-

free darts of G and all the α_i, $\forall i \in \{0, \dots, n-2\}$. α'_{n-1} is defined by following, for each n-free dart, the path of darts $\alpha_{n-1} \circ (\alpha_n \circ \alpha_{n-1})^k$ to jump over darts that do not belong to the boundary until obtaining an n-free dart. For $n = 1$ or $n = 2$, any component of G_∂ represents a boundary of G (see example in 2D in Fig. 4.20). For $n \geq 3$, any component of G_∂ represents an $(n-1)$-quasi-manifold which recovers a boundary of G.

The boundary notion is generalized in Def. 21 in order to define the i-boundary map for any n-Gmap. Given an n-Gmap G, the i-boundary map G_{∂_i} is an n-Gmap which covers the different i-boundaries of G. More precisely there is a one to one mapping between the different i-boundaries of G and the connected components of G_{∂_i}.

Definition 21 (i-boundary map) *Let $G = (D, \alpha_0, \dots, \alpha_n)$ be an n-Gmap, and $i \in \{0, \dots, n\}$. The i-boundary map $G_{\partial_i} = (D', \alpha'_0, \dots, \alpha'_n)$ is the n-Gmap defined by:*

1. *$D' = \{d \in D | d$ is i-free$\}$;*
2. *$\forall d' \in D'$:*
   - *$\forall j \in \{0, \dots, i-2, i+2, \dots, n\}$, $\alpha'_j(d') = \alpha_j(d')$;*
   - *$\alpha'_i(d') = d'$; $\alpha'_{i+1}(d') = d'$;*
   - *$\alpha'_{i-1}(d') = \alpha_{i-1} \circ (\alpha_i \circ \alpha_{i-1})^k(d')$*
     k being the smaller integer s.t. $\alpha_{i-1} \circ (\alpha_i \circ \alpha_{i-1})^k(d')$ is i-free.

Note that there are two possible cases for the definition of $\alpha'_{i-1}(d')$. The first case is when an i-free dart $d'' \neq d'$ is reached when following the path of dart $\alpha_{i-1} \circ (\alpha_i \circ \alpha_{i-1})^k(d')$. In this case, $\alpha'_{i-1}(d') = d''$. The second case is when an $(i-1)$-free dart is reached when following the path $\alpha_{i-1} \circ (\alpha_i \circ \alpha_{i-1})^k(d')$. In this case, the path goes forward until obtaining d' which is the first i-free dart of the path, and $\alpha'_{i-1}(d') = d'$. Note also that, by definition, all the darts in an i-boundary map are i-free and $(i+1)$-free.

Note also that the generalized i-boundary map has the same dimension than the initial n-Gmap. This is required for $i < n$ in order to associate correctly during the closure operation (cf. Section 6.1 page 185) darts of initial n-Gmap with darts of i-boundary map. However this is not required for $i = n$ as there is no dimension higher than i.

This generalized definition of the i-boundary map can be used to retreive the classical boundary of an n-Gmap G, when all involutions except α_n are without fixed points. First G_{∂_n} is computed, then its dimension is decreased (by definition each dart d in G_{∂_n} is such that $\alpha_n(d) = d$), producing G_∂.

Definition 22 (boundary map) *Let G be an n-Gmap without boundary except possibly for dimension n. The $(n-1)$-Gmap of the boundaries is $G_\partial = G^{-}_{\partial_n}$.*

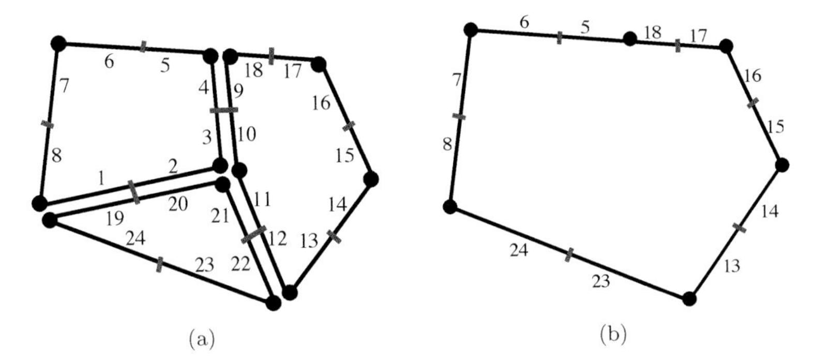

FIGURE 4.20
(a) 2-Gmap G with 2-boundary.
(b) The corresponding 2-boundary map $G_{\partial_2} = (D', \alpha'_0, \alpha'_1, \alpha'_2 = \mathrm{Id}_D)$. 1-Gmap $G_\partial = G^-_{\partial_2} = (D', \alpha'_0, \alpha'_1)$ corresponds to a closed curve, i.e. to the boundary of the surface represented by G.

Figure 4.20 shows an example of a 2-Gmap with a 2-boundary and its corresponding 2-boundary map. A second example is given in Fig. 4.21 that shows a 3-Gmap with a 3-boundary and the corresponding 3-boundary map. Note that in both cases, we get the $(n-1)$-Gmap of the boundaries by decreasing their dimension. The third example given in Fig. 4.22 shows a case where we obtain d' for the definition of $\alpha'_{i-1}(d')$ (for darts 1, 6, 12 and 15).

Figure 4.23 shows the 1-boundary map of the 2-Gmap given in Fig. 4.22(a) (which has 1-boundaries and 2-boundaries) and in Fig. 4.24 an example of 1-boundary map for a 3-Gmap.

Given an n-Gmap G with i-boundaries, it is then possible to "fill" all i-boundaries by linking the n-Gmap with a copy of its i-boundary map G_{∂_i}. More precisely, each dart of G belonging to a boundary is linked by α_i with its corresponding dart of the copy of G_{∂_i}: this produces the *i-closure* of the n-Gmap detailed in Section 6.1.

4.5.2 Duality

Duality is a classical notion that, for geometrical objects, is the link between two objects having similar structure, but regarding their dimensions in reverse order. A classical example is a planar graph embedded in a plane, which describes vertices, edges and faces. Its dual graph, embedded also in the plane, is a planar graph where each vertex (resp. edge, face) corresponds to a face (resp. edge, vertex) of the initial graph. Another classical example is the Delaunay subdivision of a set of points which is the dual of the Voronoï subdivision of the same set of points.

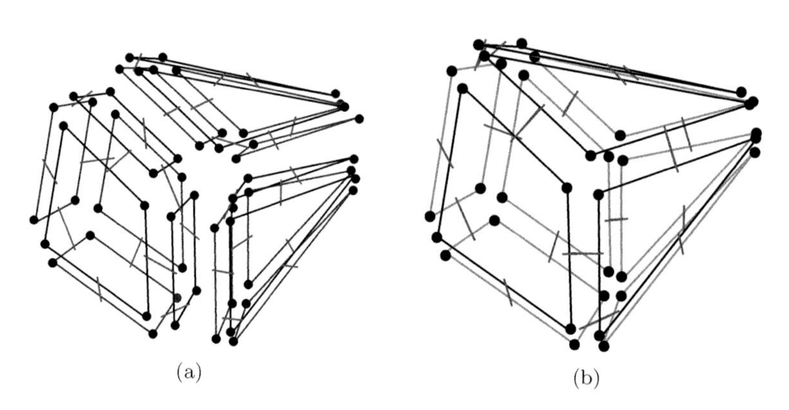

FIGURE 4.21
(a) 3-Gmap G with 3-boundary (three volumes two by two adjacent).
(b) The corresponding 3-boundary map $G_{\partial_3} = (D', \alpha'_0, \alpha'_1, \alpha'_2, \alpha'_3 = \mathrm{Id}_D)$. 2-Gmap $G_\partial = G^-_{\partial_3} = (D', \alpha'_0, \alpha'_1, \alpha'_2)$ corresponds to a surface, which is the boundary of the 3-dimensional manifold represented by G.

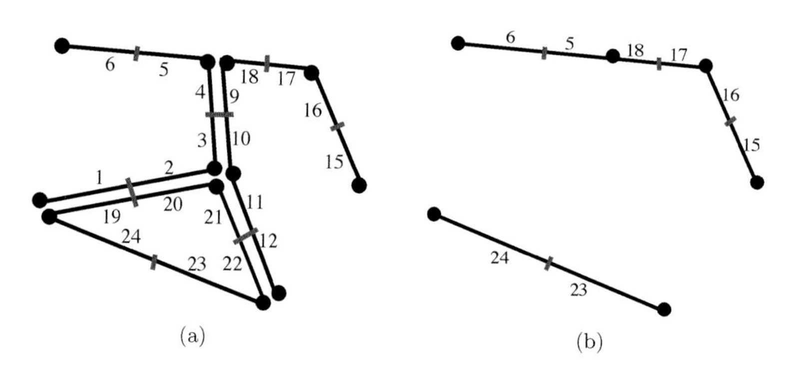

FIGURE 4.22
(a) 2-Gmap G with 1-boundaries and 2-boundaries.
(b) The corresponding 2-boundary map $G_{\partial_2} = (D', \alpha'_0, \alpha'_1, \alpha'_2 = \mathrm{Id}_D)$. For example, darts 6 and 23 are 1-free in G_{∂_2} since $6 = \alpha_1 \circ (\alpha_2 \circ \alpha_1)^0(6)$ and $\alpha_1 \circ (\alpha_2 \circ \alpha_1)^2(23) = 23$ are 2-free in G. G_{∂_1} is given in Fig. 4.23.

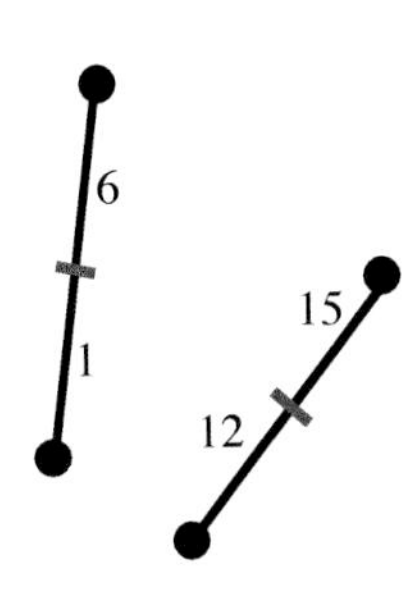

FIGURE 4.23
1-boundary map G_{∂_1} of the 2-Gmap given in Fig. 4.22(a) composed with four darts describing two isolated edges.

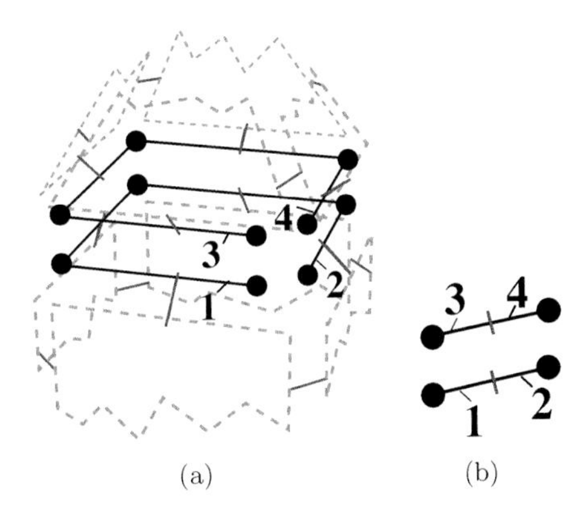

FIGURE 4.24
(a) Partial representation of 3-Gmap G. Zoom on the face separating two volumes which has a 1-boundary.
(b) The corresponding 1-boundary map $G_{\partial_1} = (D', \alpha'_0, \alpha'_1 = \mathrm{Id}_D, \alpha'_2 = \mathrm{Id}_D, \alpha'_3)$. Darts 1 and 2 (resp. 3 and 4) are 0-sewn in G_{∂_1}, and darts 1 and 3 (resp. 2 and 4) are 3-sewn in G_{∂_1}.

For n-Gmaps, defining the dual consists in reversing the order of the α involutions (see Def. 23).

Definition 23 (Dual n-Gmap) *Let $G = (D, \alpha_0, \ldots, \alpha_n)$ be an n-Gmap. The dual n-Gmap of G is $G^* = (D, \alpha_n, \ldots, \alpha_0)$.*

By definition of cells in n-Gmaps, i-cells are transformed into $(n-i)$-cells. Thanks to this definition, it is straightforward to prove that $G^{**} = G$.

An example of a 2-Gmap $G = (D, \alpha_0, \alpha_1, \alpha_2)$ without boundary is presented in Fig. 4.25, and also its corresponding dual 2-Gmap $G^* = (D, \alpha_2, \alpha_1, \alpha_0)$. G is composed by seven vertices, nine edges and four faces, while G^* is composed by four vertices, nine edges and seven faces.

As illustrated in Fig. 4.26, the dual map definition is valid for n-Gmaps with boundaries. In this example, 2-Gmap $G = (D, \alpha_0, \alpha_1, \alpha_2)$ has boundaries in all dimensions. Its dual is 2-Gmap $G^* = (D, \alpha_2, \alpha_1, \alpha_0)$. G is composed by five vertices, five edges and three faces, while G^* is composed by three vertices, five edges and five faces.

4.5.3 Orientability

n-Gmaps can describe the topology of orientable and nonorientable quasi-manifolds (cf. chapter 3). Definition 24 allows to test if a given n-Gmap is

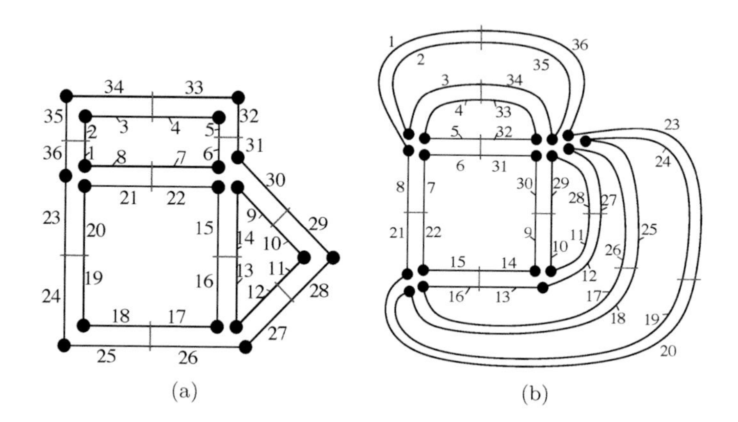

FIGURE 4.25
(a) 2-Gmap $G = (D, \alpha_0, \alpha_1, \alpha_2)$ without boundary.
(b) The dual 2-Gmap $G^* = (D, \alpha_2, \alpha_1, \alpha_0)$. The vertex located down left (resp. down right, up left, up right) corresponds to the square face down (resp. the triangle, the square face up, the external face) in Fig. 4.25(a).

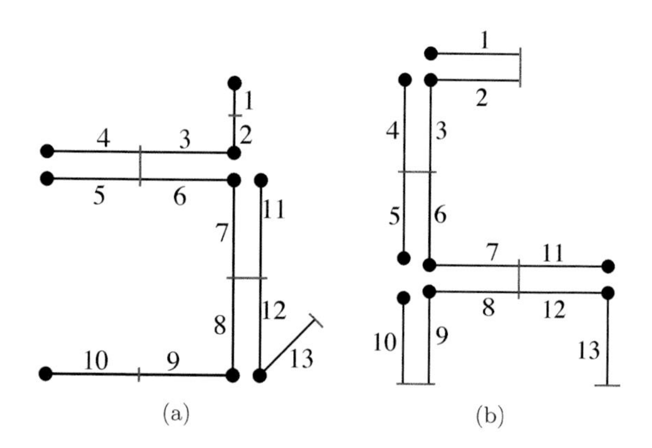

FIGURE 4.26
(a) 2-Gmap $G = (D, \alpha_0, \alpha_1, \alpha_2)$ having boundaries in all dimensions.
(b) The dual 2-Gmap $G^* = (D, \alpha_2, \alpha_1, \alpha_0)$.

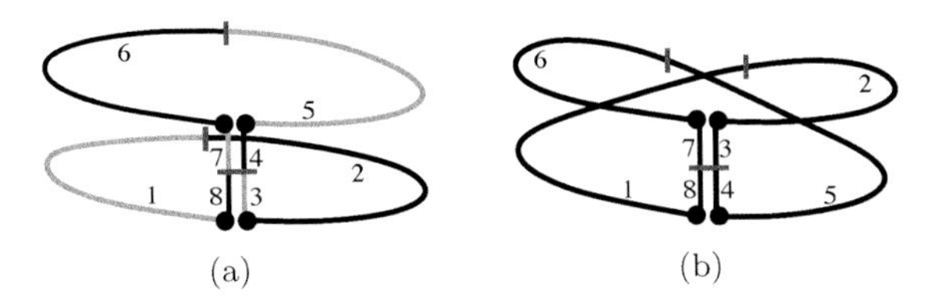

FIGURE 4.27
Orientable and nonorientable 2-Gmaps.
(a) An orientable 2-Gmap. E_1 is the set of darts in black, and E_2 is the set of darts in gray. Any pair of darts d, d' such that $d = \alpha_i(d')$ belong to two different sets, $\forall i \in \{0, 1, 2\}$.
(b) A nonorientable 2-Gmap. It is not possible to define two sets E_1 and E_2 satisfying the orientable condition.

orientable or not. It is orientable if the set of darts can be split in two such that any involution links two darts belonging to two different sets.

Definition 24 (Orientable n-Gmap) *An n-Gmap $G = (D, \alpha_0, \ldots, \alpha_n)$ is* orientable *if $D = E_1 \cup E_2$, such that $E_1 \cap E_2 = \varnothing$, and, $\forall d \in D$, $\forall i \in \{0, \ldots, n\}$, d is not i-free $\Rightarrow$ d and $\alpha_i(d)$ do not belong to the same set E_1 or E_2. G is* nonorientable *otherwise.*

Two 2-Gmaps are shown in Fig. 4.27, the first one being orientable while the second one not. In the first case, two sets of darts can be defined, which satisfy the orientable condition. In the second case, this is not possible: for instance, if $E_1 = \{1, 3, 5, 7\}$ and $E_2 = \{2, 4, 6, 8\}$, these two sets satisfy the orientable condition for α_0 and α_1, but not for α_2 because $\alpha_2(3) = 7$, and darts 3 and 7 both belong to E_1. And any other pair of sets E_1 and E_2 does not satisfy the orientable condition for α_0 or α_1. We will see in Section 5.5.2 page 178 how to compute an orientation of any orientable n-Gmap.

The orientable condition can also be defined in a similar way for any i-cell (and more generally for any orbit): cf. Def. 25. The only difference with the definition of orientable n-Gmap is to consider here only the darts of the given cell, and all involutions except α_i.

Definition 25 (Orientable i-cell) *An i-cell c in an n-Gmap $G = (D, \alpha_0, \ldots, \alpha_n)$ is* orientable *if $c = E_1 \cup E_2$, such that $E_1 \cap E_2 = \varnothing$, and $\forall d \in c$, $\forall j \in \{0, \ldots, n\}$, $j \neq i$, d is not j-free $\Rightarrow$ d and $\alpha_j(d)$ do not belong to the same set E_1 or E_2. c is* nonorientable *otherwise.*

If an n-Gmap is orientable, then all its i-cells are orientable for all $i \in \{0, \ldots, n\}$ (and more generally all its orbits are orientable). However the reverse is not true. An n-Gmap can be nonorientable even if all its cells are orientable as we can see in Fig. 4.27(b). Indeed, this 2-Gmap is nonorientable, while all its cells are orientable. For example face $c_2(1)$ is orientable, as α_2 is not taken into account, but only α_0 and α_1.

When an n-Gmap is orientable, the two subsets E_1 and E_2 correspond to the two orientations of the corresponding quasi-manifold, and these two orientations are inverse from each other (see Fig. 4.27(a)). When all involutions are without fixed points, sets E_1 and E_2 are equal to the two orbits for $\{\alpha_1 \circ \alpha_0, \ldots, \alpha_n \circ \alpha_0\}$ (cf. Section 5.5.2 where the links between n-maps and n-Gmaps are studied).

4.5.4 Classification of 2-Gmaps

Thanks to the definitions given in this chapter, given a connected 2-Gmap G having no 0 nor 1-boundary, it is possible to compute the different topological characteristics of G introduced in Section 2.3 for the classification of paper surfaces:

- $b(G)$ is the number of 2-boundaries of G, i.e. the number of connected components of G_∂;
- $\chi(G)$ is the Euler-Poincaré characteristic of G:
$$\chi(G) = -\#\langle\alpha_2\rangle + \#\langle\alpha_0, \alpha_1\rangle + \#\langle\alpha_0, \alpha_2\rangle + \#\langle\alpha_1, \alpha_2\rangle$$
$\#\langle\rangle$ denotes the number of orbits, see Section 8.3.2.1 page 329 for details on this characteristic;
- $q(G)$ is the orientability factor of G defined by:
$$q(G) = \begin{cases} 0 & \text{if G is orientable;} \\ 1 & \text{if G is nonorientable and } (b(G) + q(G)) \text{ is odd;} \\ 2 & \text{otherwise.} \end{cases}$$
- $g(G)$ is the genus of G defined by $g(G) = 1 - (b(G) + \chi(G) + q(G))/2$.

These characteristics make it possible to define a classification of connected 2-Gmaps by the triple $(b(G), q(G), g(G))$ related to the classification of the associated surfaces. Note that we can classify nonconnected 2-Gmaps by classifying each of its connected component. The following array gives some characteristics of usual surfaces:

Type of surface	$(b(G), q(G), g(G))$
Sphere	$(0, 0, 0)$
Torus	$(0, 0, 1)$
Double torus	$(0, 0, 2)$
Disk	$(1, 0, 0)$
Annulus	$(2, 0, 0)$
Projective plane	$(0, 1, 0)$
Klein bottle	$(0, 2, 0)$

Note that the number of n-boundaries, the Euler-Poincaré characteristic and the orientability notion are defined for any n-Gmaps whatever their dimension is (cf. Section 8.3.2.1 page 329 for Euler-Poincaré characteristic), but for $n > 2$, they do not provide a classification.

5

n-maps

An n-map is a combinatorial data structure allowing to describe an n-dimensional oriented quasi-manifold with or without boundary. The main difference with n-Gmaps introduced in the previous chapter is the fact that n-maps cannot describe nonorientable quasi-manifolds. The main interest of n-maps comparing to n-Gmaps is to use twice less darts for representing orientable quasi-manifolds[1]. The main drawback is the "inhomogeneity" of the definition, which often involves more complex algorithms. We structure this chapter as for n-Gmaps, in order to emphasize the similarities and the differences between the two data structures. n-maps are defined in Section 5.1, as the basic notions of cells, incidence and adjacency relations between the cells. Some basic operations allowing to modify existing n-maps are presented in Section 5.2. These operations allow to add/remove darts, increase/decrease the dimension of an n-map, merge/split n-maps; the sew/unsew operations allow to identify cells. We show in Section 5.3 that these operations make a small basis of operations allowing to build any n-map. Moreover, it is pointed out that it is possible to take multi-incidence between cells into account, and some specific configurations are illustrated, related to multi-incidence such as dangling cells and folded cells. A possible data structure for implementing n-maps is proposed in Section 5.4, and also some algorithms allowing to develop a computer software handling n-maps. Some additional notions related to n-maps are presented in Section 5.5. The relations between n-maps and n-Gmaps are studied in Section 5.5.2.

5.1 Basic Definitions

As seen in chapter 3, an n-map is defined by a set of darts on which permutations and involutions act, which satisfy some constraints. This leads to the following definition of n-maps.

Definition 26 (n-map) *An n-dimensional combinatorial map, or n-map, with $0 \leq n$, is an $(n+1)$-tuple $M = (D, \beta_1, \ldots, \beta_n)$ where:*

[1] When an orientable quasi-manifold is represented, the corresponding n-Gmap represents the two possible orientations, but the corresponding n-map represents only one orientation.

1. *D is a finite set of darts;*
2. *β_1 is a partial permutation on D; let β_0 denote β_1^{-1};*
3. *$\forall i \in \{2, \ldots, n\}$: β_i is a partial involution on D;*
4. *$\forall i \in \{0, \ldots, n-2\}$, $\forall j \in \{3, \ldots, n\}$, such that $i + 2 \leq j$, $\beta_i \circ \beta_j$ is a partial involution.*

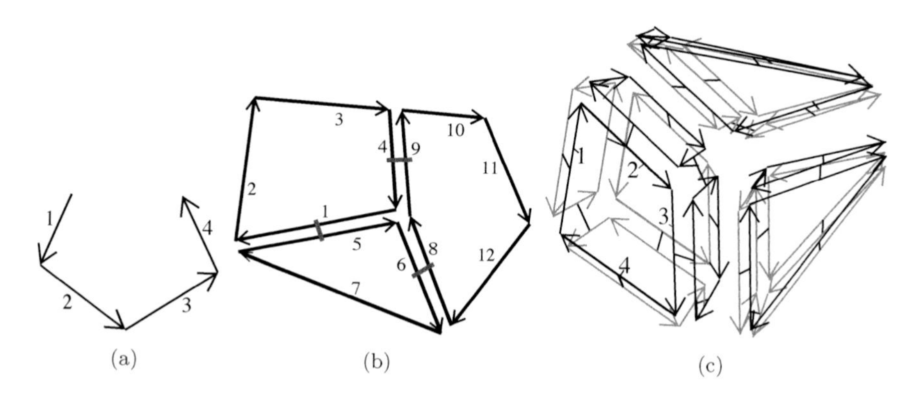

FIGURE 5.1
Examples of 1-map, 2-map and 3-map.
(a) A 1-map (D, β_1).
(b) A 2-map (D, β_1, β_2).
(c) A 3-map $(D, \beta_1, \beta_2, \beta_3)$.

A 0-map (D) is a set of darts representing isolated vertices, a 1-map (D, β_1) represents the structure of oriented polygonal curves (cf. Fig. 5.1(a)), a 2-map (D, β_1, β_2) represents the structure of oriented subdivided surfaces (cf. Fig. 5.1(b)), a 3-map $(D, \beta_1, \beta_2, \beta_3)$ represents the structure of assemblies of oriented volumes (cf. Fig. 5.1(c))...

$\beta_0 = \beta_1^{-1}$ is a partial permutation. For $i \in \{2, \ldots, n\}$, β_i is a partial involution, and thus $\beta_i = \beta_i^{-1}$.

Since all β relations are *partial* relations, darts can be linked with $\varnothing$: this corresponds to the fact that n-maps can describe objects with boundaries. Dart d is *i-free* if it is linked with $\varnothing$ by β_i, and it is *i-sewn* otherwise (cf. Def. 27 and example in Fig. 5.2).

Definition 27 (i-free, i-sewn) *Let $M = (D, \beta_1, \ldots, \beta_n)$ be an n-map, $d \in D$, and $i \in \{0, \ldots, n\}$:*

- *d is i-free if $\beta_i(d) = \varnothing$;*
- *d is i-sewn with dart $d_2 \in D$ if $\beta_i(d) = d_2$.*

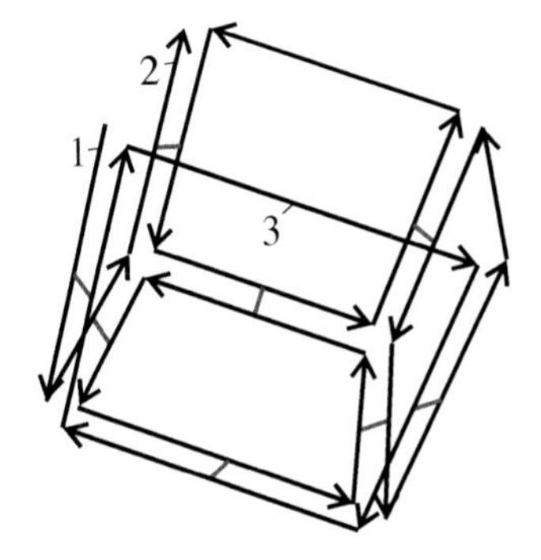

FIGURE 5.2
A 2-map with 1-boundary and 2-boundary.
For example dart 1 is 0-free, i.e. $\beta_0(1) = \varnothing$. Thus dart 0 belongs to an edge which has no previous edge in the same face.
Dart 2 is 1-free, i.e. $\beta_1(2) = \varnothing$. Thus dart 2 belongs to an edge which has no next edge in the same face.
Dart 3 is 2-free, i.e. $\beta_2(3) = \varnothing$, so it belongs to an edge which is incident to only one face.

In all generality, boundaries can exist in all dimensions; in practical applications, nD objects are often without boundary, except for dimension n, i.e. there are no free darts except for β_n.

Note that a 0-boundary (i.e. an edge incident to only one vertex, as for instance an edge cut by a plane) corresponds to a dart free for all $i \in \{1, \ldots, n\}$. Indeed, as soon as a dart d_1 is i-sewn with a dart d_2, for $i \in \{1, \ldots, n\}$, the edge containing dart d_1 is incident to two vertices: the vertex containing dart d_1, and the vertex containing dart d_2 (cf. definition of cells below).

As seen in the intuitive presentation, an n-map can contain different connected components. The set of darts belonging to the same connected component than a given dart can be retrieved, thanks to Def. 28. This is the set of darts of the orbit made of all the permutations.

Definition 28 (connected component) *Let $M = (D, \beta_1, \ldots, \beta_n)$ be an n-map, and $d \in D$. $\langle\beta_1, \ldots, \beta_n\rangle(d)$ is the connected component containing d.*

As said before, according to the context, an orbit denotes either a set of darts, or this set of darts together with the corresponding β relations. Thanks to this definition, it can be directly tested if a given n-map is connected, i.e. made of only one connected component, simply by testing if the connected component containing a dart is the set of all the darts of the n-map (cf. Def. 29).

Definition 29 (connected n-map) *Let $M = (D, \beta_1, \ldots, \beta_n)$ be an n-map. M is connected if $\langle\beta_1, \ldots, \beta_n\rangle(d)$ is equal to D, with $d \in D$.*

As for graph theory, the notion of connexity is related to the notion of path given in Def. 30.

Definition 30 (path) *Let $M = (D, \beta_1, \ldots, \beta_n)$ be an n-map and $d, d' \in D$. A path between d and d' is a sequence of darts $(d_1, \ldots, d_k)$ such that $d_1 = d$, $d_k = d'$ and $\forall i \in \{2, \ldots, k\}$, $d_i = \beta_{k_i}(d_{i-1}) \in D$, with $k_i \in \{0, \ldots, n\}$.*

Given any pair of darts d and d' that belong to the same connected component, a path exists between d and d'. Reciprocally, if d and d' do not belong to the same connected component, there is no path between d and d'.

In an n-map, a dart corresponds to an $(n+1)$-tuple of oriented cells of different dimensions (vertex, edge, face, volume ...), each pair of cells in this tuple being incident. Moreover, the vertex is the origin of the oriented edge. In other words, a dart describes a "small part" of each oriented cell (this explains why cells are not the basic elements of n-maps). A consequence is the fact that very local notions and operations can be defined: for instance, if an edge is a loop, its two extremities can be distinguished, even if it is incident to only one vertex. More generally, n-maps naturally take multi-incidence into account, and this can be useful for representing free-form objects.

As we have seen in the intuitive presentation, a cell containing a given dart corresponds to some specific orbit (cf. Def. 31).

Definition 31 (i-cell) *Let $M = (D, \beta_1, \ldots, \beta_n)$ be an n-map, $d \in D$ and $i \in \{0, \ldots, n\}$. The i-dimensional cell (or i-cell) containing d is $c_i(d)$:*

- *if $i = 0$, $c_0(d) = \langle \{\beta_j \circ \beta_k | \forall j, k \in \{1, \ldots, n\},\ j < k\} \rangle (d)$;*
- *otherwise, $c_i(d) = \langle \beta_1, \ldots, \beta_{i-1}, \beta_{i+1}, \ldots, \beta_n \rangle (d)$.*

The cells can equivalently be defined as the connected components of the $(n-1)$-map of cells defined in Def. 32. For $i = 0$, this property is only satisfied for n-maps without 1-boundary. Indeed, in such a map, the 0-cell definition is equal to the simplified form $\langle \beta_2 \circ \beta_0, \ldots, \beta_n \circ \beta_0 \rangle$ while this is not true for n-maps with 1-boundary (as shown for the example given in Fig. 3.16 page 68).

Definition 32 ($(n-1)$-map of i-cells) *Let $M = (D, \beta_1, \ldots, \beta_n)$ be an n-map, with $n \geq 1$, and $i \in \{0, \ldots, n\}$. The $(n-1)$-map of i-cells is M_{c_i}:*

- *if $i = 0$, $M_{c_0} = (D, \beta_2 \circ \beta_0, \ldots, \beta_n \circ \beta_0)$;*
- *otherwise, $M_{c_i} = (D, \beta_1, \ldots, \beta_{i-1}, \beta_{i+1}, \ldots, \beta_n)$.*

Thanks to this definition of cells, the incidence and adjacency relations between cells can be directly retrieved (cf. Def. 33).

Definition 33 (incidence and adjacency) *Let $M = (D, \beta_1, \ldots, \beta_n)$ be an n-map, $d, d' \in D$, and $i, j \in \{0, \ldots, n\}$, $i \neq j$:*

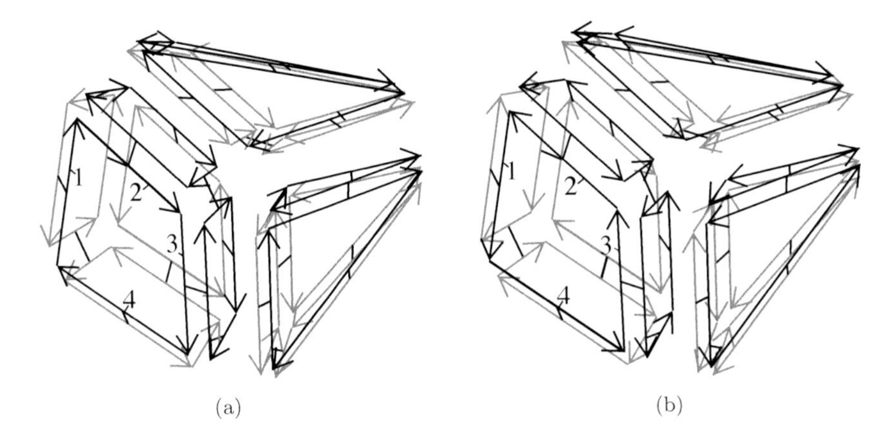

FIGURE 5.3
(a) 3-map M. For example, $\beta_1(1) = 2$, $\beta_1(2) = 3$, $\beta_1(3) = 4$ and $\beta_1(4) = 1$.
(b) The inverse 3-map M^{-1}. For example, $\beta_1(1) = 4$, $\beta_1(4) = 3$, $\beta_1(3) = 2$ and $\beta_1(2) = 1$.

- *cells* $c_i(d)$ *and* $c_j(d')$ *are incident iff* $c_i(d) \cap c_j(d') \neq \emptyset$*;*
- *cells* $c_i(d)$ *and* $c_i(d')$ *are adjacent iff two darts* d_1 *and* d_2 *exist, such that* $d_1 \in c_i(d)$, $d_2 \in c_i(d')$*, and:*
  - *if* $i = 0$*:* $d_1 = \beta_k(d_2)$ *with* $k \in \{0, \ldots, n\}$*;*
  - *if* $i > 0$*:* $d_1 = \beta_i(d_2)$ *or* $d_2 = \beta_i(d_1)$*.*

Note that there are two different cases for the adjacency definition: a first one for 0-cells and a second one for i-cells when $i > 0$. Two vertices are incident if two darts exist in the two vertices which are linked by any β_k. For $i > 0$, the two darts must be linked by β_i. Note that the second condition of the adjacency definition ($d_2 = \beta_i(d_1)$) is required for $i = 1$, but not if i is greater than 1, because in this case β_i is an involution. Note also that, according to this definition, a cell is never incident to itself, since i must be different from j; but it can be adjacent to itself: for instance a loop is adjacent to itself, and more generally cells which are multi-incident to other cells: cf. Section 5.3.

An n-map M describes an nD oriented quasi-manifold. The *inverse* of M describes the same quasi-manifold, but with the reverse orientation (cf. Def. 34 and example in Fig. 5.3).

Definition 34 (inverse) *Let* $M = (D, \beta_1, \ldots, \beta_n)$ *be an* n*-map. The* n*-map inverse of* M *is* $M^{-1} = (D, \beta_1^{-1}, \beta_2, \ldots, \beta_n)$*.*

It can be tested if two n-maps describe the same subdivision, through the *isomorphism* notion. Two n-maps are isomorphic if there is a one-to-one mapping between their darts which preserves all relations (cf. Def. 35).

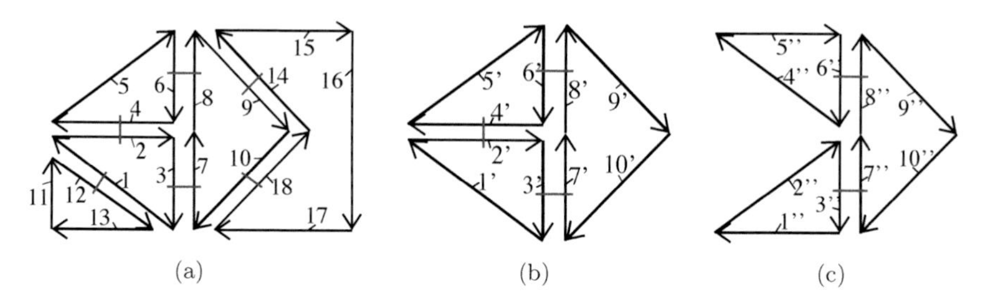

FIGURE 5.4
Submap isomorphism.
(a) 2-map M.
(b) 2-map M' is a submap of M as it is obtained from M by deleting darts 11 to 18.
(c) 2-map M'' is not isomorphic to a submap of M, as the injection f that respectively matches darts $1''$ to $10''$ with darts 1 to 10 does not verify the conditions of Def. 36: dart $4''$ is 2-free and it is matched with dart 4 which is 2-sewn with dart 2 which is itself matched (i.e. $f^{-1}(2) = 2''$).

Definition 35 (isomorphism) *Two n-maps $M = (D, \beta_1, \ldots, \beta_n)$ and $M' = (D', \beta'_1, \ldots, \beta'_n)$ are isomorphic if there exists an* isomorphism *mapping D onto D', i.e. a one-to-one mapping $f : D \cup \{\varnothing\} \to D' \cup \{\varnothing\}$ such that $f(\varnothing) = \varnothing$, and $\forall d \in D$, $\forall i \in \{1, \ldots, n\}$, $f(\beta_i(d)) = \beta'_i(f(d))$.*

We can also test if an n-map M' is included in a second n-map M, i.e. if there exists a part of M being isomorphic to M'. M' is subisomorphic to M if an injective mapping exists between the darts of M' onto the darts of M which preserves all the relations inside the concerned part of M (cf. Def. 36 and example in Fig. 5.4).

Definition 36 (subisomorphism) *An n-map $M' = (D', \beta'_1, \ldots, \beta'_n)$ is iso-*morphic to a submap *of $M = (D, \beta_1, \ldots, \beta_n)$ if there exists a* subisomorphism *between D' and D, i.e. an injective mapping $f : D' \cup \{\varnothing\} \to D \cup \{\varnothing\}$ such that $f(\varnothing) = \varnothing$, and $\forall d \in D'$, $\forall i \in \{1, \ldots, n\}$:*

- *if d is not i-free, then $f(\beta'_i(d)) = \beta_i(f(d))$;*
- *otherwise, either $f(d)$ is i-free, or $\forall d_k \in D', f(d_k) \neq \beta_i(f(d)))$.*

5.2 Basic Operations

In the previous section, several basic definitions concerning n-maps have been introduced. Now we present some basic operations allowing to modify existing

n-maps. Note that many high level operations have been defined, sometimes based on these basic ones, allowing to construct and modify n-maps. Even if the basic operations introduced in this section can be directly used to interactively construct an object, their interest is mainly theoretical: they provide a small basis of operations allowing to build any n-map. The high level operations usually handle orbits, as cells for instance, although basic operations handle directly the darts and the β functions. All these operations take an n-map as input, and produce another n-map which is the result of the operation.

5.2.1 Basic Tools

The first basic tool consists in adding a new dart in an n-map. To guaranty the validity of the n-map, the new dart is isolated, i.e. $\forall i \in \{0, \dots, n\}$, $\beta_i(d) = \varnothing$. This operation can be used several times to add any number of isolated darts in the n-map.

Definition 37 (add isolated dart) *Let $M = (D, \beta_1, \dots, \beta_n)$ be an n-map. The n-map obtained from M by adding an isolated dart is $M_{+d} = (D \cup \{d\}, \beta'_1, \dots, \beta'_n)$; $\forall i \in \{1, \dots, n\}$, β'_i is defined by:*

- $\forall e \in D$, $\beta'_i(e) = \beta_i(e)$;
- $\beta'_i(d) = \varnothing$.

The inverse operation given in Def. 38 consists in removing an isolated dart.

Definition 38 (remove isolated dart) *Let $M = (D, \beta_1, \dots, \beta_n)$ be an n-map, and let $d \in D$ be an isolated dart, i.e. $\forall i \in \{0, \dots, n\}$, $\beta_i(d) = \varnothing$. The n-map obtained from M by removing d is $M_{-d} = (D \setminus \{d\}, \beta'_1, \dots, \beta'_n)$; $\forall i \in \{1, \dots, n\}$, β'_i is the restriction of β_i to the set of remaining darts, i.e. $\beta'_i = \beta_{i|D\setminus\{d\}}$.*

Since only isolated darts are removed, the result of the operation is obviously an n-map. Note that any dart can be removed, but it is necessary to isolate it by applying (maybe several times) the unsew operation (cf. Section 5.2.3 for the definition of the unsew operation).

The operation which increases the dimension of an n-map is straightforward: cf. Def. 39. Increasing the dimension of an $(n-1)$-map M by one leads to create one n-cell for each connected component of M. For any i in $\{0, \dots, n-1\}$, no i-cell is modified by this operation.

Definition 39 (increase dimension) *Let $M = (D, \beta_1, \dots, \beta_{n-1})$ be an $(n-1)$-map. The n-map obtained from M by increasing its dimension is $M^+ = (D, \beta_1, \dots, \beta_n)$, where β_n is defined by: $\forall d \in D$, $\beta_n(d) = \varnothing$.*

The operation which decreases the dimension of an n-map is defined in Def. 40. This operation is the inverse of the increase dimension operation. It leads to the disappearance of all the n-cells of M.

Definition 40 (decrease dimension) *Let $M = (D, \beta_1, \ldots, \beta_n)$ be an n-map such that $\forall d \in D$, $\beta_n(d) = \varnothing$. The n-map obtained from M by decreasing its dimension is $M^- = (D, \beta_1, \ldots, \beta_{n-1})$.*

Note that this operation is restricted to an n-map which has all its darts n-free, in order to get the inverse of the operation which increases the dimension. However, it is possible to first n-unsew all the non n-free darts, and then to decrease the dimension of the resulting n-map.

Two distinct n-maps can be merged into a same n-map, which connected components are thus the connected components of the initial n-maps (cf. Def. 41). Its number of cells (resp. of connected components) is the sum of the numbers of cells (resp. of connected components) of the two initial n-maps.

Definition 41 (merge) *Let $M = (D, \beta_1, \ldots, \beta_n)$ and $M' = (D', \beta'_1, \ldots, \beta'_n)$ be two n-maps such that $D \cap D' = \emptyset$. The n-map obtained by merging M and M' is $M \cup M' = (D'', \beta''_1, \ldots, \beta''_n)$ defined by:*

- $D'' = D \cup D'$;
- $\forall d \in D''$, $\forall i \in \{1, \ldots, n\}$, $\beta''_i(d) = \begin{cases} \beta_i(d) & \text{if } d \in D; \\ \beta'_i(d) & \text{otherwise.} \end{cases}$

The reverse operation consists in splitting a given n-map in two. It is based on the restrict operation defined in Def. 42: it takes as input a given n-map and a subset of darts which correspond to a set of connected components, and build the n-map restricted to this set of darts.

Definition 42 (restrict) *Let $M = (D, \beta_1, \ldots, \beta_n)$ be an n-map, and let $D' \subseteq D$ be such that $\forall d \in D'$, $\forall i \in \{0, \ldots, n\}$, $\beta_i(d) \in D'$. The n-map obtained by restricting M to D' is $M_{|D'} = (D', \beta'_1, \ldots, \beta'_n)$, such that $\forall i \in \{1, \ldots, n\}$, $\beta'_i = \beta_{i|D'}$.*

Note that the restrict operation is only possible if the given set of darts D' corresponds to a set of connected components. Otherwise, some darts would be linked with darts that do not belong to D', contradicting the permutation and involution definitions, and thus the definition of n-maps. However, the restrict operation can be applied to any set of darts, simply by applying first the required unsew operations for isolating this set of darts.

The split operation, given in Def. 43, takes as input an n-map M and a subset of darts D', and produces two n-maps by using twice the restrict operation: a first time for D' and the second time for $D \setminus D'$.

Definition 43 (split) *Let $M = (D, \beta_1, \ldots, \beta_n)$ be an n-map, and let $D' \subseteq D$ be such that $\forall d \in D'$, $\forall i \in \{0, \ldots, n\}$, $\beta_i(d) \in D'$. The two n-maps obtained by splitting M along D' are $M_{s1}(D') = M_{|D'}$ and $M_{s2}(D') = M_{|D \setminus D'}$.*

It is straightforward to prove that each operation is the inverse of its corresponding operation. For that, it suffices to show that, starting from M, and applying an operation and then the corresponding inverse operation, the resulting n-map M' is isomorphic to M. For each operation, it is easy to choose the parameters of the operations which lead to this conclusion.

5.2.2 Sew Operations

The i-sew operation allows to "link" some i-free darts by β_i, for a given i. In order to satisfy the "composition constraint" of the n-map definition (i.e. $\beta_i \circ \beta_j$ is a partial involution for $|i-j| \geq 2$), we often need to put in relation not only two darts but several darts: for instance, if two darts are linked by β_3 in a 3-map, it is necessary to link also the darts belonging to the two faces containing these two darts. For the i-sew operation, we choose to put in relation the minimal number of darts such that the result of the operation is an n-map.

In fact, the i-sew operation implies some identifications of cells: for instance, when two darts are sewn by β_2 in a 2-map, the corresponding edges are identified into a single edge: this identification can involve the identification of the vertices incident to the initial edges.

For this reason, it is not possible to apply the i-sew operation to any set of darts: the cells to identify must have the same structure, i.e. they have to be isomorphic to each other. Indeed, when two cells have different structures, we do not know how to identify them. For example it is not possible to sew a cube and a tetrahedron since we do not know how to identify a square face with a triangle face. So, before defining the sew operation, it is necessary to be able to test if a sew is possible. The property of "being i-sewable" is detailed in Def. 44.

Definition 44 (i-sewable) *Let $M = (D, \beta_1, \ldots, \beta_n)$ be an n-map, $i \in \{1, \ldots, n\}$, $d, d' \in D$. Let o_d denote $\langle \beta_1, \ldots, \beta_{i-2}, \beta_{i+2}, \ldots, \beta_n \rangle (d)$, and $o_{d'}$ denote $\langle \beta_1, \ldots, \beta_{i-2}, \beta_{i+2}, \ldots, \beta_n \rangle (d')$. d is i-sewable with d' if $\beta_i(d) = \varnothing$, $\beta_i^{-1}(d') = \varnothing$ and an isomorphism*[2] *f exists between o_d and the inverse of $o_{d'}$, such that $f(d) = d'$. Moreover, if $o_d = o_{d'}$ then $f = f^{-1}$.*

Considering the whole orbits $\langle \beta_1, \ldots, \beta_{i-2}, \beta_{i+2}, \ldots, \beta_n \rangle (d)$ and $\langle \beta_1, \ldots, \beta_{i-2}, \beta_{i+2}, \ldots, \beta_n \rangle (d')$ is required in order to satisfy, after the sew, the condition that $\beta_i \circ \beta_j$ is a partial involution when $|i-j| \geq 2$. In fact, the i-sew operation will "link" together all the darts of o_d with the darts of $o_{d'}$. d must be i-free and d' must be i^{-1}-free (to consider correctly the case when $i = 1$), so all the darts in o_d and $o_{d'}$ are also free since $\beta_i \circ \beta_j$ are partial

[2] f is an isomorphism between o_d and the inverse of $o_{d'}$ iff:

$$\forall e \in o_d, \forall j \in \{1, \ldots, i-2, i+2, \ldots, n\}, f(\beta_j(e)) = \beta_j^{-1}(f(e))$$

involutions when $|i - j| \geq 2$. Note that a dart can be i-sewn with itself; more generally, it is possible that $o_d = o_{d'}$: the corresponding sew will create *folded* cells (see Section 5.3).

Note that any pair of free darts is always i-sewable in any 1-map and in any 2-map; for these cases, orbits o_d and $o_{d'}$ contain only one dart, and the set $\{1, \ldots, i-2, i+2, \ldots, n\}$ is empty. So, no precondition is required for sewing darts in 1-maps and in 2-maps. Similarly, it is always possible to 2-sew free darts in 3-maps.

The notion of "being i-sewable" is used to define the i-sew operation in Def. 45. When applying the i-sew operation, all the darts in o_d and $o_{d'}$ are linked by β_i. This implies the identification of the two $(i-1)$-cells containing darts d and d'. Moreover, several other cells can be identified i.e. the j-cells incident to the two identified $(i-1)$-cells, with $j \in \{0, \ldots, i-2, i+1, \ldots, n\}$ (cf. below for some examples).

Definition 45 (i-sew) *Let $M = (D, \beta_1, \ldots, \beta_n)$ be an n-map, $n > 0$, $i \in \{2, \ldots, n\}$, and let $d, d' \in D$ be such that d and d' are i-sewable by the isomorphism f of Def. 44. The n-map obtained from M by the i-sewing of d and d' is $M_i\text{-}sew(d, d') = (D, \beta_1, \ldots, \beta_{i-1}, \beta'_i, \beta_{i+1}, \ldots, \beta_n)$, where β'_i is defined by: $\forall e \in D$,*

- *if $e \in o_d$, $\beta'_i(e) = f(e)$;*
- *if $o_d \neq o_{d'}$ and $e \in o_{d'}$, $\beta'_i(e) = f^{-1}(e)$;*
- *otherwise, $\beta'_i(e) = \beta_i(e)$.*

Given an n-map and two i-sewable darts, the i-sew operation consists simply in linking all the darts in o_d with their "equivalent" darts in $o_{d'}$. More precisely, β'_i is directly given by β_i and isomorphism f. So, the i-sew of d and d' produces the same n-map than the i-sew of any dart d'' of o_d and the corresponding dart $f(d'')$ of $o_{d'}$.

As often for n-maps, there is a special case for the dimension 1, described in Def. 46; this is due to the fact that β_1 is a partial permutation, as other β's are partial involutions. More precisely, there are two differences with the generic definition for other dimensions: first, β_1 being a partial permutation, the inverse relation has not to be defined; second, given two darts e and $e' = f(e)$, it is necessary to define $\beta_1(e) = e'$ if e has the same orientation than d, and $\beta_1(e') = e$ otherwise (see example in Fig. 5.6) to obtain a correct orientation of all the linked darts. e has the same orientation than d if $e \in \langle\{\beta_i \circ \beta_j | \forall i, j \in \{3, \ldots, n\}, i < j\}\rangle(d)$.

Definition 46 (1-sew) *Let $M = (D, \beta_1, \ldots, \beta_n)$ be an n-map,and let $d, d' \in D$ be such that d and d' are 1-sewable by the isomorphism f of Def. 44. Let o_1 denote $\langle\{\beta_i \circ \beta_j | \forall i, j \in \{3, \ldots, n\}, i < j\}\rangle(d)$. The n-map obtained from M by the 1-sewing of d and d' is $M_1\text{-}sew(d, d') = (D, \beta'_1, \beta_2, \ldots, \beta_n)$, where β'_1 is defined by: $\forall e \in D$,*

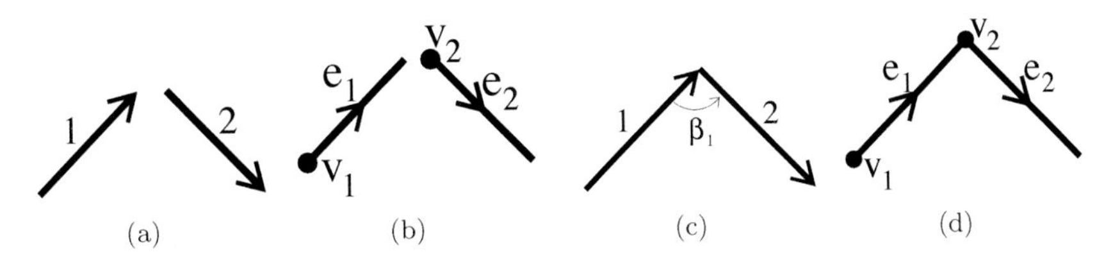

FIGURE 5.5
1-sew operation in a 1-map.
(a) 1-map $M = (D, \beta_1 = \varnothing)$ contains two isolated darts.
(b) The corresponding 1D quasi-manifold having two oriented edges (e_1 and e_2) and two represented vertices (v_1 and v_2).
(c) 1-map M_1-sew$(1, 2)$: now, $\beta_1(1) = 2$.
(d) The corresponding 1D quasi-manifold having two oriented edges and two represented vertice.

- *if* $e \in o_d$*: if* $e \in o_1$ *then* $\beta_1'(e) = f(e)$,
  else $\beta_1'(f(e)) = e$;
- *otherwise,* $\beta_1'(e) = \beta_1(e)$.

An example of 1-sew operation in a 1-map is shown in Fig. 5.5. Two isolated darts are linked by β_1. Note that in this case there is no identification of represented cells (since we are in the special case where an edge has only one of its vertex represented). As said above, the 1-sew operation in 1D concerns only the two given darts.

A second example of 1-sew operation is represented in Fig. 5.6, but now in a 3-map. In the initial 3-map shown in Fig. 5.6(a), the face $\{1, \ldots, 8\}$ between the two volumes has a 1-boundary. Indeed, darts 4 and 8 are 1-free. The 1-sew operation is applied to darts 4 and 1, producing the 3-map shown in Fig. 5.5(b): dart 4 is 1-sewn with dart 1, and dart 8 is 1-sewn with dart 5. Indeed, $o_4 = \langle\beta_3\rangle(4) = \{4, 5\}$, $o_1 = \langle\beta_3\rangle(1) = \{1, 8\}$ and the isomorphism f between o_4 and o_1 is such that $f(4) = 1$ and $f(5) = 8$. $\beta_1(4) = f(4) = 1$ because $4 \in \langle\rangle(4)$; $\beta_1(8) = \beta_1(f(5)) = 5$ because $5 \notin \langle\rangle(4)$. Note that the same result is received if the 1-sew operation is applied to darts 8 and 5.

An example of 2-sew operation in a 2-map is presented in Fig. 5.7. The two 1-cells containing darts 3 and 4 are identified, involving the identification of their two extremities. The 2-sew operation is always possible for any pair of 2-free darts in a 2-map, since in this case, o_d and $o_{d'}$ contain only one dart each.

For 3-maps, the 2-sew operation is similar as there is no additional constraint. Indeed, orbits o_d and $o_{d'}$ contain only dart d and dart d' respectively. Thus it is always possible to 2-sew any pair of 2-free darts. Note that the 2-sew operation can lead to the identification of two volumes, if the two 2-sewn darts did not belong to the same volume before applying the operation.

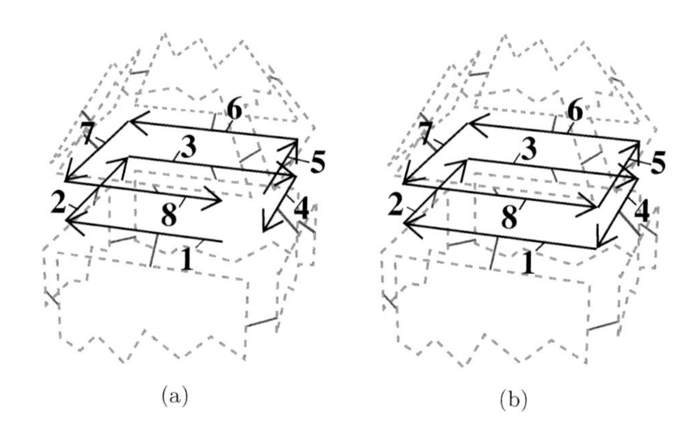

FIGURE 5.6
1-sew operation in a 3-map.
(a) Partial representation of 3-map M. Zoom on a face separating two volumes, which has a 1-boundary.
(b) 3-map $M_{1\text{-sew}}(4, 1)$: now, $\beta_1(4) = 1$ and $\beta_1(8) = 5$.

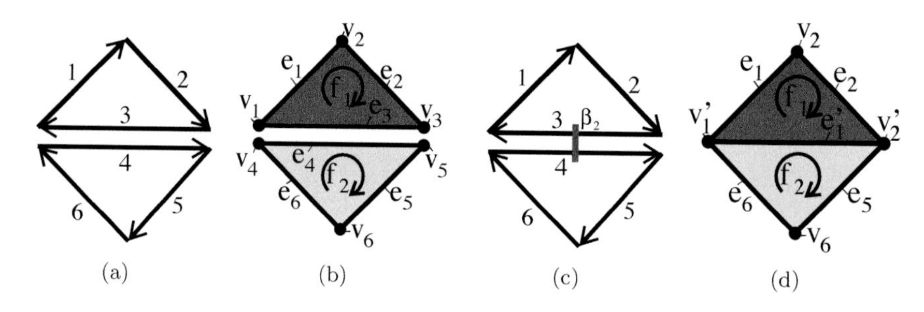

FIGURE 5.7
2-sew operation in a 2-map.
(a) 2-map $M = (D, \beta_1, \beta_2 = \varnothing)$ contains six darts.
(b) The corresponding 2D quasi-manifold has two oriented faces, six edges and six vertices.
(c) 2-map $M_{2\text{-sew}}(3, 4)$: $\beta_2(3) = 4$ (and $\beta_2(4) = 3$).
(d) The corresponding 2D quasi-manifold. Edges e_3 and e_4 are identified into edge e'_1; vertices v_1 and v_4 (resp. v_3 and v_5) are identified into vertex v'_1 (resp. v'_2).

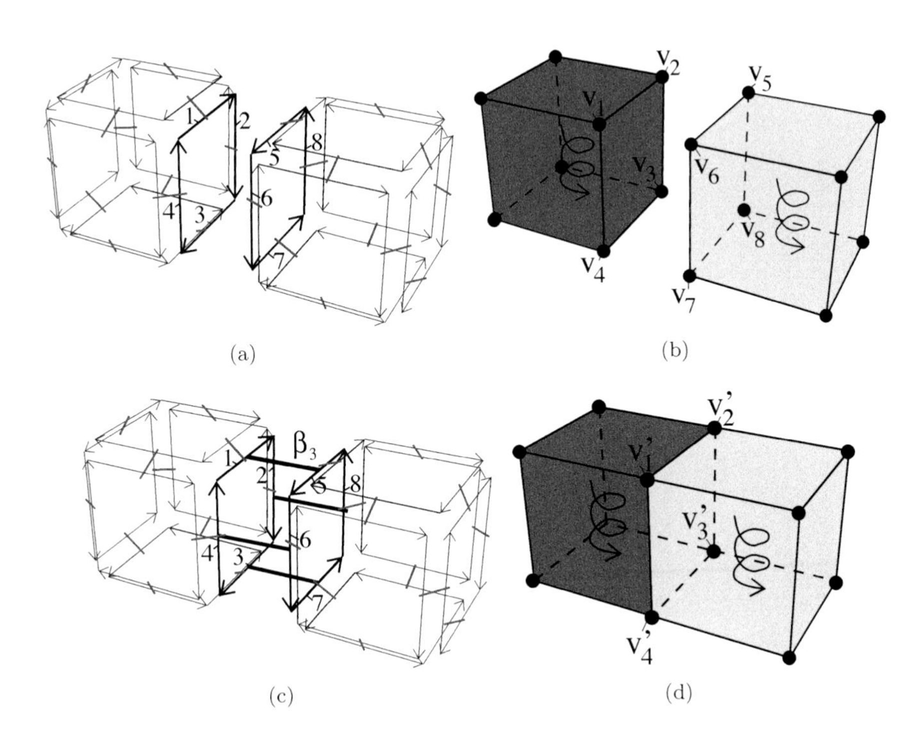

FIGURE 5.8
3-sew operation in a 3-map.
(a) 3-map $M = (D, \beta_1, \beta_2, \beta_3 = \varnothing)$ contains 48 darts.
(b) The corresponding 3D quasi-manifold has two oriented volumes, twelve oriented faces, twenty-four oriented edges and sixteen vertices.
(c) 3-map $G_{3\text{-sew}}(1, 5)$: $\beta_3(1) = 5$, $\beta_3(2) = 8$, $\beta_3(3) = 7$ and $\beta_3(4) = 6$ (and conversely).
(d) The corresponding 3D quasi-manifold. Let denote a cell by a sequence of its incident vertices. Faces (v_1,v_2,v_3,v_4) and (v_6,v_5,v_8,v_7) are identified into one face (v_1',v_2',v_3',v_4'). Edges (v_1,v_2) and (v_6,v_5) (resp. (v_2,v_3) and (v_5,v_8), (v_3,v_4) and (v_8,v_7), (v_4,v_1) and (v_7,v_6)) are identified into edge (v_1',v_2') (resp. (v_2',v_3'), (v_3',v_4'), (v_4',v_1')). Vertices v_1 and v_6 (resp. v_2 and v_5, v_3 and v_8, v_4 and v_7) are identified into vertex v_1' (resp. v_2', v_3', v_4').

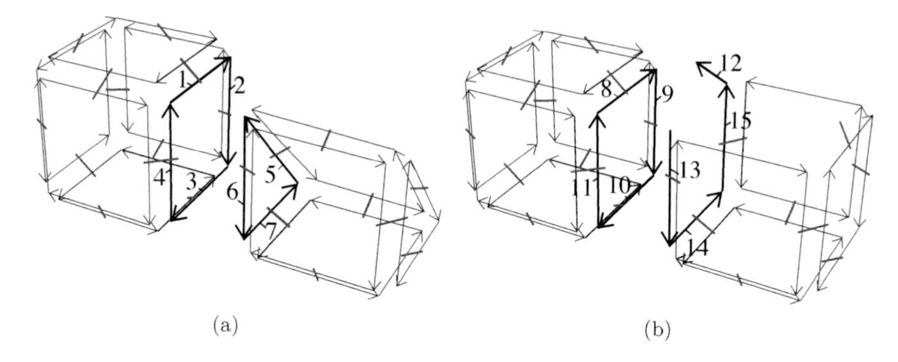

FIGURE 5.9
Two examples of non 3-sewable darts in a 3-map.
(a) Darts 1 and 5 are not 3-sewable. It is not possible to define a one to one mapping, since the two 2-cells do not have the same number of darts.
(b) Darts 8 and 12 are not 3-sewable. Let f be the one to one mapping defined by $f(8) = 12$, $f(9) = 15$, $f(10) = 14$ and $f(11) = 13$. $\beta_1(f(8)) = \beta_1(12) = \varnothing$ while $f(\beta_1^{-1}(8)) = f(11) = 13$. This is not an isomorphism.

An example of 3-sew operation in a 3-map is presented in Fig. 5.8. Two 2-cells are identified by linking two by two by β_3 all the darts belonging to the two initial faces. This identification of faces involves the identification of the boundaries of the two faces (each pair of edges and each pair of vertices of the two faces in relation by the isomorphism f are identified two by two). Such 3-sew operation is not always possible: the two initial faces must be 3-sewable, i.e. they must have the same structure. In this example, it is easy to verify that darts 1 and 5 are 3-sewable. $o_1 = \{1, 2, 3, 4\}$ and $o_2 = \{5, 6, 7, 8\}$ are isomorphic, the isomorphism f is defined by $f(1) = 5$, $f(2) = 8$, $f(3) = 7$ and $f(4) = 6$, and for each dart $e \in o_1$, $\beta_1(f(e)) = f(\beta_1^{-1}(e))$.

The 3-sew operation cannot be applied to the two examples depicted in Fig. 5.9. In the first example, darts 1 and 5 are not 3-sewable since faces $c_2(1)$ and $c_2(5)$ do not have the same number of darts. Thus no one to one mapping exists between o_1 and o_5. In the second example, the two faces $c_2(8)$ and $c_2(12)$ do not have the same structure: the first one has no 1-boundary while the second one has one 1-boundary. Thus no isomorphism exists between o_8 and o_{12}.

5.2.3 Unsew Operations

The i-unsew operation is defined in Def. 47 (in Def. 48 for the specific case of the 1-unsew operation); it is the inverse of the i-sew operation. It is always possible to i-unsew non i-free darts.

Definition 47 (i-unsew) *Let $M = (D, \beta_1, \ldots, \beta_n)$ be an n-map, $i \in$*

$\{2,\ldots,n\}$ and let $d \in D$ be a non i-free dart. The n-map obtained from M by i-unsewing d and $\beta_i(d)$ is $M_i\text{-}unsew(d) = (D, \beta_1, \ldots, \beta_{i-1}, \beta'_i, \beta_{i+1}, \ldots, \beta_n)$, where β'_i is defined by: $\forall e \in D$,

- *if $e \in \langle \beta_1, \ldots, \beta_{i-2}, \beta_i, \beta_{i+2}, \ldots, \beta_n \rangle(d)$, $\beta'_i(e) = \varnothing$;*
- *otherwise, $\beta'_i(e) = \beta_i(e)$.*

The i-unsew operation consists in setting β_i to $\varnothing$ for each dart of the orbit $\langle \beta_1, \ldots, \beta_{i-2}, \beta_i, \beta_{i+2}, \ldots, \beta_n \rangle(d)$. So, the i-unsew applied to any dart of this orbit produces the same result.

For the 1-unsew operation given in Def. 48, two different orbits must be considered: first $o_1 = \langle \{\beta_i \circ \beta_j | \forall i, j \in \{3, \ldots, n\},\ i < j\} \rangle(d)$ is all the darts of the first edge having the same orientation than d, second $o_2 = \langle \beta_3, \ldots, \beta_n \rangle(\beta_1(d)) \setminus \langle \{\beta_i \circ \beta_j | \forall i, j \in \{3, \ldots, n\},\ i < j\} \rangle(\beta_1(d))$ is all the darts of the second edge having the opposite orientation than d. Each dart in o_1 and in o_2 must be linked by β_1 with $\varnothing$, all the other relations are not modified. In case of a folded edge, $o_1 \cup o_2$ is the set of all the darts of the folded edge.

Definition 48 (1-unsew) *Let $M = (D, \beta_1, \ldots, \beta_n)$ be an n-map and $d \in D$ be a non 1-free dart. Let o_1 denote $\langle \{\beta_i \circ \beta_j | \forall i, j \in \{3, \ldots, n\},\ i < j\} \rangle(d)$ and o_2 denote $\langle \beta_3, \ldots, \beta_n \rangle(\beta_1(d)) \setminus \langle \{\beta_i \circ \beta_j | \forall i, j \in \{3, \ldots, n\},\ i < j\} \rangle(\beta_1(d))$. The n-map obtained from M by 1-unsewing d and $\beta_1(d)$ is $M_1\text{-}unsew(d) = (D, \beta'_1, \beta_2, \ldots, \beta_n)$, where β'_1 is defined by: $\forall e \in D$,*

- *if $e \in o_1 \cup o_2$: $\beta'_1(e) = \varnothing$;*
- *otherwise: $\beta'_1(e) = \beta_1(e)$.*

It can be proven that i-sew and i-unsew are inverse from each other: applying i-sew to two i-free darts d and d', and then applying i-unsew to d, produces the original n-map; applying i-unsew to dart d, which is non i-free, and then applying i-sew to d and d', such that d and d' were originally linked by β_i, produces the original n-map. As a consequence, we can see examples of unsew operations by looking at the examples of sew operations and regarding figures in reverse direction. For example, considering the 3-map given in Fig. 5.6(b) and applying the 1-unsew operation to dart 4, the 3-map represented in Fig. 5.6(a) is obtained. As another example, starting from the 3-map given in Fig. 5.8(c), the 3-map given in Fig. 5.8(a) is obtained by applying the 3-unsew operation to dart 1.

5.3 Completeness, Multi-Incidence

Any n-map can be built thanks to the basic operations, by following one of the two basic construction methods. The first one consists in applying an iterative creation process: the dimension of the n-map is progressively increased, starting from 1 until the dimension of the object; at each step, some darts are sewn for the new dimension. The second one consists in starting directly with the dimension of the object, in adding some isolated darts, and in sewing some darts for any dimension. Note that, as for the basic operations, the main interest of these construction methods is mainly theoretical: usually, high level operations are used for practical applications.

5.3.1 Construction by Increasing Dimensions

The first construction method is the inverse of the decomposition process described for the intuitive presentation of n-maps (cf. Section 3.1). The process is initialized in 1D with a set of darts having β_1 equal to $\varnothing$. This set of darts describes a set of isolated oriented edges. Then some darts are 1-sewn, in order to glue the corresponding edges together. The 1-map describes the structure of oriented polygonal lines. The dimension of the map is increased by adding β_2, initialized to $\varnothing$. The 2-map describes the structure of isolated oriented faces, whose boundaries are the previous oriented polygonal lines. Now, some darts are 2-sewn in order to identify some edges and maybe their boundaries. So, the faces are glued along their boundary edges. The 2-map describes the topology of oriented subdivided surfaces, and so on.

This process can be generalized for any dimension. For that, the two following operations are iteratively applied:

- increase the dimension by adding β_n initialized to $\varnothing$;
- n-sew some darts in order to identify some $(n-1)$-cells (and their boundaries). Remember that the identification of two $(n-1)$-cells is possible only if they have the same structure.

An example showing the whole process of a 2D object construction is presented in Fig. 5.10. The initial 1-map contains only isolated darts (see (a)), corresponding to isolated oriented edges (see (e)). Seven 1-sew operations are applied in order to form oriented polygonal lines by gluing isolated edges (for example, 1-sewing darts 1 and 2, or darts 2 and 3). The 1-map shown in (b) corresponds to the subdivision given in (f). Now the dimension of the 1-map is increased in order to obtain the 2-map shown in (c), where β_2 is equal to $\varnothing$. This operation transforms all the oriented polygonal lines into oriented faces (see (g)). Lastly, the 2-sew operation is applied to darts 1 and 5, producing the 2-map shown in (d); it corresponds to the subdivided 2D object shown in (h).

A second example showing the construction of a 3D object is presented in Fig. 5.11. The first steps are similar to the previous 2D example. Starting from a 1-map made of isolated darts (see (a)), 1-sew operations are applied; then the dimension of the 1-map is increased (see (b)). 2-sew operations are applied, then the dimension of the 2-map is increased (see (c)). The last step

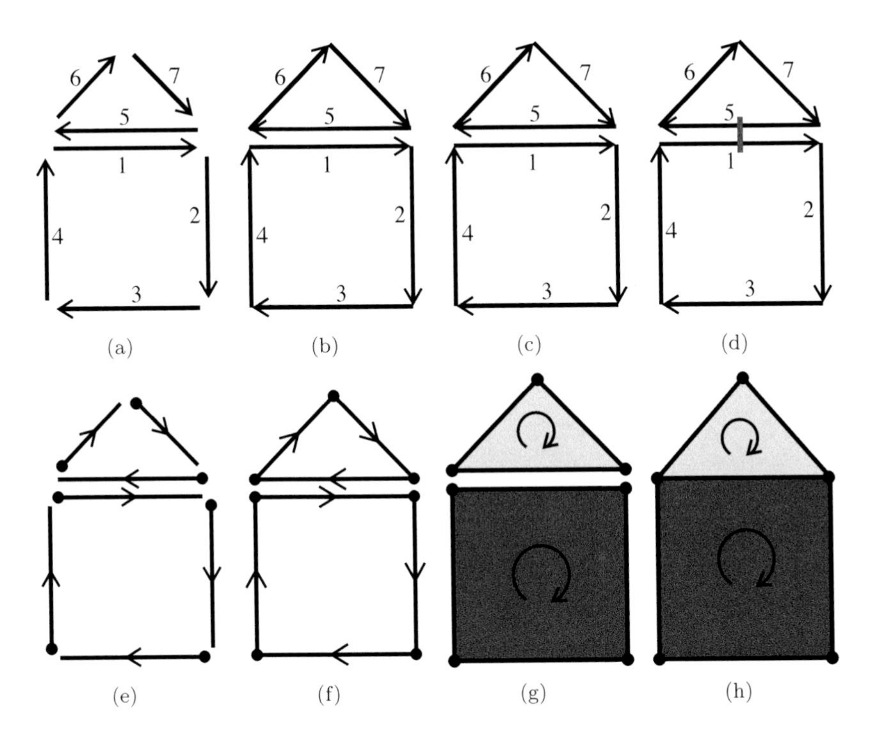

FIGURE 5.10
Construction of a 2D object by using the first method. The first line shows the different n-maps, and the second line the corresponding objects.
(a) 1-map $(D, \beta_1 = \varnothing)$ corresponds to (e), a set of isolated oriented edges.
(b) 1-map (D, β_1) obtained after seven 1-sew operations. This 1-map corresponds to the two oriented polygonal lines shown in (f).
(c) 2-map $(D, \beta_1, \beta_2 = \varnothing)$ corresponds to the two oriented faces shown in (g).
(d) 2-map (D, β_1, β_2) obtained after one 2-sew operation. In (h) the corresponding 2D subdivided object.

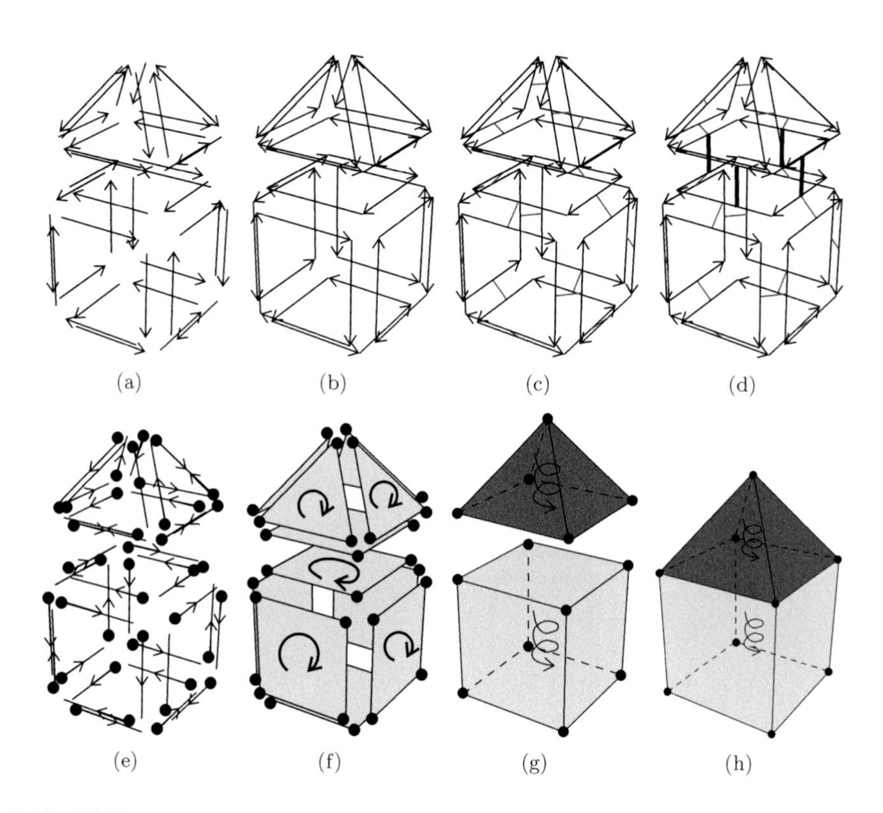

FIGURE 5.11
Construction of a 3D object by using the first method (partial representation). The first line shows the different n-maps, and the second line the corresponding objects.
(a) 1-map $(D, \beta_1 = \varnothing)$ corresponds to (e), a set of isolated oriented edges.
(b) 2-map $(D, \beta_1, \beta_2 = \varnothing)$ obtained after forty 1-sew operations and one increase dimension operation. This 2-map corresponds to eleven oriented faces, shown in (f).
(c) 3-map $(D, \beta_1, \beta_2, \beta_3 = \varnothing)$ obtained after twenty 2-sew operations and one increase dimension operation. This 3-map corresponds to the 3D subdivided object shown in (g) containing two volumes.
(d) 3-map $(D, \beta_1, \beta_2, \beta_3)$ obtained after one 3-sew operation, and (h) the corresponding 3D subdivided object.

corresponds to the 3-sew operation, producing the 3-map shown in (d): it corresponds to the 3D subdivided object shown in (h).

The inverse operations can be applied in order to destroy a given n-map: this corresponds exactly to the decomposition process described for the intuitive presentation of n-maps (cf. Section 3.1). First, n-unsew operations are applied until having all the darts n-free, then the dimension of the n-map is decreased. This process is iterated until obtaining in 1D a set of darts on which β_1 is equal to $\varnothing$. As examples of this "destruction process", we can consider the examples given for the construction process in reverse direction (cf. Figs. 5.10 and 5.11), and apply i-unsew (resp. decrease dimension) operations instead of i-sew (resp. increase dimension) operations.

5.3.2 Construction Directly in a Given Dimension

The second basic construction method consists in working directly in dimension n. Starting with an empty n-map, the operation which adds an isolated dart is applied as many times as necessary. Next, any i-sew operation can be applied, for $i \in \{1, \ldots, n\}$ (contrary to the previous construction where only k-sew operations can be applied to a k-map).

Fig. 5.12 represents an example of the construction of the object depicted in Fig. 5.10, applying this second construction method. The main differences with the first one are:

- the initial map is an n-map having its final dimension, and containing isolated darts (here a 2-map, see Fig. 5.12(a));
- different sew operations can be applied, in any order; for example, 1-sew operations can be first applied (see Fig. 5.12(b)), and then the 2-sew operation; conversely, 2-sew operations can be first applied (see Fig. 5.12(c)), then 1-sew operations. In both cases, the same 2-map is obtained: cf. Fig. 5.12(d).

A second construction of the object depicted in Fig. 5.11 is presented in Fig. 5.13. The initial 3-map contains a set of isolated darts (see (a)). Then the sew operations can be applied in any order. For example, the same order than in Fig. 5.11 can be followed: start by 1-sew operations, then 2-sew operations and finish by 3-sew operation. The 3-maps obtained after each step are similar to the ones in Fig. 5.11(a), (b), (c) and (d), the only difference being the dimension of the maps which is always 3 along the whole construction process. For this reason, the corresponding cellular objects are different since each dart describes always a part of a vertex, an edge, a face and a volume (contrary to the objects shown in Fig. 5.11).

Another order is chosen in Fig. 5.13. First, 3-sew operations are applied; the resulting 3-map is shown in (b). Then, 1-sew operations are applied, producing the 3-map shown in (c). The four 1-sew operations between darts that are not 3-free imply the modification of β_1 for the darts linked by β_3 (in order

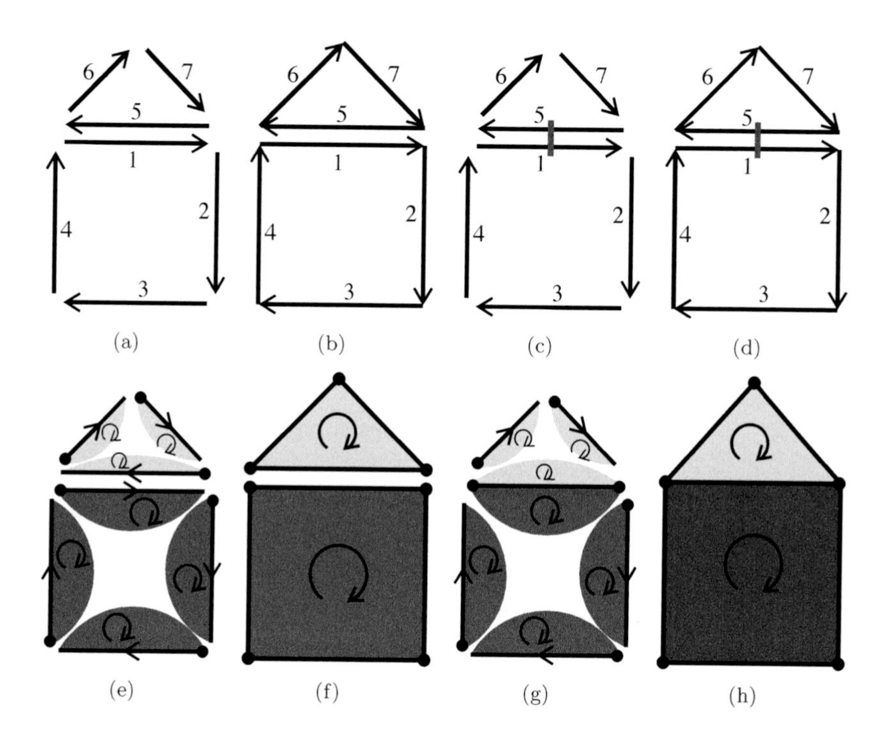

FIGURE 5.12
Construction of a 2D object by applying the second method. The first line shows the different 2-maps, and the second line the corresponding objects.
(a) 2-map $(D, \beta_1 = \varnothing, \beta_2 = \varnothing)$ corresponds to (e), a set of isolated oriented faces with boundaries.
(b) 2-map $(D, \beta_1, \beta_2 = \varnothing)$ obtained from (a) after seven 1-sew operations; it corresponds to the object shown in (f).
(c) 2-map $(D, \beta_1 = \varnothing, \beta_2)$ obtained from (a) after one 2-sew operation; it corresponds to the object shown in (g).
(d) 2-map (D, β_1, β_2) corresponds to the 2D oriented subdivided object shown in (h). This 2-map is either obtained from (b) after one 2-sew operation, or obtained from (c) after seven 1-sew operations.

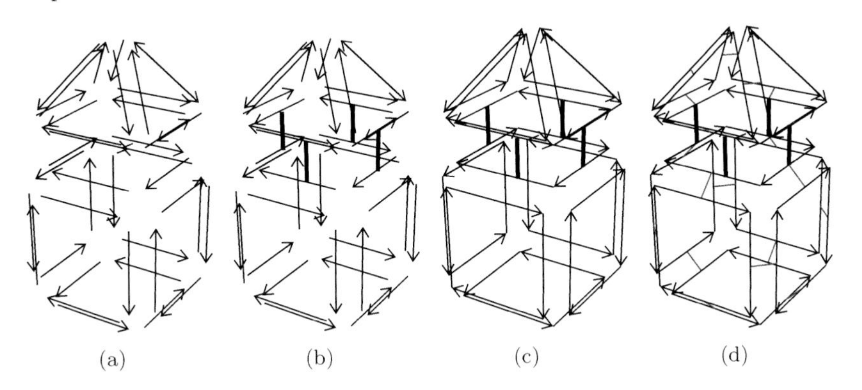

FIGURE 5.13
Construction of a 3D object by applying the second method (partial representation). In this example, the different 3-maps are drawn, but not the corresponding objects.
(a) 3-map $(D, \beta_1 = \varnothing, \beta_2 = \varnothing, \beta_3 = \varnothing)$, i.e. a set of isolated darts.
(b) 3-map obtained after four 3-sew operations.
(c) 3-map obtained after thirty-six 1-sew operations.
(d) 3-map obtained after twenty 2-sew operations.

to satisfy the constraint that $\beta_1 \circ \beta_3$ is a partial involution). Lastly, 2-sew operations are applied, producing the 3-map shown in (d).

The inverse operations can be applied in order to destroy a given n-map. Different i-unsew operations can be applied in any order, until all the darts are isolated; then, all these isolated darts are removed, producing an empty n-map. As example of these destruction operations, consider the examples given for the construction operations in reverse direction (cf. Figs. 5.12 and 5.13) and apply i-unsew operations instead of i-sew operations.

At last, note that for practical applications, working directly in the required dimension is more flexible, since sew (and/or unsew) operations can be applied in any order.

5.3.3 Completeness

Any n-map can be constructed thanks to the two construction methods.

Theorem 3 (Completeness of the first method for n-map construction or destruction) *Any n-map can be transformed into a set of isolated darts in dimension 0, by a sequence of unsew and decrease dimension operations, unsew being always applied for the highest possible dimension. Conversely, any n-map can be constructed starting from a set of isolated darts in dimension 0, by using only the increase dimension and sew operations, sew being always applied for the highest possible dimension.*

The proof of this theorem is easy. First, it is proved that any n-map can be decomposed into a set of darts in dimension 0 by applying a sequence of decrease dimension and unsew operations. Let M be an n-map. The n-unsew operation is applied to any non n-free dart. This operation can always be applied to a non n-free dart, and it increases at least by two the number of n-free darts. Thus, at the end, all the darts of the resulting n-map are n-free. Next, the decrease dimension operation can be applied, producing an $(n-1)$-map. The process can be reiterated until obtaining a 0-map composed by a set of isolated darts. In order to prove the second part of the theorem, let us consider the previous operation sequence in reverse order, and replace each operation by its inverse operation (decrease dimension by increase dimension and i-unsew by i-sew). This sequence transforms the 0-map into M.

A similar theorem and proof can be stated for the second construction method.

Theorem 4 (Completeness of the second method for n-map construction or destruction) *Any n-map can be transformed into an empty n-map by a sequence of i-unsew and remove isolated darts operations. Conversely, any n-map can be constructed starting from an empty n-map by applying the add isolated dart and i-sew operations.*

The proof of the first part of this theorem is based upon the fact that an i-unsew operation can be applied to any non i-free dart, increasing at least by two the number of i-free darts. And when a dart is isolated, it can be removed. We get thus a sequence of operations which at last destructs the n-map; in order to get the initial n-map, the inverse operations can be applied in reverse order, proving thus the theorem.

5.3.4 Multi-Incidence

Any n-map can be constructed by the basic construction operations; possible configurations are now studied. Indeed, there are two cases of sew operations: the first one is the identification of two distinct cells; the second one is the identification of one cell with itself. For practical applications, identification is (almost) always applied to two distinct cells.

5.3.4.1 Identification of Two Distinct Cells

This standard case has already been illustrated in all the examples given in Section 5.2.2. But note that an important property of n-maps is the fact that it is possible to represent subdivisions containing multi-incident cells. It is thus possible to describe complex subdivisions using few cells. For instance, it is possible to construct a torus starting with a single four-sided face, leading to the minimal subdivision of the torus. The first part of the construction consists in applying the 2-sew operation to two darts which belong to the same square face, but not to two consecutive edges. In the example of Fig. 5.14 darts 2

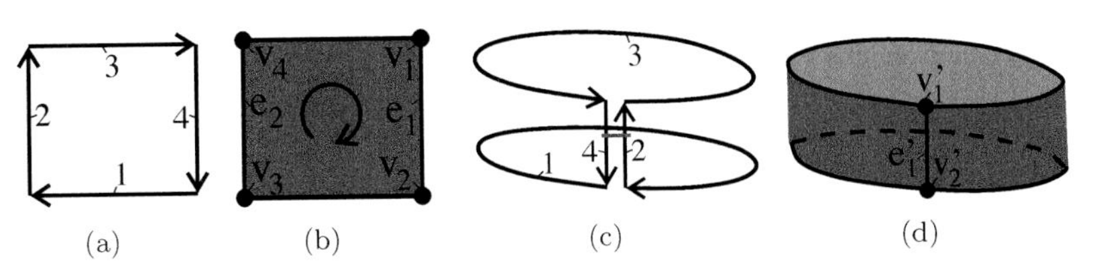

FIGURE 5.14
2-sew operation in a 2-map, creating multi-incidence.
(a) 2-map $M = (D, \beta_1, \beta_2 = \varnothing)$ contains four darts.
(b) The corresponding 2D quasi-manifold has one oriented face, four edges and four vertices.
(c) 2-map M_2-sew$(2, 4)$ describes a cylinder.
(d) The corresponding 2D quasi-manifold. Edges e_1 and e_2 are identified into edge e'_1; vertices v_1 and v_4 (resp. v_2 and v_3) are identified into vertex v'_1 (resp. v'_2). Edge e'_1 is incident twice to the face.

and 4 are 2-sewn. This folds the face on itself, producing a cylinder. In this cylinder, the face is incident twice to the edge containing dart 2, since 2 and $4 = \beta_2(2)$ both belong to the edge and to the face. In order to get a torus, darts 1 and 3 are 2-sewn. The two edges, which make the boundary of the cylinder, are thus identified into one edge, which is also incident twice to the face.

A dangling cell corresponds to a particular case of multi-incidence. A 2D example is depicted in Fig. 5.15: two darts which belong to two consecutive edges of a same face are 2-sewn. This creates a dangling edge, i.e. an edge which is incident twice to a face (since darts 1 and $\beta_2(1) = 2$ belong to the same face and to the same edge), and which has one of its two vertices incident only to this edge (since $\beta_2(1) = 2 = \beta_1(1)$). Dangling edges can be created in any dimension. In 3D, a similar construction can create dangling faces by applying the 3-sew operation. More generally, dangling cells can be constructed in any dimension.

5.3.4.2 Identification of a Cell With Itself

The second possible case of the sew operation consists in identifying a cell with itself, if at least an automorphism different from the identity exists, i.e. an isomorphism mapping the cell onto itself. This leads to a folded cell. Such identifications are illustrated in Figs. 5.16 and 5.17. In the first example, the 2-sew operation is applied in a 2-map between a dart and itself. Indeed, this is allowed by the definitions of "being sewable" and sewing operation; as shown in Fig. 5.16(c), this results into a 2-map, corresponding to the 2D quasi-manifold shown in Fig. 5.16(d). Note that edge e'_1, which results from the identification of edge e_1 with itself, is incident to one face and to one

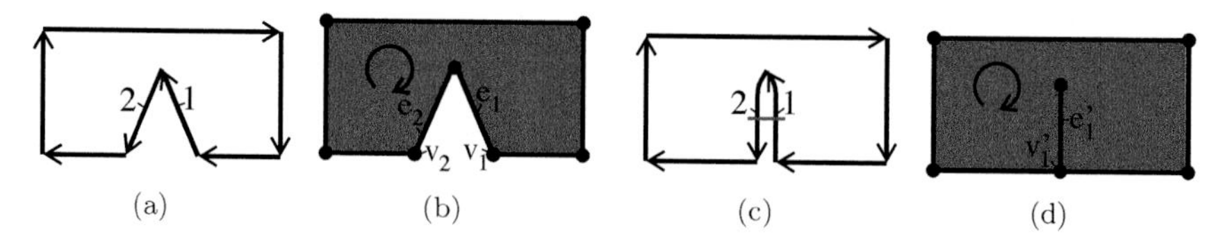

FIGURE 5.15
2-sew operation in a 2-map, creating a dangling edge.
(a) 2-map $M = (D, \beta_1, \beta_2 = \varnothing)$ contains seven darts.
(b) The corresponding 2D quasi-manifold has one oriented face, seven edges and seven vertices.
(c) 2-map $M_{2\text{-sew}}(1, 2)$.
(d) The corresponding 2D quasi-manifold. Edges e_1 and e_2 are identified into edge e'_1; vertices v_1 and v_2 are identified into vertex v'_1. Edge e'_1 is dangling.

vertex: this is not a dangling edge as in the example of Fig. 5.15; note also that it is not a loop, since a loop which does not belong to a boundary is incident to two different faces.

A similar example is presented in Fig. 5.17. The 3-sew operation is applied in a 3-map. The initial 3-map represents a cube, and a dart of a face (dart 1) is 3-sewn with the dart belonging to the same face but to the opposite edge (dart 3). This is possible because the face has four automorphisms, and this association between darts 1 and 3 corresponds to one of these automorphims. After the 3-sew, $\beta_3(1) = 3$, $\beta_3(2) = 2$, $\beta_3(3) = 1$ and $\beta_3(4) = 4$. The resulting 3-map has a folded face and two folded edges. As for the previous case in 2D, the face is not a dangling face since its boundary is made of three edges, one edge being incident to two vertices and the other edges being folded ones. The original face (v_1, v_2, v_3, v_4), is transformed into a face incident to two vertices v'_1 and v'_2. Edges (v_1, v_2) and (v_4, v_3) are identified into edge (v'_1, v'_2); edge (v_1, v_4) is folded into an edge incident to vertex v'_1 and edge (v_2, v_3) is folded into an edge incident to vertex v'_2. Note that, as four automorphims map the face onto itself, four different identifications of the cell with itself can be done, each one leading to a different 3-map. But in each case, the resulting face is folded.

This specific identification is generally not applied for usual applications. It can be easily avoided, either in a constructive way, by adding a constraint to the definition of the sew operation, or directly by adding a constraint to the definition of n-maps. In the first case, the i-sew operation is restricted in order to take as input two darts belonging to two distinct orbits $\langle \beta_1, \ldots, \beta_{i-2}, \beta_{i+2}, \ldots, \beta_n \rangle$. In the second case, a constraint is added to the definition of n-maps, which requires that any dart $d \notin \langle \beta_1, \ldots, \beta_{i-2}, \beta_{i+2}, \ldots, \beta_n \rangle(\beta_i(d))$. Note that it is easy to test if the constraint is satisfied, by searching such darts during a traversal of the n-map.

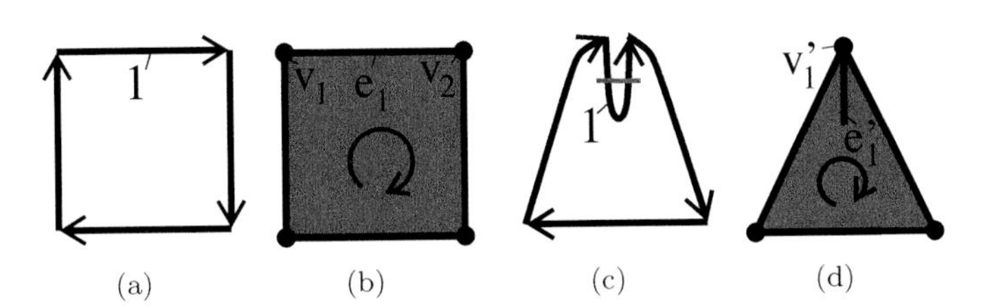

FIGURE 5.16
2-sew operation in a 2-map, creating a folded cell.
(a) 2-map $M = (D, \beta_1, \beta_2 = \varnothing)$ contains four darts.
(b) The corresponding 2D quasi-manifold has one oriented face, four edges and four vertices.
(c) 2-map M2-sew$(1, 1)$.
(d) The corresponding 2D quasi-manifold. Edge e_1 is folded on itself. It is incident to only one vertex v_1', which is the result of the identification of v_1 and v_2, although it is not a loop.

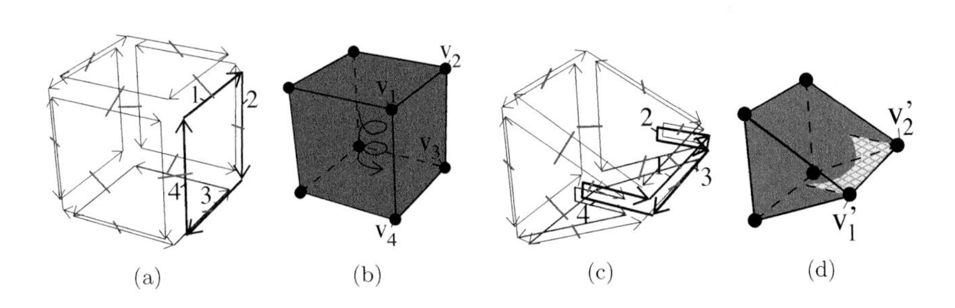

FIGURE 5.17
3-sew operation in a 3-map, creating folded cells.
(a) 3-map $M = (D, \beta_1, \beta_2, \beta_3 = \varnothing)$ contains twenty-four darts.
(b) The corresponding 3D quasi-manifold has one oriented volume, six faces, twelve edges and eight vertices.
(c) 3-map M3-sew$(1, 3)$.
(d) The corresponding 3D quasi-manifold. Face (v_1, v_2, v_3, v_4) is folded on itself. It is incident twice to edge (v_1', v_2'), which is the result of the identification of edges (v_1, v_2) and (v_4, v_3). Moreover, the two edges containing darts 2 and 4 are folded, as they are incident twice to the same vertex.

5.4 Data Structures, Iterators and Algorithms

In this section, data structures and algorithms are defined in order to handle n-maps in computer softwares. We choose to use an algorithmic language to facilitate the coding in different programming languages. To simplify notations, we do not make the distinction in algorithms between a pointer to an element and the element itself.

The objective here is to describe one way to encode darts and relations. We choose a solution based on structures with pointers, but it is easy to modify the algorithms in order to deal with other possibilities. Note that only the combinatorial part of the data structure is described here. You can see chapter 7 for two examples showing how is it possible to represent the geometrical part. The solution presented here is the one used for example in the `Combinatorial Maps` package of CGAL [61].

5.4.1 Data Structures

The basic element of the definition of n-maps is the dart. Thus the main data structure is the one which represents the darts. In dimension n, a dart structure contains an array `Betas` of $n+1$ pointers for representing the different β_i links. We choose here to explicitly encode β_0 relations for efficiency reasons. Indeed, in algorithms, it is often needed to retrieve the two neighbor darts in a same face of a given dart. This operation is performed in constant time if β_0 is encoded, while it is needed to iterate through the face otherwise (or even worse, to iterate through all the darts of the n-map if the considered face has a 1-boundary).

Listing 5.1
`Dart` data structure for n-maps

```
structure Dart
{
  pointer to Dart Betas[n+1] ;
  Boolean Marks[NB_MARKS] ;
} ;
```

In addition to the $n+1$ pointers, a dart structure contains an array `Marks` of Boolean which serves to mark darts. These marks are used in several algorithms in order to test in constant time if a dart is already processed. They also can be used to mark some part of an n-map having a specific property. The number of marks is defined by a constant number, `NB_MARKS`, which can be fixed by the users depending on their requirements. Generally, 8 is enough for usual applications. In the algorithms, marks satisfy an important invariant: all darts are always unmarked for unused marks. This invariant allows us to avoid initialization of marks in algorithms, and it is often more efficient to unmark only the marked darts at the end of each algorithm instead of unmarking all the darts.

The nMap structure is mainly composed by a set of darts Darts, and a set of integer FreeMarks allowing to manage Boolean marks. Note that the dimension n of the nMap could be a global constant, a template argument or a parameter of the function that creates an n-map. null_dart is a specific dart which is used to describe free darts. As said in the definition of n-maps, a dart d is i-free if $\beta_i(d) = \varnothing$. Here using a dart for $\varnothing$ instead of a specific value (such as NULL) allows us to use null_dart as a sentinel during algorithms (cf. for example Algorithm 29). Note that this specific dart does not belong to the set of darts of the n-map.

Listing 5.2
nMap data structure

```
structure nMap
{
  set of Dart Darts;
  set of integer FreeMarks;
  Dart null_dart;
};
```

An nMap has a set of free marks, i.e. marks that are not in use. When creating a new nMap, we add in the set of free marks all the integers between 0 and NB_MARKS − 1, which correspond to all the available marks (see Algorithm 22). Moreover we initialize the special dart null_dart to be linked with itself for all β_i. This will allows us to compose several β links without previously testing the special case of free darts. Lastly all the marks of null_dart are initialized to false.

Algorithm 22: createNMap(n): create a new n-map

```
   Input: n: the dimension of the n-map.
   Output: An empty new n-map.
1  cm ← a new nMap;
2  cm.Darts ← ∅;
3  cm.FreeMarks ← ∅;
4  for i ← 0 to n do
5  |  cm.null_dart.Betas[i] ← cm.null_dart;
6  for i ← 0 to NB_MARKS − 1 do
7  |  add i in cm.FreeMarks;
8  |  cm.null_dart.Marks[i] ← false;
9  return cm;
```

When an algorithm needs a mark, we just get an index from this set and remove it from this set (cf. Algorithm 23). As said above, marks satisfy the invariant: all darts of `cm` are unmarked for this new mark. We can use this mark during the algorithm, and at the end, unmark all the marked darts and free the mark by adding it in the set of free marks (cf. Algorithm 24). Before to free a mark, one must ensure that all the darts of `cm` are unmarked for this mark. This test is not done in Algorithm 24 because it is costly, but it can be added for debugging purposes.

Algorithm 23: `reserveMarkNMap(cm)`: reserve a new mark for n-maps

Input: `cm`: an n-map.
Output: The index of a new reserved mark;
-1 if there is no more available mark.

1 **if** `cm.FreeMarks` *is empty* **then return** *-1*;
2 `i` $\leftarrow$ an element of `cm.FreeMarks`;
3 remove `i` from `cm.FreeMarks`;
4 **return** `i`;

Algorithm 24: `freeMarkNMap(cm,i)`: free a reserved mark for n-maps

Input: `cm`: an n-map having all its darts unmarked for mark `i`;
`i`: the index of a reserved mark.
Result: Free the reserved mark `i`.

1 add `i` into `cm.FreeMarks`;

We can test if a mark is free or reserved by testing if it belongs to `FreeMarks` or not. Moreover, given a reserved mark `i`, we can test if a dart `d` is marked or not by looking directly at `d.Marks[i]` (see Algorithm 25), and we can mark/unmark this dart by setting `d.Marks[i]` to true/false (see Algorithms 26 and 27).

Algorithm 25: `isMarkedNMap(d,i)`: test if a dart is marked for n-maps

Input: `d`: a dart;
`i`: the index of a reserved mark.
Output: True iff `d` is marked for mark `i`.

1 **return** `d.Marks[i]`;

Lastly, Algorithm 28 describes the function that gives, for a given index i, the index of the inverse permutation of β_i, denoted $inv(i)$.

The complexity of all these algorithms is in constant time, because only atomic operations are applied, and no loop exists, except in Algorithm 22; but in this case `n` and `NB_MARKS` are constant values. A stack is used to store the set of free marks, allowing constant time access to a new free mark and constant time insertion and removal of elements.

Note that it is possible to add in the `nMap` data structure an array of Boolean masks which define, for each mark, the values corresponding to

Algorithm 26: `markNMap(d,i)`: mark a dart for n-maps

Input: d: a dart;
 i: the index of a reserved mark.
Result: Mark dart d for mark i.

1 d.Marks[i] $\leftarrow$ true;

Algorithm 27: `unmarkNMap(d,i)`: unmark a dart for n-maps

Input: d: a dart;
 i: the index of a reserved mark.
Result: Unmark dart d for mark i.

1 d.Marks[i] $\leftarrow$ false;

marked/unmarked. Thanks to these masks we are able to reverse in constant time the value of a mark for all the darts of the n-map. This is very efficient to unmark all the darts of the n-map when we know that they are all marked.

5.4.2 Iterators

As we have seen before, cells are defined thanks to the orbit notion. Thus is it important to be able to iterate through the darts belonging to a given orbit of an n-map. Hopefully, this can be achieved easily thanks to Algorithm 29. This algorithm is generic since it allows to run through any orbit $\langle\beta_{\mathtt{i}_1},\ldots,\beta_{\mathtt{i}_k}\rangle(\mathtt{d})$ for any valid sequence $(\mathtt{i}_1,\ldots,\mathtt{i}_k)$ and any dart d. We say that the sequence $(\mathtt{i}_1,\ldots,\mathtt{i}_k)$ is valid if:

- $\forall j \in \{1,\ldots,k\}$: $\mathtt{i}_j \in \{0,\ldots,n\}$;
- $\forall j,l \in \{1,\ldots,k\}$: $j \neq l \Rightarrow \mathtt{i}_j \neq \mathtt{i}_l$;
- $0 \in (\mathtt{i}_1,\ldots,\mathtt{i}_k) \Leftrightarrow 1 \in (\mathtt{i}_1,\ldots,\mathtt{i}_k)$.

A *sequence* of integers is used, because the order of the integers can be taken into account in algorithms (cf. for example Algorithm 43 page 172).

Note that, by definition of orbits, we need to consider functions $\beta_{\mathtt{i}_\mathtt{j}}$ and their inverses $\beta_{\mathtt{i}_\mathtt{j}}^{-1}$. When $\mathtt{i}_\mathtt{j} > 1$, $\beta_{\mathtt{i}_\mathtt{j}}$ is a partial involution and thus $\beta_{\mathtt{i}_\mathtt{j}}^{-1} =$

Algorithm 28: `inv(i)`: return the inverse index of i for n-maps

Input: $\mathtt{i} \in \{0,\ldots,n\}$.
Output: The inverse index of i.

1 **if** i $=0$ **then return** *1*;
2 **else if** i $=1$ **then return** *0*;
3 **return** i;

β_{i_j}. For $i_j = 0$ and $i_j = 1$, we ensure by the third validity condition that inv(i_j) is present in the sequence, and thus we are sure that $\beta_{i_j}^{-1}$ will also be considered.

Algorithm 29: Generic iterator for n-maps

Input: cm: an n-map;
d ∈ cm.Darts: a dart;
($i_1, \ldots, i_k$): a *valid* sequence of integers between 0 and n.
Result: Run through all the darts of $\langle\beta_{i_1}, \ldots, \beta_{i_k}\rangle$(d).

```
1  ma ← reserveMarkNMap(cm);
2  P ← an empty stack of pointer to Dart;
3  push(P,d);
4  markNMap(d,ma);
5  markNMap(cm.null_dart,ma);
6  while P is not empty do
7      cur ← top(P);
8      // process dart cur
9      pop(P);
10     for j ← 1 to k do
11         if not isMarkedNMap(cur.Betas[i_j],ma) then
12             markNMap(cur.Betas[i_j],ma);
13             push(P,cur.Betas[i_j]);
14 unmark all marked darts and cm.null_dart for ma;
15 freeMarkNMap(cm,ma);
```

Algorithm 29 makes a breadth first traversal. For that we use a stack of darts to process. We take a dart, cur, which belongs to $\langle\beta_{i_1}, \ldots, \beta_{i_k}\rangle$(d), and we process it. Then for each i_j in the sequence, we know that β_{i_j}(cur) belongs also to $\langle\beta_{i_1}, \ldots, \beta_{i_k}\rangle$(d). Thus we need to process it and for that it is enough to add it in the stack of dart to process. Before to do so, we test if this dart is marked. Indeed, if it is the case, this dart was already processed. Otherwise, we mark it in order to not reconsider it later and push it in the stack.

At the end of the algorithm, we must unmark all the marked darts in order to satisfy the property that all the darts are always unmarked for nonused marks. This can be achieved easily by reusing the same algorithm to iterate onto these darts and reversing the role of marked and unmarked darts.

The complexity of this algorithm is linear in the number of darts of $\langle\beta_{i_1}, \ldots, \beta_{i_k}\rangle$(d), because each dart is considered exactly twice (a first time during the main loop, a second time during the unmarking step), and all operations are atomic ones.

Algorithm 29 can for example be used with the sequence $(0, \ldots, i-1, i+1, \ldots, n)$ in order to iterate through all darts of the i-cell $c_i(d)$ containing a given dart, for any $i \in \{1, \ldots, n\}$. For instance, the sequence $(0, 1, 2)$ is used to iterate through all the darts of the volume containing a given dart in a 3-map.

Note that this generic algorithm can be optimized for several specific cases by using the properties of n-maps. For example, the use of stack P can be avoided, when iterating through the darts of a face in a 3-map, because a linear order exists for these darts. For instance, given dart d, we apply successively β_1 in order to traverse orbit $\langle\beta_1\rangle(d)$, then apply β_3 to d, and apply again successively β_1. This optimization is based upon the fact that $\beta_1 \circ \beta_3$ is a partial involution. The traversal of other orbits can be optimized in this way for similar reasons.

The generic iterator algorithm allows to iterate through all the darts of a given orbit, and in particular through the darts of any cells containing a given dart. However, there is the special case of vertices (0-cells), since the definition of 0-cells does not follow the same principle than the definition of other cells. This is done in Algorithm 30.

Algorithm 30: Vertex iterator for n-maps

```
Input: cm: an n-map;
       d ∈ cm.Darts: a dart.
Result: Run through all the darts of c0(d).
1  ma ← reserveMarkNMap(cm);
2  P ← an empty stack of pointer to Dart;
3  push(P,d);
4  markNMap(d,ma);
5  markNMap(cm.null_dart,ma);
6  while P is not empty do
7      cur ← top(P);
8      // process dart cur
9      pop(P);
10     for i ← 1 to n − 1 do
11         for j ← i + 1 to n do
12             if not isMarkedNMap(cur.Betas[j].Betas[i],ma) then
13                 markNMap(cur.Betas[j].Betas[i],ma);
14                 push(P,cur.Betas[j].Betas[i]);
15             if not isMarkedNMap(cur.Betas[inv(i)].Betas[j],ma) then
16                 markNMap(cur.Betas[inv(i)].Betas[j],ma);
17                 push(P,cur.Betas[inv(i)].Betas[j]);
18 unmark all marked darts and cm.null_dart for ma;
19 freeMarkNMap(cm,ma);
```

The principle of this algorithm is similar to the principle of Algorithm 29. The only difference is about the considered relations. We follow here the definition of 0-cells in an n-map (cf. Def. 31) and thus we need to consider $\beta_{\mathtt{i}} \circ \beta_{\mathtt{j}}$ for any $1 \leq i < j \leq n$, and their inverses.

The complexity of Algorithm 30 is linear in the number of darts of $c_0(\mathtt{d})$ (because n is constant).

These two algorithms are a basis for several other algorithms allowing to iterate through some specific cells based on incidence and adjacency relations. We illustrate this by showing three useful examples. In the first one, detailed in Algorithm 31, we run through one dart per each i-cell of an n-map. This algorithm allows us to process each i-cell exactly once.

Algorithm 31: Iterator over one dart per each i-cell for n-maps

Input: cm: an n-map;
i $\in \{0, \ldots, n\}$.

Result: Run through one dart per each i-cell of cm.

```
1 ma ← reserveMarkNMap(cm);
2 foreach dart d ∈ cm.Darts do
3     if not isMarkedNMap(d,ma) then
4         // process dart d
5         mark all the darts in c_i(d) for ma;
6 unmark all darts for ma;
7 freeMarkNMap(cm,ma);
```

Algorithm 31 runs through all darts of the n-map, and for each nonmarked dart d, it marks the darts of the i-cell $c_{\mathtt{i}}(\mathtt{d})$ containing d. Note that these darts are marked by using Algorithm 29 with the sequence $(0, \ldots, \mathtt{i}-1, \mathtt{i}+1, \ldots, n)$ for $\mathtt{i} > 0$ or Algorithm 30 for $\mathtt{i} = 0$. To optimize the algorithm, we use in Algorithms 29 and 30 the same mark ma reserved in Algorithm 31 (indeed, as the goal of line 5 is to mark all the darts in $c_{\mathtt{i}}(\mathtt{d})$, it is useless and inefficient to use a local mark in Algorithms 29 or 30 for iterating through these darts, and to unmark these darts at the end of the iterator). At the end of Algorithm 31, we need to unmark all the marked darts. Here we are in the case where all the darts of the n-map are marked, thus we can simply negate the mask of mark ma.

The complexity of Algorithm 31 is linear in the number of darts of the n-map because each dart is considered in the first loop (line 2 of the algorithm), and thus each dart is marked exactly once in line 5.

A second example given in Algorithm 32 consists in iterating through one dart per each i-cell incident to a given j-cell for two given dimensions i and j, with $\mathtt{i} \neq \mathtt{j}$.

This algorithm is similar to the previous one, by replacing the loop over all the darts of the n-map by a loop over all the darts of $c_{\mathtt{j}}(\mathtt{d})$ (loop achieved by using Algorithms 29 or 30). Thanks to the definition of incidence in n-maps, we are sure that during this loop we discover at least one dart of each incident i-cell, and by using a mark, we ensure to process each i-cell only once. To

Algorithm 32: Iterator over one dart per each i-cell incident to a j-cell for n-maps

```
Input: cm: an n-map;
       d ∈ cm.Darts: a dart;
       i ∈ {0,...,n};
       j ∈ {0,...,n}: with i ≠ j.
Result: Run through one dart per each i-cell incident to c_j(d).
1 ma ← reserveMarkNMap(cm);
2 foreach dart e ∈ c_j(d) do
3   if not isMarkedNMap(e,ma) then
4     // process dart e
5     mark all the darts in c_i(e) for ma;
6 unmark all marked darts for ma;
7 freeMarkNMap(cm,ma);
```

unmark all the marked darts, we reuse the same algorithm by reversing the role of marked and unmarked darts.

The complexity of this algorithm is linear in the number of darts discovered during the traversal, i.e. in the number of darts of $c_j(d)$ plus the number of darts of the incident cells.

The last example given in Algorithm 33 allows to run through one dart per each i-cell adjacent to a given i-cell, for $i > 0$.

Once again we use the same principle than for the two previous examples, but here, we have to test dart $\beta_i(e)$ for each $e \in c_i(d)$. Indeed, due to the definition of adjacency relations in n-maps, we know that all adjacent i-cells to $c_i(d)$ can be reached by such a dart. For $i = 1$ it is also required to test $\beta_i^{-1}(e)$, i.e. $\beta_0(e)$. It is easy to modify this algorithm to consider the case $i = 0$ thanks to Def. 33.

The complexity of this algorithm is linear in the number of darts discovered during the traversal, i.e. in the number of darts of $c_i(d)$ plus the number of darts of the adjacent cells.

Note that we generally implement the iterators as data structures (like iterators in STL [12]) allowing to start a traversal, move to the next position, get the current element pointed by the iterator and test if the iterator has reached its end position (i.e. the traversal is finished). Some algorithms given in next sections use this type of implementation (for example Algorithms 42 and 43 page 171).

5.4.3 Basic Tools

We propose now algorithms implementing the basic tools defined in Section 5.2.

Algorithm 33: Iterator over one dart per each i-cell adjacent to an i-cell for n-maps

```
Input: cm: an n-map;
       d ∈ cm.Darts: a dart;
       i ∈ {1,...,n}.
Result: Run through one dart per each i-cell adjacent to c_i(d).
1  ma ← reserveMarkNMap(cm);
2  foreach dart e ∈ c_i(d) do
3      if not isMarkedNMap(e.Betas[i],ma) then
4          // process dart e.Betas[i]
5          mark all the darts in c_i(e.Betas[i]) for ma;
6      if i = 1 and not isMarkedNMap(e.Betas[0],ma) then
7          // process dart e.Betas[0]
8          mark all the darts in c_i(e.Betas[0]) for ma;
9  unmark all marked darts for ma;
10 freeMarkNMap(cm,ma);
```

Algorithm 34 implements a method to test if a given dart is i-free or not. It only consists in looking if $\beta_i(d)$ is equal to `null_dart` or not.

Algorithm 34: `isFreeNMap(cm,d,i)`: test if a dart is i-free for n-maps

```
Input: cm: an n-map;
       d ∈ cm.Darts: a dart;
       i ∈ {0,...,n}.
Output: True iff d is i-free.
1  return d.Betas[i] = cm.null_dart;
```

The algorithms which add isolated darts and remove isolated darts are straightforward since they only add or remove elements in the set of darts. In Algorithm 35, when we create a dart, we initialize all the β links to `cm.null_dart`, so that it is i-free in all dimensions. Moreover we initialize all its marks to false so that it is unmarked for all the possible marks.

Remember that the remove dart operation can be applied only for isolated darts. Thus Algorithm 36 takes an isolated dart as input and removes it from the set of darts. Testing if `d` is isolated is easy: we test if `d` is i-free for all $i \in \{0, \ldots, n\}$.

The time complexity of these three algorithms is constant as n and `NB_MARKS` are constants, and we can use a container of darts allowing constant time insertion and removal of elements.

Now we present the two algorithms that respectively increase and decrease the dimension of a given n-map. In practice, these operations create a new n'-map which is a copy of the initial n-map, but in the case of increase dimension

Algorithm 35: `createDartNMap(cm)`: create a new isolated dart in an n-map

Input: `cm`: an n-map.
Output: A new isolated dart added in `cm`.

```
1 Dart d ← a new Dart;
2 add d in cm.Darts;
3 for i ← 0 to n do
4   d.Betas[i] ← cm.null_dart;
5 for i ← 0 to NB_MARKS − 1 do
6   d.Marks[i] ← false;
7 return d;
```

Algorithm 36: `removeIsolatedDartNMap(cm,d)`: remove an isolated dart in an n-map

Input: `cm`: an n-map;
`d` ∈ `cm.Darts`: an isolated dart.

```
1 remove d from cm.Darts;
```

we add a new relation, and in the case of decrease dimension we remove one. Thus we start first by defining Algorithm 37, which copies a given n-map into a second n'-map, by copying only the β_i links for i between 0 and the smaller dimension between n and n'.

During the copy, we need to keep an association between each dart in the original n-map and its corresponding dart in the copy. This is done through an associative array (which can be for example a hash table or a binary search tree). Note that if $n < n'$, we do not need to initialize β_i for the new darts when $n < i \leq n'$ as this is done by the algorithm that creates a new dart. Note also that we copy the free marks, and the values of all the marks for all the darts. Depending on specific needs, we can remove these copies of marks so that `cm'` have no reserved mark.

Algorithm 38 implements the operation that increase the dimension of a given n-map: it only creates a new $(n+1)$-map `cm'` and then copies `cm` into `cm'`.

Algorithm 39 implements the operation that decrease the dimension of a given n-map: it only creates a new $(n-1)$-map `cm'` and then copies `cm` into `cm'`. Note that we require here that the given n-map has all its darts n-free. We can add a test in the algorithm to check if this precondition is satisfied. Moreover we can also remove this requirement as it is possible to decrease the dimension of an n-map even if it has some darts n-sew. In this case, we obtain the same result than if we first n-unsew all these darts, then decrease the dimension.

The complexity of these three algorithms is linear in the number of darts

Algorithm 37: copyNMap(cm,cm′): copy an n-map into a second n'-map

```
   Input: cm: an n-map;
          cm′: an n′-map.
   Result: Copy cm into cm′ until minimum(n,n′).
 1 cm′.Darts ← ∅;
 2 cm′.FreeMarks ← cm.FreeMarks;
 3 assoc ← an empty associative array between darts;
 4 assoc[cm.null_dart] ← cm′.null_dart;
 5 foreach dart d ∈ cm.Darts do
 6     assoc[d] ← createDartNMap(cm′);
 7 k ← min(n, n′);
 8 foreach dart d ∈ cm.Darts do
 9     d′ ← assoc[d];
10     for i ← 0 to k do
11         d′.Betas[i] ← assoc[d.Betas[i]];
12     for i ← 0 to NB_MARKS − 1 do
13         d′.Marks[i] ← d.Marks[i];
```

Algorithm 38: increaseDimNMap(cm): increase the dimension of an n-map

```
  Input: cm: an n-map.
  Output: The (n + 1)-map obtained from cm by increasing its
          dimension.
1 cm′ ← createNMap(n + 1);
2 copyNMap(cm,cm′);
3 return cm′;
```

Algorithm 39: decreaseDimNMap(cm): decrease the dimension of an n-map

```
  Input: cm: an n-map having all its darts n-free.
  Output: The (n − 1)-map obtained from cm by decreasing its
          dimension.
1 cm′ ← createNMap(n − 1);
2 copyNMap(cm,cm′);
3 return cm′;
```

$\#d$ of the given n-map times the complexity for accessing an element in the associative array. This access can be done in $O(\log \#d)$ for hash table or binary search tree, thus the complexity is in $O(\#d. \log \#d)$. Note that hash tables allow constant time access in average by using a correct hash function.

Now we present the two algorithms allowing to merge two given n-maps, and to restrict a given n-map to a subset of its darts. As for the two previous algorithms, these operations build a new n-map, which in the case of merge is a copy of the two initial n-maps, and in the case of restrict is a partial copy of the initial n-map. For this reason, the two algorithms use once again an associative array to link the original darts and the copy ones.

Algorithm 40 implements the merge operation of two given n-maps. This operation is always possible, as in our data structure, two different n-maps have necessarily two different sets of darts (while this could be not the case for another implementation of n-maps).

Algorithm 40: `mergeNMaps(cm,cm')`: merge two given n-maps

Input: `cm`, `cm'`: two n-maps.
Output: The n-map obtained by merging `cm` and `cm'`.

```
1  cm'' ← createNMap(n);
2  cm''.FreeMarks ← cm.FreeMarks ∩ cm'.FreeMarks;
3  assoc ← an empty associative array between darts;
4  assoc[cm.null_dart] ← cm''.null_dart;
5  assoc[cm'.null_dart] ← cm''.null_dart;
6  foreach dart d ∈ cm.Darts ∪ cm'.Darts do
7      assoc[d] ← createDartNMap(cm'');
8  foreach dart d ∈ cm.Darts ∪ cm'.Darts do
9      d' ← assoc[d];
10     for i ← 0 to n do
11         d'.Betas[i] ← assoc[d.Betas[i]];
12     for i ← 0 to NB_MARKS − 1 do
13         d'.Marks[i] ← d.Marks[i];
14 return cm'';
```

In the new n-map, we consider that a mark is free if it is free in both n-maps `cm` and `cm'`: this explains line 2 which initializes `cm''.FreeMarks` as the intersection of both sets of free marks. In the associative array, we associate with each original dart its corresponding copy, both for `cm` and for `cm'`. After having created all the darts, we only need to make the same links between the new darts than the original links in the initial n-maps. This is directly done thanks to the associative array.

Algorithm 41 implements the operation that restricts an n-map to a given subset of darts. As seen in Def. 42 page 140, this operation is only possible if the subset of darts corresponds to a set of connected components. Thus we start to test if this property is satisfied, and construct the restricted n-map only when it is the case. Note that we could propose different variants, for example that unsew all the nonfree darts linked with a dart that does not belong to D′, or a nonsafe version that does not make the test and assume the user gives always a set of connected components...

Algorithm 41: `restrictNMap(cm,D′)`: restrict an n-map to a subset of darts

Input: cm: an n-map;
  D′ ⊆ cm.Darts: a subset of darts.
Output: The n-map obtained by restricting cm to D′.

```
1  foreach dart d ∈ D′ do
2      for i ← 0 to n do
3          if d.Betas[i] ∉ D′ then error;
4  cm′ ← createNMap(n);
5  cm′.FreeMarks ← cm.FreeMarks;
6  assoc ← an empty associative array between darts;
7  assoc[cm.null_dart] ← cm′.null_dart;
8  foreach dart d ∈ D′ do
9      assoc[d] ← createDartNMap(cm′);
10 foreach dart d ∈ D′ do
11     d′ ← assoc[d];
12     for i ← 0 to n do
13         d′.Betas[i] ← assoc[d.Betas[i]];
14     for i ← 0 to NB_MARKS − 1 do
15         d′.Marks[i] ← d.Marks[i];
16 return cm′;
```

The complexity of the merging operation is linear in the number of darts of the two given n-maps ($\#d_1$ and $\#d_2$) times the complexity for accessing an element in the associative array, i.e. in $O((\#d_1 + \#d_2).\log(\#d_1 + \#d_2))$. The complexity of the restrict operation is linear in the number of darts in D′ times $\log |D'|$ (this corresponds to the cost of the copy), plus the cost of the initial test (checking whether the subset of darts corresponds to a set of connected components: lines $1-3$). The test complexity is $O(|D'|.\log |D'|)$ if D′ is a container where searching an element is achieved in log. However this test can be done in $O(|D'|)$ by first marking all the darts in D′, then reiterate through all these darts and just verify if for each dart d, $\alpha_i(d)$ is marked. The global complexity of Algorithm 41 is thus in $O(|D'|.\log |D'|)$.

5.4.4 Sew/Unsew Operations

Now we present the i-sew operation and its inverse which is the i-unsew operation. As seen in Section 5.2.2 when introducing the sew operation, it is not always possible to i-sew two darts: they must be i-sewable. Algorithm 42 allows to test if two given darts are i-sewable or not. This algorithm follows directly the i-sewable definition given in Def. 44 page 141. Its main principle is to run simultaneously through the two orbits o_d and $o_{d'}$, to progressively build the isomorphism while testing it is indeed an isomorphism.

Algorithm 42: `sewableNMap(cm,d,d',i)`: test if two darts are i-sewable for n-maps

```
   Input: cm: an n-map;
          d, d' ∈ cm.Darts: two darts;
          i ∈ {1,...,n}.
   Output: True iff d is i-sewable with d'.
 1 if not isFreeNMap(d,i) or not isFreeNMap(d',inv(i)) then
 2 |  return false;
 3 if d = d' then return true;
 4 if i = 1 or i = 2 then
 5 |  S ← (i+2,...,n); it ← generic iterator(cm,d,S);
 6 |  it' ← generic iterator(cm,d',S);
 7 else
 8 |  S ← (0,1,...,i−2,i+2,...,n); it ← generic iterator(cm,d,S);
 9 |  it' ← generic iterator(cm,d',(1,0,2,...,i−2,i+2,...,n));
10 f ← an empty associative array between darts;
11 f[cm.null_dart] ← cm.null_dart;
12 while it is not to its end and it' is not to its end do
13 |  f[it] ← it';
14 |  foreach j ∈ S do
15 |  |  if it.Betas[i] is defined in f and
   |  |  f[it.Betas[i]] ≠ it'.Betas[inv(i)] then
16 |  |  |  return false;
17 |  advance it to its next position;
18 |  advance it' to its next position;
19 if it is not to its end or it' is not to its end then
20 |  return false;
21 return true;
```

Firstly in line 1, we verify that `d` is `i`-free and that `d` is `inv(i)`-free. If it is not the case, the two darts are not `i`-sewable. Secondly if `d` = `d'` they are always sewable. Thirdly we create two generic iterators that correspond to the two orbits $o_d = \langle\beta_1,\ldots,\beta_{i-2},\beta_{i+2},\ldots,\beta_n\rangle(\mathtt{d})$ and $o_{d'} = \langle\beta_1,\ldots,\beta_{i-2},\beta_{i+2},\ldots,\beta_n\rangle(\mathtt{d}')$. Note that if $\mathtt{i} > 2$ we use two different sequences, one starting with $(0,1)$ and the second one starting with $(1,0)$. This allows to iterate through the two orbits in reverse directions, and thus to follow the i-sewable definition. During the main loop, for each pair of darts pointed by `it` and `it'`, we consider all the darts in the neighborhood of `it` that are already considered. Indeed, we already have built the isomorphism between these darts through the association array `f`, and we need to test if the local property of the `i`-sewable operation is satisfied (i.e. $\mathtt{f}(\beta_{\mathtt{j}}(\mathtt{it})) = \beta_{\mathtt{j}}^{-1}(\mathtt{it}')$, with $\mathtt{f}(\mathtt{it}) = \mathtt{it}'$). When it is not the case, we directly return false as we know that the two darts `d` and `d'` are not i-sewable. Otherwise we continue the main loop until finishing at least one iterator. If only one iterator has finished, the two darts are not `i`-sewable (the two orbits do not have the same number of darts); otherwise, they are `i`-sewable, and we return true.

The complexity of the `i`-sewable algorithm is linear in the number of darts $\#d$ in the smaller orbits among o_d and $o_{d'}$ times $\log \#d$. Indeed, we stop the main loop as soon as one iterator has reached its end position, and all operations are atomic except the access to an element in the associative array which is in log, and the inner loop (line 14) is bounded by n which is a constant.

The `i`-sew operation is implemented by Algorithm 43, for $\mathtt{i} \in \{2,\ldots,n\}$. This algorithm follows directly the definition of the `i`-sew operation (see Def. 45 page 142). Mainly, we run through the two orbits $o_d = \langle\beta_1,\ldots,\beta_{i-2},\beta_{i+2},\ldots,\beta_n\rangle(\mathtt{d})$ and $o_{d'} = \langle\beta_1,\ldots,\beta_{i-2},\beta_{i+2},\ldots,\beta_n\rangle(\mathtt{d}')$, and put in relation two by two all pairs of darts by $\beta_{\mathtt{i}}$.

Algorithm 43: `sewNMap(cm,d,d',i)`: `i`-sew two darts for n-maps

Input: `cm`: an n-map;
  `d`, `d'` $\in$ `cm.Darts`: two `i`-sewable darts;
  $\mathtt{i} \in \{2,\ldots,n\}$.
Result: `i`-sew darts `d` and `d'`.

```
1  if i = 2 then
2  |   it ← generic iterator(cm,d,(4,...,n));
3  |   it' ← generic iterator(cm,d',(4,...,n));
4  else
5  |   it ← generic iterator(cm,d,(0,1,...,i−2,i+2,...,n));
6  |   it' ← generic iterator(cm,d',(1,0,2,...,i−2,i+2,...,n));
7  while it is not to its end do
8  |   it.Betas[i] ← it';
9  |   it'.Betas[i] ← it;
10 |   advance it to its next position;
11 |   advance it' to its next position;
```

The key point of this algorithm is the use once again of the two generic iterators. If $\mathtt{i} > 2$, we use the two sequences $(0, 1, \ldots, \mathtt{i}-2, \mathtt{i}+2, \ldots, n)$ and $(1, 0, 2, \ldots, \mathtt{i}-2, \mathtt{i}+2, \ldots, n)$ to iterate through the two orbits by using the same β links, except for β_0 and β_1 which are used in reverse order (if $\mathtt{i} = 2$ the two orbits are considered in the same order as they do not contain β_1). Since we know that `d` and `d'` are `i`-sewable, and since the generic iterator uses the indexes given in the sequence in the same order, at each step of the while loop, the two iterators point to two darts in relation by isomorphism f of Def. 45. For this reason, we only need to put the two pointed darts in relation by $\beta_{\mathtt{i}}$, then move the two iterators to their next positions. Note that the relation is done in both directions as $\beta_{\mathtt{i}}$ is a partial involution (because $\mathtt{i} \in \{2, \ldots, n\}$). The algorithm ends when the two iterators reach their end positions.

Algorithm 44 shows the special case of 1-sew operation. Its principle is similar to the previous algorithm. The only difference is that there are two cases depending if $\mathtt{it} \in \langle\{\beta_i \circ \beta_j | \forall i, j \in \{3, \ldots, n\}, i < j\}\rangle(\mathtt{d})$ or not (as seen in Def. 46 page 142).

Algorithm 44: `1sewNMap(cm,d,d')`: 1-sew two darts for n-maps

Input: `cm`: an n-map;
`d`, `d'` $\in$ `cm.Darts`: two 1-sewable darts.
Result: 1-sew darts `d` and `d'`.

```
1 it ← generic iterator(cm,d,(3,...,n));
2 it' ← generic iterator(cm,d',(3,...,n));
3 while it is not to its end do
4     if it ∈ ⟨{β_i ∘ β_j |∀i,j ∈ {3,...,n}, i < j}⟩(d) then i ← 1;
5     else i ← 0;
6     it.Betas[i] ← it';
7     it'.Betas[inv(i)] ← it;
8     advance it to its next position;
9     advance it' to its next position;
```

The complexity of the two sew algorithms is linear in the number of darts in the orbit $\langle\beta_1, \ldots, \beta_{i-2}, \beta_{i+2}, \ldots, \beta_n\rangle(\mathtt{d})$. In the 1-sew algorithm, all the darts in $\langle\{\beta_i \circ \beta_j | \forall i, j \in \{3, \ldots, n\}, i < j\}\rangle(\mathtt{d})$ are marked. Testing if `it` belongs to this orbit is done in constant time by testing if `it` is marked.

The `i`-unsew operation is given in Algorithm 45, for $\mathtt{i} \in \{2, \ldots, n\}$. As for the `i`-sew algorithm, it follows directly the i-unsew definition (see Def. 47 page 146). It runs through all the darts of the orbit $\langle\beta_1, \ldots, \beta_{i-2}, \beta_{i+2}, \ldots, \beta_n\rangle(\mathtt{d})$, and we `i`-sew darts $\beta_i(\mathtt{e})$ and `e` to $\varnothing$.

Note in this algorithm that we use the fact that an i-free dart is linked with the sentinel `cm.null_dart` to avoid to test this special case.

Lastly, Algorithm 46 gives the 1-unsew operation. As seen in the definition (Def. 48 page 147), there are two orbits to consider: $o_1 = \langle\{\beta_i \circ \beta_j | \forall i, j \in$

Algorithm 45: `unsewNMap(cm,d,i)`: i-unsew two darts for n-maps

Input: `cm`: an n-map;
`d` $\in$ `cm.Darts`: a non `i`-free dart;
`i` $\in \{2, \ldots, n\}$.
Result: `i`-unsew darts `d` and β_i(`d`).

```
1 foreach dart e ∈ ⟨β1, ..., βi−2, βi+2, ..., βn⟩(d) do
2 |   (e.Betas[i]).Betas[i] ← cm.null_dart;
3 |   e.Betas[i] ← cm.null_dart;
```

$\{3, \ldots, n\}, i < j\}\rangle$(d) and $o_2 = \langle\beta_3, \ldots, \beta_n\rangle(\beta_1(d)) \setminus \langle\{\beta_i \circ \beta_j | \forall i, j \in \{3, \ldots, n\}, i < j\}\rangle(\beta_1(\texttt{d}))$.

Algorithm 46: `1unsewNMap(cm,d)`: 1-unsew two darts for n-maps

Input: `cm`: an n-map;
`d` $\in$ `cm.Darts`: a non 1-free dart.
Result: 1-unsew darts `d` and β_1(`d`).

```
1 foreach dart e ∈ ⟨β3, ..., βn⟩(d) do
2 |   if e ∈ ⟨{βi ∘ βj | ∀i, j ∈ {3, ..., n}, i < j}⟩(d) then
3 |   |   (e.Betas[1]).Betas[0] ← cm.null_dart;
4 |   |   e.Betas[1] ← cm.null_dart;
5 foreach dart e ∈ ⟨β3, ..., βn⟩(d.Betas[1]) do
6 |   if e ∉ ⟨{βi ∘ βj | ∀i, j ∈ {3, ..., n}, i < j}⟩(d.Betas[1]) then
7 |   |   (e.Betas[1]).Betas[0] ← cm.null_dart;
8 |   |   e.Betas[1] ← cm.null_dart;
```

The complexity of the two unsew algorithms is linear in the number of darts of the traversed orbit $\langle\beta_1, \ldots, \beta_{i-2}, \beta_{i+2}, \ldots, \beta_n\rangle$(d). For the 1-unsew operation, the darts belonging to $\langle\{\beta_i \circ \beta_j | \forall i, j \in \{3, \ldots, n\}, i < j\}\rangle$(d) and to $\langle\{\beta_i \circ \beta_j | \forall i, j \in \{3, \ldots, n\}, i < j\}\rangle(\beta_1(\texttt{d}))$ are marked. Test if a dart belongs to these orbits or not can thus be done in constant time.

5.5 Complements

In this section, some additional definitions and properties are studied: the notion of boundary map; the relations between n-Gmaps and n-maps; a classification of 2-maps, corresponding to the classification of surfaces; the duality notion.

Algorithms corresponding to these new notions are not described, but they are not complicated to retrieve. For the boundary map and the dual map, the principle is similar to the copy of an n-map, while modifying the β links according to the definitions.

5.5.1 Boundary Map

For many applications, n-maps are such that there are no free darts except for β_n, i.e. no dart d exists, such that $\beta_i(d) = \varnothing$, for $i < n$: so, the boundary of each n-cell corresponds to an $(n-1)$-quasi-manifold without boundary, but the n-dimensional quasi-manifold corresponding to the n-map can have boundaries, which correspond to the connected components of the $(n-1)$-map of the boundaries.

More precisely, let us consider an n-map M without boundary except for dimension n. M_∂ is the $(n-1)$-map of the boundaries of M; it corresponds thus to an $(n-1)$-quasi-manifold without boundary. It is defined by keeping all the n-free darts of M and all the β_i, $\forall i \in \{1, \ldots, n-2\}$. β'_{n-1} is defined by following the path of darts $\beta_{n-1} \circ (\beta_n \circ \beta_{n-1})^k$ to jump over darts that do not belong to the boundary until obtaining an n-free dart. For $n = 1$ or $n = 2$, any component of M_∂ represents a boundary of the quasi-manifold associated with M (see example in 2D in Fig. 5.18). For $n \geq 3$, any component of M_∂ represents an $(n-1)$-quasi-manifold which recovers a boundary of the quasi-manifold associated with M.

The boundary notion is generalized in Def. 49 in order to define the i-boundary map. Given an n-map M, the i-boundary map M_{∂_i} is an n-map corresponding to the parts of M incident to i-free darts.

Definition 49 (i-boundary map) *Let $M = (D, \beta_1, \ldots, \beta_n)$ be an n-map, and $i \in \{2, \ldots, n\}$. The i-boundary map $M_{\partial_i} = (D', \beta'_1, \ldots, \beta'_n)$ is the n-map defined by:*

1. *$D' = \{d \in D | d$ is i-free$\}$;*
2. *$\forall d' \in D'$:*
    - *$\forall j \in \{1, \ldots, i-2, i+2, \ldots, n\}$, $\beta'_j(d') = \beta_j(d')$;*
    - *$\beta'_i(d') = \varnothing$; $\beta'_{i+1}(d') = \varnothing$;*
    - *$\beta'_{i-1}(d') = \beta_{i-1} \circ (\beta_i \circ \beta_{i-1})^k(d')$*
      k being the smaller integer s.t. $\beta_{i-1} \circ (\beta_i \circ \beta_{i-1})^k(d')$ is i-free.

Note that there are two possible cases for the definition of $\beta'_{i-1}(d')$. The first case is when an i-free dart d'' is reached when following the path of dart $\beta_{i-1} \circ (\beta_i \circ \beta_{i-1})^k(d')$. In this case, $\beta'_{i-1}(d') = d''$. The second case occurs when $\varnothing$ is reached when following the path. This second case occurs if a dart of the path $\beta_{i-1} \circ (\beta_i \circ \beta_{i-1})^k(d')$ is $(i-1)$-free; thus, $\beta'_{i-1}(d') = \varnothing$, since $\varnothing$ is i-free. Note also that by definition all the darts in an i-boundary map are i-free and $(i+1)$-free.

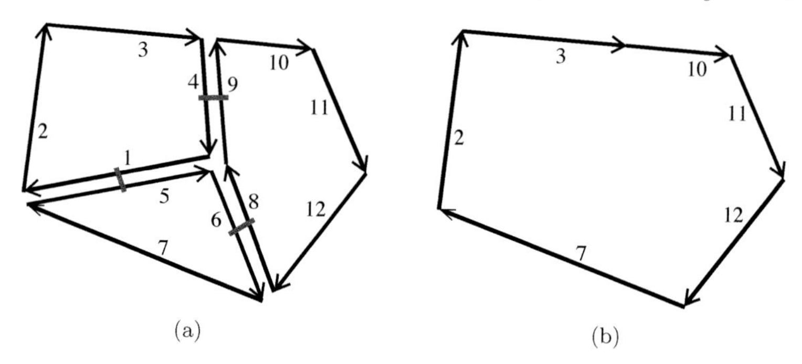

FIGURE 5.18
(a) 2-map M with 2-boundary.
(b) The corresponding 2-boundary map $M_{\partial_2} = (D', \beta'_1, \beta'_2 = \varnothing)$. 1-map $M_\partial = (D', \beta'_1)$ corresponds to a closed curve, i.e. to the boundary of the surface represented by M.

Note also that the i-boundary map has the same dimension than the initial n-map. This is required for $i < n$ in order to associate correctly during the closure operation (cf. Section 6.1 page 185) darts of initial n-map with darts of i-boundary map. However this is not required for $i = n$ as there is no dimension higher than i.

The i-boundary map can be used to retreive the classical boundary: when an n-map is without boundary except for dimension n, M_{∂_n} is computed, then its dimension is decreased (by definition, each dart d in M_{∂_n} is such that $\beta_n(d) = \varnothing$).

Definition 50 (boundary map) *Let M be an n-map without boundary except possibly for dimension n. The $(n-1)$-map boundary map $M_\partial = M^-_{\partial_n}$.*

Figure 5.18 shows an example of a 2-map with a 2-boundary and its corresponding 2-boundary map. A second example is given in Fig. 5.19, showing a 3-map with a 3-boundary and the corresponding 3-boundary map. Note that in both cases, the $(n-1)$-map of the boundaries is obtained by decreasing their dimension. The third example given in Fig. 5.20 shows a case where we obtain $\varnothing$ for the definition of $\beta'_{i-1}(d')$ (for darts 7 and 11).

There is a special case for the 1-boundary map, defined in Def. 51.

Definition 51 (1-boundary map) *Let $M = (D, \beta_1, \ldots, \beta_n)$ be an n-map. The 1-boundary map $M_{\partial_1} = (D', \beta'_1, \ldots, \beta'_n)$ is the n-map defined by:*

1. *$D' = \{d \in D | d$ is 1-free$\}$;*

2. *$\forall d' \in D'$:*

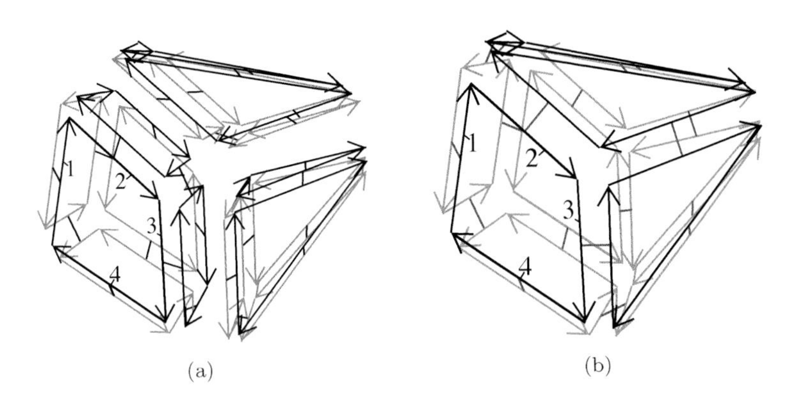

FIGURE 5.19
(a) 3-map M with 3-boundary (three volumes two by two adjacent).
(b) The corresponding 3-boundary map $M_{\partial_3} = (D', \beta'_1, \beta'_2, \beta'_3 = \varnothing)$. 2-map $M_\partial = (D', \beta'_1, \beta'_2)$ corresponds to a surface, which is the boundary of the 3-dimensional manifold represented by M.

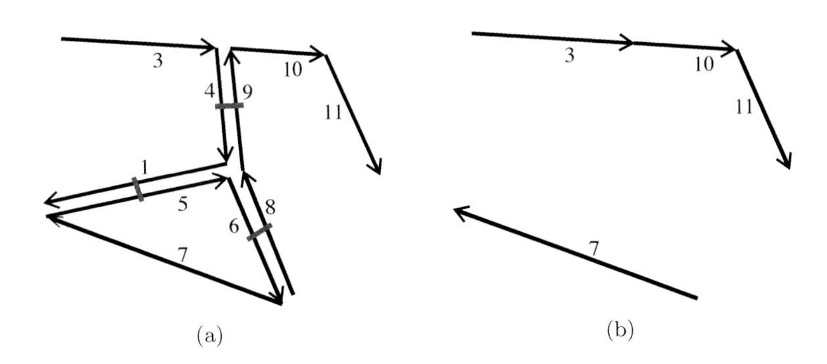

FIGURE 5.20
(a) 2-map M with 1-boundaries and 2-boundaries.
(b) The corresponding 2-boundary map $M_{\partial_2} = (D', \beta'_1, \beta'_2 = \varnothing)$. Darts 7 and 11 are 1-free in M_{∂_2} since $\beta_1 \circ (\beta_2 \circ \beta_1)^1(7) = \varnothing$ and $\beta_1 \circ (\beta_2 \circ \beta_1)^0(11) = \varnothing$ in M. M_{∂_1} is given in Fig. 5.21.

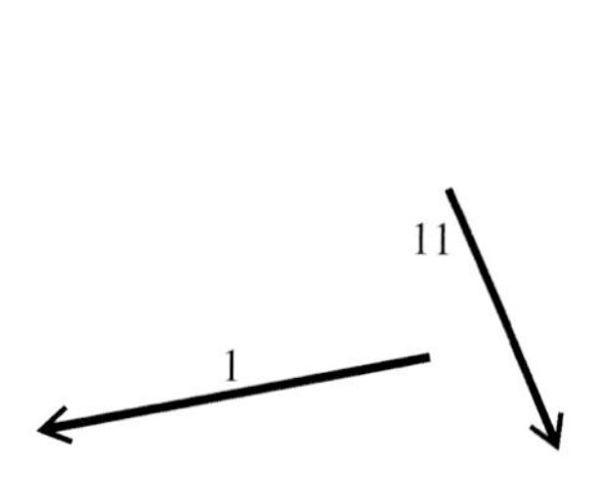

FIGURE 5.21
1-boundary map M_{∂_1} of the 2-map given in Fig. 5.20(a). M_{∂_1} is composed by two isolated darts.

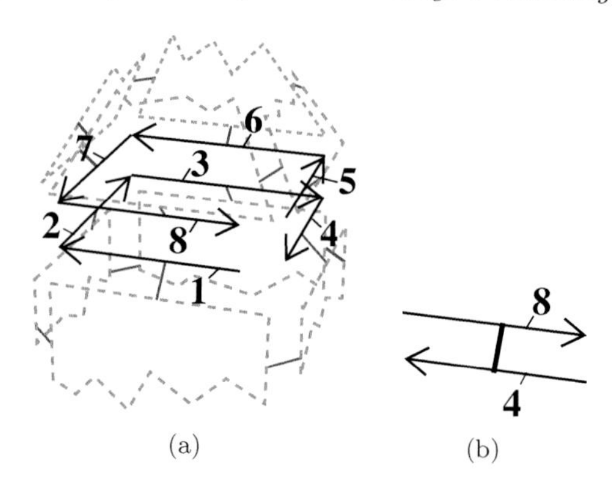

FIGURE 5.22
(a) Partial representation of 3-map M. Zoom on a face separating two volumes which has a 1-boundary.
(b) The corresponding 1-boundary map $M_{\partial_1} = (D', \beta'_1 = \varnothing, \beta'_2 = \varnothing, \beta'_3)$. Darts 4 and 8 are 3-sewn in M_{∂_1}.

- *$\forall j \in \{3, \ldots, n\}$, $\beta'_j(d') = \beta_j(d'')$, where $d'' = \beta_0^k(d')$, k being the smaller integer s.t. $\beta_0^k(d')$ is 0-free;*
- *$\beta'_1(d') = \emptyset$; $\beta'_2(d') = \emptyset$.*

The main difference with the definition of the i-boundary map for $i \in \{2, \ldots, n\}$ concerns the definition of $\beta'_j(d')$. Instead of being directly equal to $\beta_j(d)$, we need to retrieve $\beta_j(d'')$, where d'' the second extremity of the $\langle\beta_1\rangle$ orbit. This is due to the fact that β_1 is not an involution, and also to satisfy the constraint of the n-maps definition (i.e. $\beta_1 \circ \beta_j$ is a partial involution for any $j \in \{3, \ldots, n\}$). To retrieve the second extremity of the $\langle\beta_1\rangle$ orbit, β_0 is applied, starting from dart d' as many times as necessary until obtaining a 0-free dart. Note that all the darts in a 1-boundary map are 1-free and 2-free (see two examples in Figs. 5.21 and 5.22).

Given an n-map M with i-boundaries, it is then possible to "fill in" all i-boundaries by linking the n-map with a copy of its i-boundary map M_{∂_i}. More precisely, each dart of M belonging to a boundary is linked by β_i with its corresponding dart of the copy of M_{∂_i}: this produces the *i-closure* of the n-map detailed in Section 6.1.

5.5.2 Links between n-maps and n-Gmaps

Let $G = (D, \alpha_0, \ldots, \alpha_n)$ be an n-Gmap, and let Q be its associated quasi-manifold. G is orientable (resp. connected, without boundary) iff Q is orientable (resp. connected, without boundary): cf. chapter 8. Let Q be orientable, connected and without boundary (and so is G). Q can be oriented in two ways, inverse each from the other. Similarly, G "contains" these two inverse orientations; more formally, it contains two orbits E_1 and E_2 for $\{\alpha_1 \circ \alpha_0, \ldots, \alpha_n \circ \alpha_0\}$. Note that, for any i, $0 \leq i \leq n$, for any dart d of E_1 (resp. E_2), $\alpha_i(d)$ belongs to E_2 (resp. E_1). E_1 and E_2 correspond to the two possible orientations of G (and Q).

More precisely, for any i, $1 \leq i \leq n$, let β_i denote $\alpha_i \circ \alpha_0$. $O(G)$, the n-map of the orientations of G, is defined as $(D, \beta_1, \ldots, \beta_n)$. It is an n-map, which contains two connected components $M_1 = (E_1, \beta_{1_{|E_1}}, \ldots, \beta_{n_{|E_1}})$ and $M_2 = (E_2, \beta_{1_{|E_2}}, \ldots, \beta_{n_{|E_2}})$, and M_1 and M_2 are inverse each from the other. M_1 (resp. M_2) *is* an orientation of G, and corresponds thus to an orientation of Q.

Conversely, given two connected n-maps M_1 and M_2 which are inverse to each other, it is possible to construct an n-Gmap G, such that $O(G)$, its n-map of the orientations, contains two connected components isomorphic to M_1 and to M_2.

At last, note that:

- if G is connected without boundary, G is not orientable iff $O(G)$, its n-map of the orientations, contains one connected component. In this case, the oriented quasi-manifold associated with $O(G)$ is the recovering quasi-manifold of Q (as for instance, a projective plane can be recovered by an oriented sphere, or a Klein bottle can be recovered by an oriented torus);

- if G is connected with boundaries, its n-map of the orientations contains one connected component.

These relations between n-Gmaps and n-maps can be generalized.

Definition 52 (transformation of an n-map into n-Gmap) *Let $M = (D, \beta_1, \ldots, \beta_n)$ be an n-map, where $D = \{d_1, \ldots, d_k\}$. The corresponding n-Gmap is $G = (D \cup D', \alpha_0, \ldots, \alpha_n)$, where $D' = \{d'_1, \ldots, d'_k\}$ is such that $D \cap D' = \emptyset$, and $\forall i \in \{1, \ldots, k\}$:*

- $\alpha_0(d_i) = d'_i$ *and* $\alpha_0(d'_i) = d_i$*;*
- *if d_i is 0-free in M then $\alpha_1(d_i) = d_i$ else $\alpha_1(d_i) = \alpha_0(\beta_0(d_i))$;*
- *if d_i is 1-free in M then $\alpha_1(d'_i) = d'_i$ else $\alpha_1(d'_i) = \beta_1(d_i)$;*
- *$\forall j \in \{2, \ldots, n\}$, if d_i is j-free in M then $\alpha_j(d_i) = d_i$ and $\alpha_j(d'_i) = d'_i$, else $\alpha_j(d_i) = \alpha_0(\beta_j(d_i))$ and $\alpha_j(d'_i) = \beta_j(d_i)$.*

An intuitive explanation of the relation between M and G is the following. To transform an n-map M into a corresponding n-Gmap G, each dart d_i of M is associated with an inverse dart d'_i in G. An example of conversion of a

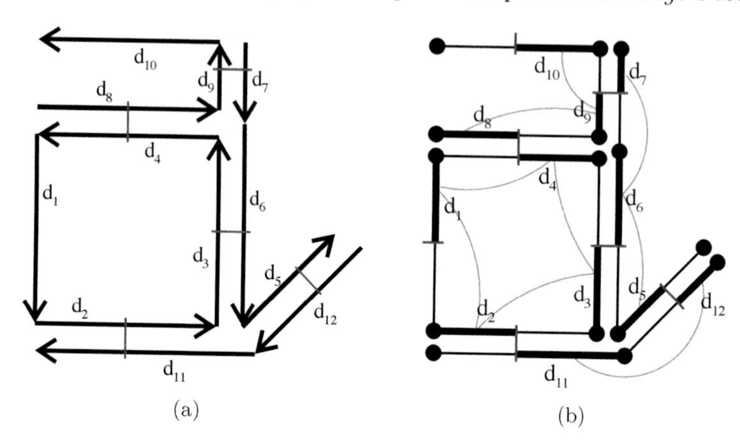

FIGURE 5.23
Conversion between a 2-map and a 2-Gmap.
(a) A 2-map M.
(b) The corresponding 2-Gmap G.

2-map into a corresponding 2-Gmap is presented in Fig. 5.23. Black darts of G are the initial darts of M, while gray darts are new darts.

Each dart d_i in M describes in G a part of the same cells, the opposite dart being d_i'; these two darts are thus 0-sewn together: so, α_0 is an involution without fixed points. In order to define other involutions α_i, it is necessary to distinguish when darts are i-free or not. Moreover, there is a special case for α_1 since β_1 is a permutation, i.e. to take into account the orientation of M.

The reverse conversion, defined in Def. 53, allows to transform an orientable n-Gmap into a corresponding n-map describing the same orientable quasi-manifold. More precisely, an orientable n-Gmap has two possible orientations, thus two n-maps correspond to these two possible orientations, each n-map being the inverse of the second one.

Definition 53 (Orientation of an n-Gmap) *Let $G = (D, \alpha_0, \ldots, \alpha_n)$ be an orientable n-Gmap, E_1 and E_2 the two sets defined by the orientation. $\forall i \in \{1, \ldots, n\}$, β_i is defined by: $\forall d \in D$,*

- *if d is 0-free in G or $\alpha_0(d)$ is i-free in G, $\beta_i(d) = \varnothing$;*
- *otherwise, $\beta_i(d) = \alpha_i \circ \alpha_0(d)$.*

n-map $M_1 = (E_1, \beta_{1|E_1}, \ldots, \beta_{n|E_1})$ is one orientation of G. The second orientation is n-map $M_2 = (E_2, \beta_{1|E_2}, \ldots, \beta_{n|E_2})$.

As example, reconsider Fig. 5.23 in the reverse order. Starting from the 2-Gmap G given in (b), it can be transformed into the corresponding 2-map M in (a), which is one orientation of G. Black darts belong to this first orientation, gray darts belong to the second orientation.

At last, let G be an orientable n-Gmap, and let M be one of its orientation. Note that G has twice the number of darts of M. Thus, in order to describe an orientable quasi-manifold, the memory space occupation is the main advantage of an n-map. However, the main advantage of n-Gmaps is related to their homogeneous definitions, leading to generic algorithms without special cases, which are thus simpler to conceive.

5.5.3 Classification of 2-maps

Given a connected 2-map M having no 0 nor 1-boundary, it is possible to compute the different topological characteristics of M introduced in Section 2.3 for the classification of paper surfaces:

- $b(M)$ is the number of 2-boundaries of M, i.e. the number of connected components of M_∂;
- $\chi(M)$ is the Euler-Poincaré characteristic of M:
  $$\chi(M) = \#\langle\beta_2 \circ \beta_0\rangle - \#\langle\beta_2\rangle + \#\langle\beta_1\rangle + |\{d|\beta_2(d) = d\}|$$
  $\#\langle\rangle$ denotes the number of orbits, see Section 8.3.2.2 page 331 for details on this characteristic;
- $q(M)$ is the orientability factor of M: $q(M) = 0$, since a 2-map is always orientable;
- $g(M)$ is the genus of M defined by $g(M) = 1 - (b(G) + \chi(G))/2$.

These characteristics make it possible to define a classification of connected 2-maps by the pair $(b(G), g(G))$ related to the classification of the associated surfaces. Note that nonconnected 2-maps can be classified, by classifying each of their connected components. The following array gives some characteristics of usual surfaces:

Type of surface	$(b(G), g(G))$
Sphere	$(0,0)$
Torus	$(0,1)$
Double torus	$(0,2)$
Disk	$(1,0)$
Annulus	$(2,0)$

Note that the number of n-boundaries and the Euler-Poincaré characteristic are defined for any n-maps whatever their dimensions is (cf. Section 8.3.2.2 page 331 for Euler-Poincaré characteristic); but for $n > 2$, no complete classification exists.

5.5.4 Duality

Remember that duality is a classical notion which makes the link between two objects having a similar structure, but regarding their dimensions in reverse order.

The definition of duality for n-maps can be easily deduced from the definition of duality for n-Gmaps and from the links between n-Gmaps and n-maps. Let $G = (D, \alpha_0, \ldots, \alpha_n)$ be an orientable n-Gmap without boundary, and let $O(G) = (D, \beta_1, \ldots, \beta_n)$ be its n-map of the orientations; so, for any i, $1 \le i \le n$, $\beta_i = \alpha_i \circ \alpha_0$. Let $G^* = (D, \alpha_n, \ldots, \alpha_0)$ be the dual of G.

In order to satisfy the relations between the associated quasi-manifolds, it is clear that $O(G)^*$, the dual of $O(G)$, has to be the n-map of the orientations of G^*, i.e. $O(G^*)$. The definition of $O(G)^*$ can thus be easily deduced:

- let $n = 1$. $G = (D, \alpha_0, \alpha_1)$ and $O(G) = (D, \beta_1 = \alpha_1 \circ \alpha_0)$; $G^* = (D, \alpha_1, \alpha_0)$, and $O(G^*) = (D, \beta_1^{-1} = \alpha_0 \circ \alpha_1)$;
- let $n > 1$; then for any $i > 1$, $\beta_i = \alpha_i \circ \alpha_0$ is an involution, i.e. $\beta_i = \beta_i^{-1}$. $G = (D, \alpha_0, \ldots, \alpha_n)$, and $O(G) = (D, \beta_1 = \alpha_1 \circ \alpha_0, \ldots, \beta_n = \alpha_n \circ \alpha_0)$. $G^* = (D, \alpha_n, \ldots, \alpha_0)$, and $O(G^*) = (D, \beta_1^* = \alpha_{n-1} \circ \alpha_n, \ldots, \beta_{n-1}^* = \alpha_1 \circ \alpha_n, \beta_n^* = \alpha_0 \circ \alpha_n)$. $\beta_n^* = \beta_n^{-1} = \beta_n$. For $1 \le i \le n-1$, $\beta_i^* = \alpha_{n-i} \circ \alpha_n = \alpha_{n-i} \circ \alpha_0 \circ \alpha_0 \circ \alpha_n = \beta_{n-i} \circ \beta_n^{-1} = \beta_{n-i} \circ \beta_n$.

This leads to the following definition.

Definition 54 (dual) *Let $M = (D, \beta_1, \ldots, \beta_n)$ be an n-map without n-boundary. The dual n-map of M is M^* defined by:*

- *if $n = 1$, $M^* = (D, \beta_1^{-1})$;*
- *otherwise, $M^* = (D, \beta_{n-1} \circ \beta_n, \ldots, \beta_1 \circ \beta_n, \beta_n)$.*

To ensure that the composition $\beta_i \circ \beta_n$ can be done without loss of any relation, the n-map must be without n-boundary. This ensures that $\beta_n \neq \varnothing$ and thus that composing β_i with β_n gives $\varnothing$ only when $\beta_i = \varnothing$. This property is required to guaranty the important property that $M^{**} = M$. For all other dimensions $i \neq n$, the n-map can have i-boundary.

By definition of cells, i-cells are transformed into $(n-i)$-cells by duality. Thanks to this definition, it is straightforward to prove that $M^{**} = M$.

An example of a 2-map $M = (D, \beta_1, \beta_2)$ without boundary is presented in Fig. 5.24, as its corresponding dual 2-map $M^* = (D, \beta_1 \circ \beta_2, \beta_2)$. M is composed by seven vertices, nine edges and four faces, and M^* is composed by four vertices, nine edges and seven faces.

As illustrated in Fig. 5.25, the dual definition is still valid for n-maps with boundaries. In this example, the 2-map $M = (D, \beta_1, \beta_2)$ is without 2-boundary, but with 1-boundaries. Its dual is 2-map $M^* = (D, \beta_1 \circ \beta_2, \beta_2)$ is also without 2-boundary, but with 1-boundaries. M is composed by six vertices, five edges and five faces, and M^* is composed by five vertices, five edges and six faces.

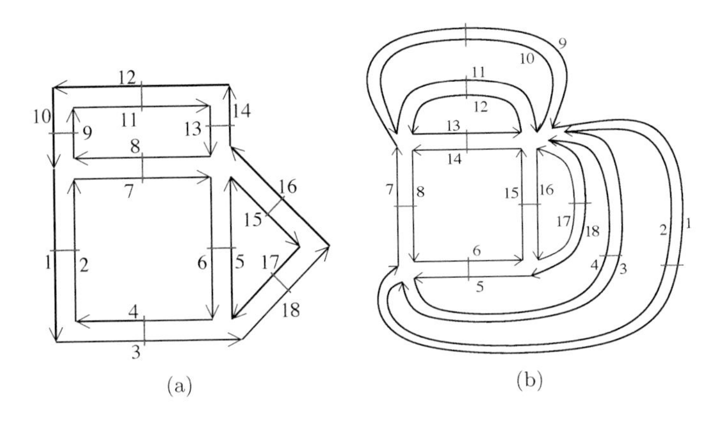

FIGURE 5.24
(a) 2-map $M = (D, \beta_1, \beta_2)$ without boundary.
(b) The dual 2-map $M^* = (D, \beta_1 \circ \beta_2, \beta_2)$. The vertex located down left (resp. down right, up left, up right) corresponds to the square face down (resp. the triangle, the square face up, the external face) in Fig. 5.24(a).

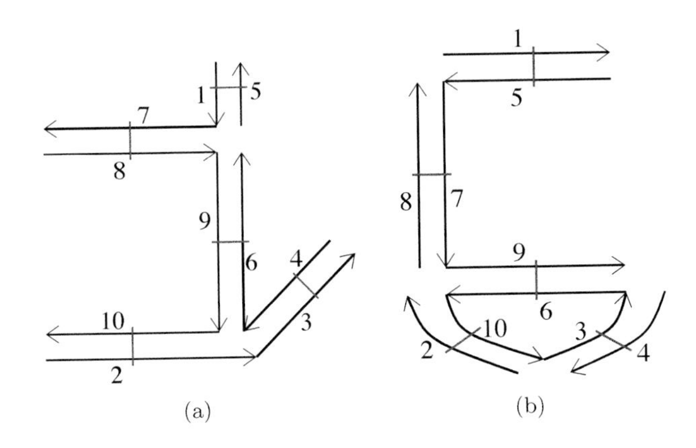

FIGURE 5.25
(a) 2-map $M = (D, \beta_1, \beta_2)$ is without 2-boundary, but with 1-boundaries.
(b) The dual 2-map $M^* = (D, \beta_1 \circ \beta_2, \beta_2)$.

6

Operations

In this chapter, some operations allowing to modify n-maps and n-Gmaps are defined. The corresponding algorithms handle the data structures introduced in Section 4.4 page 108 and Section 5.4 page 158. The first operation given in Section 6.1 is the *closure* operation, allowing to fill boundaries. The second and third operations given in Section 6.2 and Section 6.3 allow to *remove* and *contract* an i-cell, by merging the two incident $(i+1)$- or $(i-1)$-cells. For these three operations, the definitions and algorithms are stated for both n-maps and n-Gmaps, pointing out the differences and allowing to understand how to define an operation on n-maps by adapting the corresponding definition on n-Gmaps.

Then, the *insertion* and the *expansion* operations are defined in Section 6.4 and in Section 6.5; they are the two inverse operations of the removal and contraction. The *chamfering* operation is defined in Section 6.6, the *extrusion* operation in Section 6.7, and the *triangulation* operation in Section 6.8. For these five operations, the definitions are stated only for n-Gmaps; the transposition for n-maps is left to the reader.

Note that all these operations are defined here in the most generic way. The only constraints are required in order to guaranty the validity of the operation, i.e. to ensure that a valid n-Gmap or n-map is obtained. In order to satisfy some specific needs depending on a given application, it is often required to add more restrictive constraints (like for example the preservation of the connectivity for the removal operation, or the preservation of the genus of a surface for the insertion operation). These constraints can be tested in a preliminary step before applying the generic operation. Note also that, depending on the application, additional specific properties make it possible to optimize some operations.

6.1 Closure

The *closure* operation allows to fill boundaries. As seen in the two previous chapters, boundaries can exist in any dimension. Thus, we define the i-closure operation, which fills i-dimensional boundaries. This operation is based on the definition of the i-boundary map. Intuitively, each connected component of the i-boundary map allows to close one boundary by identifying its darts with the corresponding i-free darts of the n-Gmap or the n-map.

6.1.1 For n-Gmaps

The i-closure operation for an n-Gmap G is defined in Def. 55. It consists in creating new darts, one for each i-free dart in G. Then, the new darts are linked between them by α_j, $\forall j \in \{0, \ldots, n\}$, $j \neq i$, by using the corresponding links in the i-boundary map G_{∂_i} (cf. Def. 21 page 124); each new dart is linked by α_i with its corresponding dart in G.

Definition 55 (i-closure) *Let $G = (D, \alpha_0, \ldots, \alpha_n)$ be an n-Gmap, $i \in \{0, \ldots, n\}$, and $G_{\partial_i} = (D', \alpha'_0, \ldots, \alpha'_n)$ be the corresponding i-boundary map. The i-closure of G is the n-Gmap $G_{i\text{-}close} = (D'', \alpha''_0, \ldots, \alpha''_n)$ defined by:*

1. *$D'' = D \cup \{d''_1, \ldots, d''_{|D'|}\}$, where $\{d''_1, \ldots, d''_{|D'|}\}$ is a set of new darts, such that a bijection f maps D' onto $\{d''_1, \ldots, d''_{|D'|}\}$;*
2. *$\forall j \in \{0, \ldots, n\}$ s.t. $j \neq i$:*
   - *$\forall d \in D$, $\alpha''_j(d) = \alpha_j(d)$;*
   - *$\forall d' \in D'$, $\alpha''_j(f(d')) = f(\alpha'_j(d'))$.*
3. *$\forall d \in D \setminus D'$, $\alpha''_i(d) = \alpha_i(d)$;*
4. *$\forall d' \in D'$, $\alpha''_i(d) = f(d)$, $\alpha''_i(f(d)) = d$.*

The links between $D'' \setminus D$ (i.e. the new darts) and D' (i.e. the darts of i-boundaries of G) are defined through the bijection f. Note that any bijection between the two sets of darts can be considered, as the order of the new darts has no importance. Note also that $D' \subseteq D$, by definition of boundary map: this explains why new darts are added to D and linked with D'.

The 1-closure operation is illustrated in Fig. 6.1. Starting from the 1-Gmap with boundary depicted in (a), the 1-Gmap shown in (b) is obtained by applying the 1-closure operation; note that this 1-Gmap has no boundary. In the initial 1-Gmap, there are two 1-free darts (numbered 1 and 6); thus two new darts (numbered 7 and 8) are created; moreover, $f(6)$ (resp. $f(1)$) is arbitrarily set to 7 (resp. 8). As the two darts 1 and 6 are 0-sewn in the 1-boundary map G_{∂_1}, the corresponding darts 7 and 8 are also 0-sewn in $G_{1\text{-close}}$. $\alpha''_1(1) = 8$ (resp. $\alpha''_1(6) = 7$), since $f(1) = 8$ (resp. $f(6) = 7$).

Another example of the 1-closure operation is presented in Fig. 6.2; it is here applied to a 3-Gmap. In this example, four darts, namely 1, 6, 7 and 12, are 1-free darts; thus four new darts are created, numbered 13, 14, 15 and 16, and are respectively associated with 1, 6, 7 and 12 by bijection f. As darts 1 and 6 are 0-sewn in the 1-boundary map G_{∂_1}, the corresponding darts 13 and 14 are also 0-sewn in $G_{1\text{-close}}$ (this is similar for darts 7 and 12, and the corresponding darts 15 and 16). Moreover, as 6 and 12 are 3-sewn in the

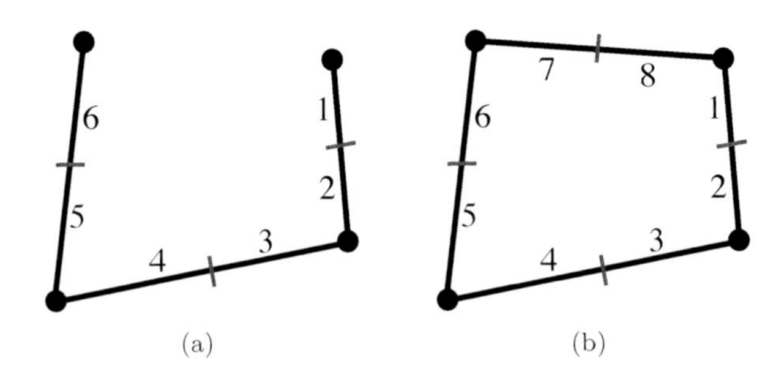

FIGURE 6.1
1-closure for 1-Gmap.
(a) A 1-Gmap G with 1-boundary.
(b) The 1-Gmap $G_{1\text{-close}}$. 7 and 8 are new darts, such that $f(6) = 7$ and $f(1) = 8$.

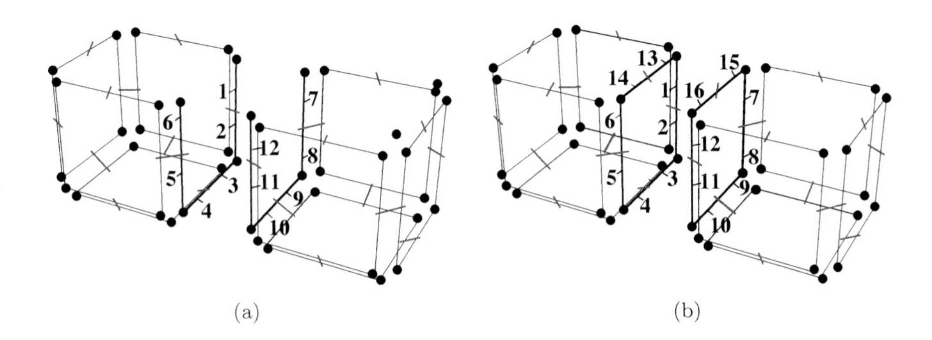

FIGURE 6.2
1-closure for 3-Gmap.
(a) A 3-Gmap G with 1-boundary, 2-boundary and 3-boundary. The four darts 1, 6, 7 and 12 are 1-free.
(b) The 3-Gmap $G_{1\text{-close}}$. Dart 13, 14, 15 and 16 are new darts, respectively associated with darts 1, 6, 7 and 12.

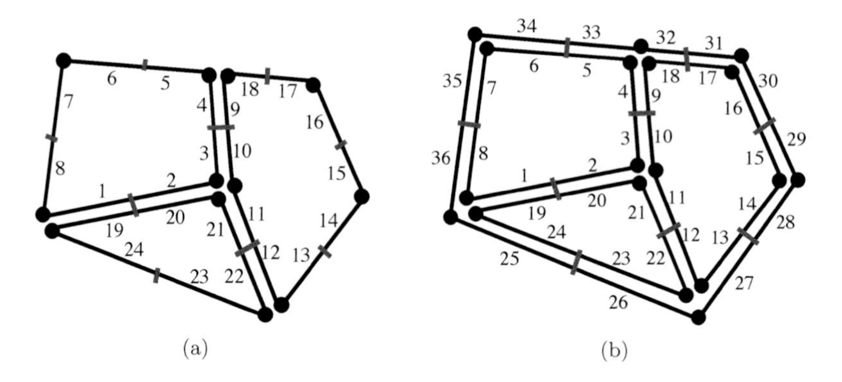

FIGURE 6.3
2-closure for 2-Gmap.
(a) A 2-Gmap G with 2-boundary. Darts $5, 6, 7, 8, 13, 14, 15, 16, 17, 18, 23, 24$ are 2-free.
(b) The 2-Gmap $G_{2\text{-close}}$. New darts are numbered from 25 to 36.

1-boundary map G_{∂_1}, 14 and 16 are also 3-sewn in $G_{1\text{-close}}$ (this is similar for 1 and 7, and the corresponding darts 13 and 15). $\alpha''_1(1) = f(1) = 13$ (resp. $\alpha''_1(6) = 14$, $\alpha''_1(7) = 15$ and $\alpha''_1(12) = 16$).

An example of 2-closure operation in a 2-Gmap is presented in Fig. 6.3. Starting from the 2-Gmap with 2-boundary depicted in (a), the 2-Gmap shown in (b), obtained by applying the 2-closure operation, has no more 2-boundary. In this example, there are twelve 2-free darts, thus twelve new darts numbered from 25 to 36 are created. These new darts are linked by α''_2 with their corresponding darts. Moreover, these new darts are linked between them by α''_0 and α''_1, according to the corresponding involutions of the 2-boundary map G_{∂_2}.

We can see in Fig. 6.4 a 2-Gmap which has 1-boundaries and 2-boundaries. In this case, the 2-closure operation produces a 2-Gmap without 2-boundary, but it still has 1-boundaries.

Thanks to the i-closure definition, it is straightforward to define a *closure operation* that removes all the boundaries in all dimensions. For that, all the possible i-closure operations are successively applied, from $i = 0$ to $i = n$. For example, starting from the 3-Gmap given in Fig. 6.2(a) (which has no 0-boundary), and applying the 1-closure operation, the 3-Gmap shown in Fig. 6.2(b) is obtained: this 3-Gmap has no 1-boundary. Then the 2-closure operation is applied, producing the 3-Gmap given in Fig. 6.5(a), which has no 2-boundary. Finally, the 3-closure operation is applied, producing the 3-Gmap in Fig. 6.5(b), which has no 3-boundary, and thus no boundary at all.

Algorithm 47 implements the i-closure operation for n-Gmaps. This algorithm follows directly the i-closure definition (see Def. 55). Note that the

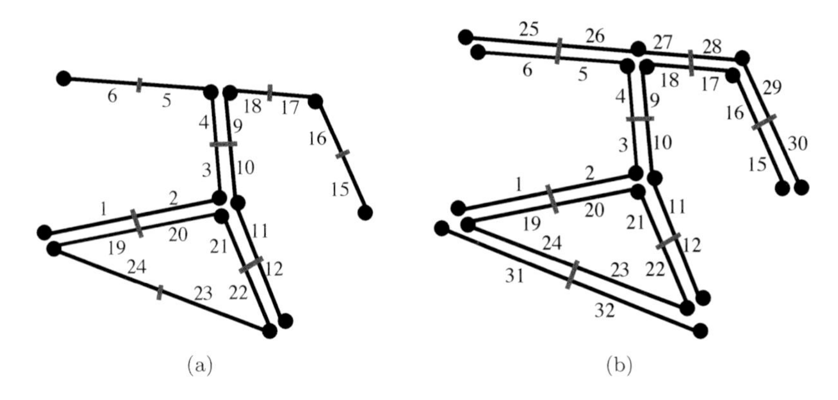

FIGURE 6.4
2-closure for 2-Gmap.
(a) A 2-Gmap G with 1-boundaries and 2-boundaries.
(b) The 2-Gmap $G_{2\text{-close}}$. New darts are numbered from 25 to 32. This 2-Gmap has no 2-boundary, but it still has 1-boundaries.

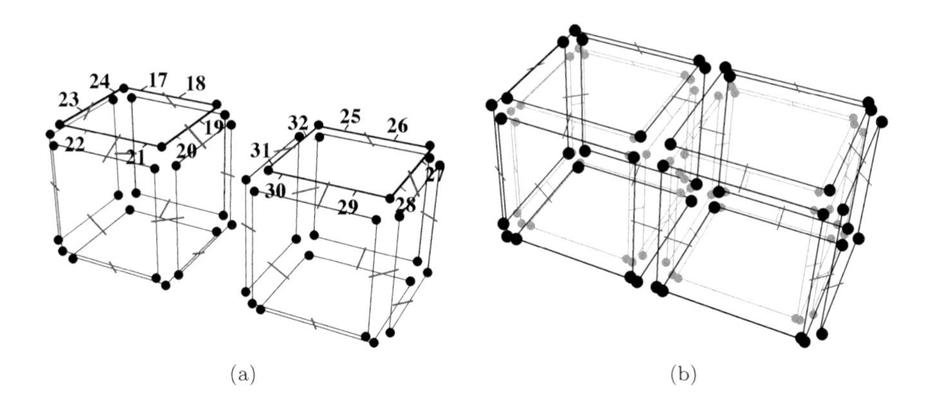

FIGURE 6.5
Closure for 3-Gmap.
(a) The 3-Gmap G' obtained from Fig. 6.2(b) by applying the 2-closure operation. Darts 17 to 32 are new darts.
(b) The 3-Gmap $G'_{3\text{-close}}$. The eighty black darts are new darts, describing the external volume enclosing the two initial cubes.

i-boundary map is not explicitly computed; nevertheless, the new darts are linked according to its definition.

Algorithm 47: `closeNGMap(gm,i)`: i-close an n-Gmap

```
   Input: gm: an n-Gmap;
          i ∈ {0, ..., n}.
   Result: gm is i-closed.
1  foreach i-free dart d ∈ gm.Darts do
2      d' ← createDartNGMap(gm);
3      d'.Alphas[i] ← d; d.Alphas[i] ← d';
4      foreach j ∈ {0, ..., i − 2, i + 2, ..., n} do
5          d'' ← d.Alphas[j];
6          if not isFreeNGMap(d,j) and not isFreeNGMap(d'',i) then
7              d'.Alphas[j] ← d''.Alphas[i];
8              d''.Alphas[i].Alphas[j] ← d';
9      if i > 0 then
10         d'' ← d.Alphas[i − 1];
11         while not isFreeNGMap(d'',i) and not
           isFreeNGMap(d''.Alphas[i],i − 1) do
12             d'' ← d''.Alphas[i].Alphas[i − 1];
13         if not isFreeNGMap(d'',i) then
14             d'.Alphas[i − 1] ← d''.Alphas[i];
15             d''.Alphas[i].Alphas[i − 1] ← d';
```

In this algorithm, the new darts are created and linked by $\alpha_{\mathtt{i}}$ with the corresponding `i`-free darts. This corresponds to item 4 of the i-closure definition. Then, the other α links are defined according to the definition of the boundary map. $\alpha_{\mathtt{i}+1}$ is initialized to the identity for all new darts (second item of the i-boundary map definition): this is a result of the `createDartNGMap` function.

For all `j` in $\{0, \ldots, \mathtt{i}-2, \mathtt{i}+2, \ldots, n\}$, $\alpha_{\mathtt{j}}$ is deduced from the definition of $\alpha_{\mathtt{j}}$ on the `i`-free dart `d`. Indeed, $\alpha_{\mathtt{i}} \circ \alpha_{\mathtt{j}}$ is an involution. As $\alpha_{\mathtt{i}}(\mathtt{d}) = \mathtt{d}'$ and $\alpha_{\mathtt{j}}(\mathtt{d}) = \mathtt{d}''$, it is necessary that $\alpha_{\mathtt{j}}(\mathtt{d}') = \alpha_{\mathtt{i}}(\mathtt{d}'')$; since $\mathtt{d}' = f(\mathtt{d})$, $\alpha_{\mathtt{j}}(\mathtt{d}') = \alpha_{\mathtt{i}}(\mathtt{d}'')$ is equivalent to $\alpha''_{\mathtt{j}}(f(\mathtt{d})) = f(\alpha'_{\mathtt{j}}(\mathtt{d}))$: this corresponds to item 2 of the i-closure definition (remember that, in this algorithm, α_i stands also for α'_i and α''_i).

Note that $\alpha_{\mathtt{j}}$ can be defined only if `d` is not `j`-free and `d''` is not `i`-free. Indeed, if `d` is `j`-free then `d'` must be also `j`-free; thus, there is nothing to do as darts are created free in all dimensions. If `d''` is `i`-free, the $\alpha_{\mathtt{j}}$ link between `d'` and $\alpha_{\mathtt{i}}(\mathtt{d}'')$ cannot yet be defined: the new dart associated with $\alpha_{\mathtt{j}}(\mathtt{d})$ is not yet created. This definition will be done in a future step of the algorithm, when dart $\alpha_{\mathtt{j}}(\mathtt{d})$ will be processed. Indeed, it will be possible to define the

α_j links when processing $\alpha_j(d)$; this is why this definition is done in both directions.

α_{i-1} is initialized by following the path of darts $\alpha_{i-1} \circ (\alpha_i \circ \alpha_{i-1})^k(d)$ until obtaining either an `i`-free dart, or a dart such that its dart associated by α_i is (`i-1`)-free. The first case corresponds to the definition of α'_{i-1} in the i-boundary map. However, in this case, it is not possible to define α_{i-1}, as the second dart concerned by this definition is not yet created: the definition will be done when the second extremity of this path will be processed. For this second dart, the first dart is no longer `i`-free, because its corresponding dart is already created and linked with it by α_i. However, the new dart is (`i-1`)-free since it was not yet (`i-1`)-linked, and this corresponds to the second case. So, $\alpha_{i-1}(d') = \alpha_i(d'')$, and reciprocally. As $d' = f(d)$ and $\alpha_i(d'') = f(\alpha_{i-1} \circ (\alpha_i \circ \alpha_{i-1})^k(d))$, this is equivalent to $\alpha''_{i-1}(f(d)) = f(\alpha_{i-1} \circ (\alpha_i \circ \alpha_{i-1})^k(d))$; this corresponds to item 2 of the i-closure definition.

The complexity of Algorithm 47 is linear in number of darts of `gm`. Indeed, each dart is considered exactly once in the main foreach loop. In this loop, each operation is atomic except the while loop (line 11) which is linear in number of darts in the path $\alpha_{i-1} \circ (\alpha_i \circ \alpha_{i-1})^k(d)$. However these darts are considered exactly twice, the first time when `d` is processed, and the second time when the second i-free dart is processed. In other words, these paths are disjoint for different pairs of i-free darts. So, the overall complexity of the algorithm is linear in number of darts of the n-Gmap.

To illustrate the behavior of this algorithm, consider the 2-Gmap given in Fig. 6.4(a), and apply the 2-closure algorithm. Darts are processed in any order, and there is nothing to do for non 2-free darts. So, for example, start with dart $d = 5$. A new dart $d' = 26$ is created, and darts 5 and 26 are 2-sewn. The first loop (line 4) considers only $j = 0$. As $d'' = 6$ is 2-free, the new dart is not 0-sewn. The second loop (line 11) allows to define α_1. Starting from $d'' = 4$, $d'' = 18$ is reached after one iteration; it is 2-free, thus α_1 is not defined for d'. Then another 2-free dart is processed, for example $d = 17$. As for the previous case, a new dart $d' = 28$ is created, darts 17 and 28 are 2-sewn, α_0 and α_1 are not defined for new dart 28. Now if dart $d = 18$ is considered, a new dart $d' = 27$ is created, and darts 18 and 27 are 2-sewn. First, as $d'' = 17$ is not 2-free, $\alpha_0(27)$ is defined as $\alpha_2(17) = 28$, and $\alpha_0(28)$ is defined as 27 (lines 7 and 8). Second, starting from $d'' = 9$, $d'' = 5$ is reached after one iteration; dart 5 is not 2-free, but $\alpha_2(5)$ is 1-free. Thus $\alpha_1(27)$ is defined as $\alpha_2(5) = 26$, and $\alpha_1(26)$ is defined as 27 (lines 14 and 15).

The principle is similar in 3D. Consider for example the 3-Gmap given in Fig. 6.5(a), and apply the 3-closure operation. If dart $d = 32$ is first processed, a new dart $d' = 32'$ is created, and darts 32 and $32'$ are 3-sewn. The first loop (line 4) considers $j \in \{0, 1\}$, but in both cases, $\alpha_j(32)$ is 3-free. The second loop (line 11) allows to define α_2. Starting from $d'' = \alpha_2(32)$, $d'' = 19$ is obtained after one iteration; it is 3-free, and thus α_2 is not defined for dart $32'$. Consider now $d = 31$; a new dart $d' = 31'$ is created, and darts 31 and $31'$ are 3-sewn. During the first loop, darts $31'$ and $32'$ are 0-sewn. Consider now

$d = 20$; a new dart $d' = 20'$ is created, and darts 20 and $20'$ are 3-sewn. No sew is done during the first loop; darts $20'$ and $31'$ are 2-sewn, as dart $d'' = 31$ is obtained, which is such that $\alpha_3(31)$ is 2-free, and which is not 3-free.

6.1.2 For n-maps

The i-closure operation, for $i \geq 2$, is defined in Def. 56 for an n-map M. As for n-Gmaps, new darts are created, one for each i-free dart in M; these darts are linked between them for β_j, $\forall j \in \{1, \ldots, n\}$, $j \neq i$, according to the corresponding links in the inverse of the i-boundary map M_{∂_i} (cf. Def. 49 page 175); each new dart is linked by β_i with its corresponding dart in M.

Definition 56 (i-closure) *Let $M = (D, \beta_1, \ldots, \beta_n)$ be an n-map, $i \in \{2, \ldots, n\}$, and $M_{\partial_i}^{-1} = (D', \beta'_1, \ldots, \beta'_n)$ be the inverse of the i-boundary map of M. The i-closure of M is the n-map $M_{i\text{-close}} = (D'', \beta''_1, \ldots, \beta''_n)$ defined by:*

1. *$D'' = D \cup \{d''_1, \ldots, d''_{|D'|}\}$, where $\{d''_1, \ldots, d''_{|D'|}\}$ is a set of new darts, such that a bijection f maps D' onto $\{d''_1, \ldots, d''_{|D'|}\}$;*
2. *$\forall j \in \{1, \ldots, n\}$ s.t. $j \neq i$:*
   - *$\forall d \in D$, $\beta''_j(d) = \beta_j(d)$;*
   - *$\forall d' \in D'$, $\beta''_j(f(d')) = f(\beta'_j(d'))$.*
3. *$\forall d \in D \setminus D'$, $\beta''_i(d) = \beta_i(d)$;*
4. *$\forall d' \in D'$, $\beta''_i(d) = f(d)$, $\beta''_i(f(d)) = d$.*

The 2-closure operation is illustrated in Fig. 6.6. Starting from the 2-map with boundary given in (a), the 2-map shown in (b), obtained by applying the 2-closure operation, has no 2-boundary. In the initial 2-map, six darts are 2-free, thus six new darts are created, numbered from 13 and 18. These new darts are linked by β''_2 with their corresponding darts. Moreover, these new darts are linked between them by β''_1, according to the corresponding links in the inverse of the 2-boundary map $M_{\partial_2}^{-1}$.

The 2-map depicted in Fig. 6.7 has 1-boundaries and 2-boundaries. In this case, the 2-closure operation produces a 2-map without 2-boundary, but it still has 1-boundaries.

The 1-closure operation is defined in Def. 57.

Definition 57 (1-closure) *Let $M = (D, \beta_1, \ldots, \beta_n)$ be an n-map, with $n > 0$, and $M_{\partial_1} = (D', \beta'_1, \ldots, \beta'_n)$ be the corresponding 1-boundary map. The 1-closure of M is the n-map $M_{1\text{-close}} = (D'', \beta''_1, \ldots, \beta''_n)$ defined by:*

1. *$D'' = D \cup \{d''_1, \ldots, d''_{|D'|}\}$, where $\{d''_1, \ldots, d''_{|D'|}\}$ is a set of new darts, such that a bijection f maps D' onto $\{d''_1, \ldots, d''_{|D'|}\}$;*

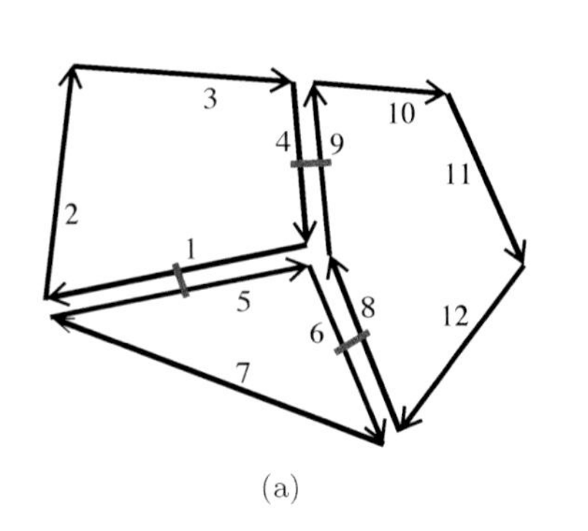

(a)

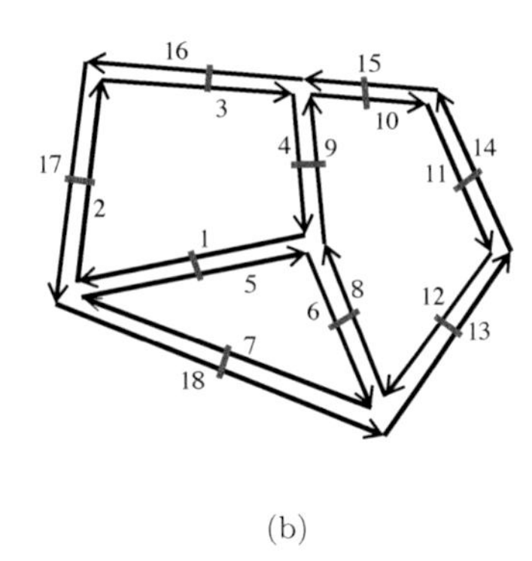

(b)

FIGURE 6.6
2-closure for 2-map.
(a) A 2-map M with 2-boundary. Darts $2, 3, 7, 10, 11, 12$ are 2-free.
(b) The 2-map $M_{2\text{-close}}$. New darts are numbered from 13 to 18.

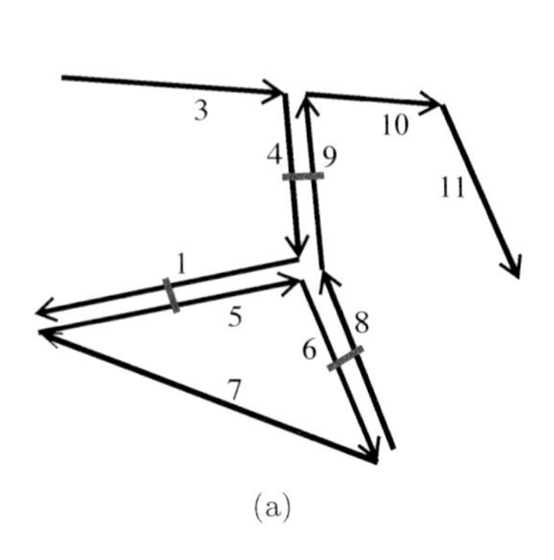

(a)

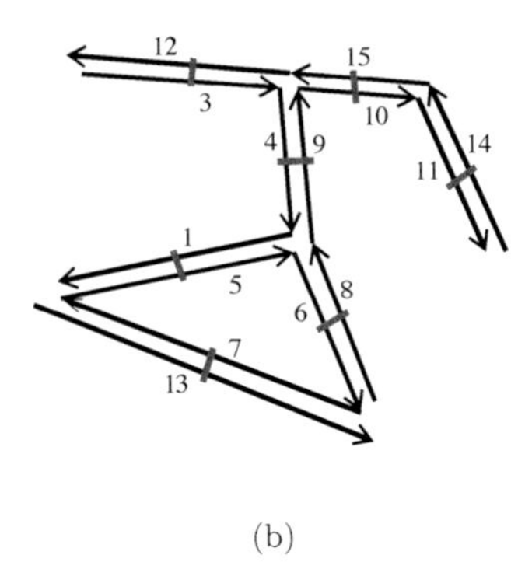

(b)

FIGURE 6.7
2-closure for 2-map.
(a) A 2-map M with 1-boundaries and 2-boundaries.
(b) The 2-map $M_{2\text{-close}}$. New darts are numbered from 12 to 15. This 2-map has no 2-boundary, while it still has 1-boundaries.

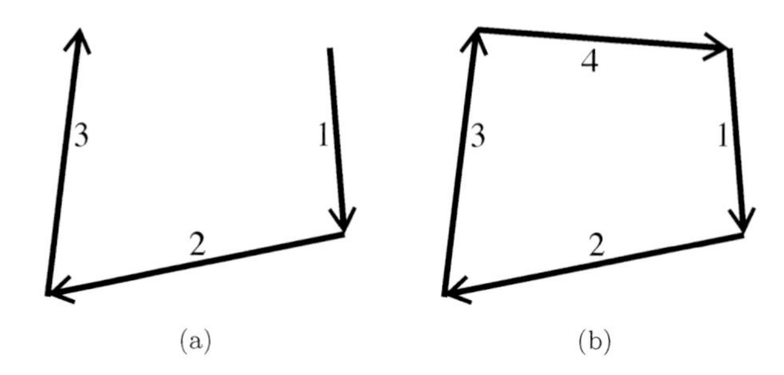

FIGURE 6.8
1-closure for 1-map.
(a) A 1-map M with 1-boundary.
(b) The 1-map $M_{1\text{-close}}$. Dart 4 is new.

2. $\forall j \in \{2, \ldots, n\}$:
   - $\forall d \in D$, $\beta''_j(d) = \beta_j(d)$;
   - $\forall d' \in D'$, $\beta''_j(f(d')) = f(\beta'_j(d'))$.
3. $\forall d \in D \setminus D'$, $\beta''_1(d) = \beta_1(d)$;
4. $\forall d' \in D'$, $\beta''_1(d) = f(d)$, $\beta''_1(f(d)) = \beta_0^k(d)$
   k being the smaller positive integer s.t. $\beta_0^k(d)$ is 0-free.

There are two differences with the definition of the i-closure for $i \geq 2$. First, the 1-boundary map M_{∂_1} is considered and not its inverse. Indeed, all darts in D' are 1-free, thus $M_{\partial_1} = M_{\partial_1}^{-1}$. Second, $\beta''_1(f(d))$ is not equal to d, since β_1 is not an involution. It is equal to $\beta_0^k(d)$, k being the smaller positive integer such that $\beta_0^k(d)$ is 0-free. Intuitively, β_1 defines polygonal lines, and β_0 is the second extremity of the open polygonal line incident to d.

The 1-closure operation is illustrated in Fig. 6.8. Starting from the 1-map with boundary depicted in (a), the 1-map without boundary shown in (b) is obtained by applying the 1-closure operation. In the initial 1-map, there is one 1-free dart (numbered 3), thus one dart (numbered 4) is created. $\beta''_1(3) = f(3) = 4$, since dart 4 is the new dart corresponding to dart 3. However, it is more complex to define $\beta''_1(4)$, since it is necessary to retrieve the 0-free dart belonging to the same polygonal line. So, $\beta_0^k(3)$ has to be computed, where k is the smaller positive integer such that $\beta_0^k(3)$ is 0-free. Thus $\beta''_1(4) = \beta_0^2(3) = 1$.

Another example of the 1-closure operation is presented in Fig. 6.9, but here for a 3-map. There are two 1-free darts numbered 3 and 4, thus two new darts (numbered 7 and 8) are created. As darts 3 and 4 are 3-sewn in the 1-boundary map M_{∂_1}, the corresponding darts 7 and 8 are also 3-sewn in $M_{1\text{-close}}$. $\beta''_1(3) = f(3) = 7$, since dart 7 corresponds to the 1-free dart 3 (similarly, $\beta''_1(4) = 8$), and $\beta''_1(7) = \beta_0^2(3) = 1$ (similarly, $\beta''_1(8) = \beta_0^2(4) = 2$).

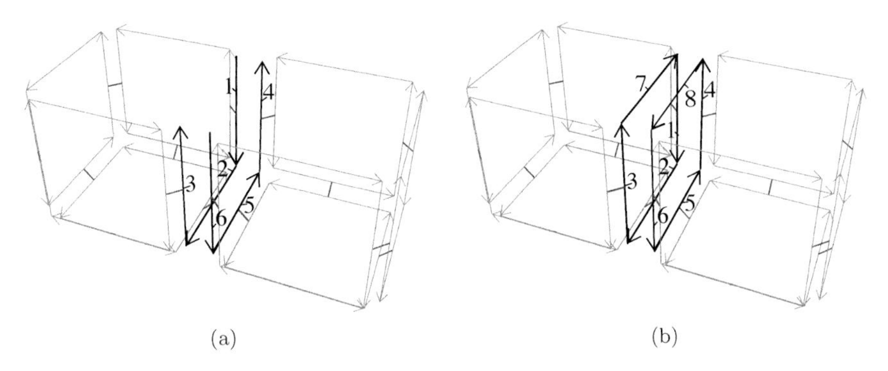

FIGURE 6.9
1-closure for 3-map.
(a) A 3-map M with 1-boundary, 2-boundary and 3-boundary. Darts 3 and 4 are 1-free.
(b) The 3-map $M_{1\text{-close}}$. Dart 7 and 8 are new darts associated with darts 3 and 4.

As for n-Gmaps, a *closure operation* can be defined for removing all boundaries for all dimensions. For that, all the possible i-closure operations are applied, from $i = 1$ to $i = n$. For example, starting from the 3-map given in Fig. 6.9(a), the 1-closure operation produces the 3-map shown in Fig. 6.9(b), which has no 1-boundary. Then, the 2-closure operation produces the 3-map given in Fig. 6.10(a), which has no 2-boundary. Finally, the 3-closure operation produces the 3-map given in Fig. 6.10(b), which has no 3-boundary, and thus no boundary at all.

Algorithm 48 implements the i-closure operation for n-maps and for $i \in \{2, \ldots, n\}$. This algorithm follows directly the i-closure definition (see Def. 56). As for Algorithm 47, the i-boundary map is not explicitly computed.

In this algorithm, the new darts are created and linked by $\beta_{\mathtt{i}}$ with the corresponding `i`-free darts (cf. item 4 of the i-closure definition). Then, the other β links are defined, according to the definition of the boundary map. $\beta_{\mathtt{i+1}}$ is initialized to $\varnothing$ for all the new darts (second item of the i-boundary map definition), by definition of the `createDartNMap` function.

All `j`, such that $|\mathtt{i} - \mathtt{j}| \geq 2$, have to be considered, thus two cases occur: if $\mathtt{i} = 2$, this is the set $\{4, \ldots, n\}$; otherwise, this is the set $\{0, 1, \ldots, \mathtt{i} - 2, \mathtt{i} + 2, \ldots, n\}$. For all `j` in this set, $\beta_{\mathtt{j}}$ is deduced from the definition of $\beta_{\mathtt{j}}$ for the `i`-free dart `d`. Indeed, $\beta_{\mathtt{i}} \circ \beta_{\mathtt{j}}$ is a partial involution. As $\beta_{\mathtt{i}}(\mathtt{d}) = \mathtt{d}'$ and $\beta_{\mathtt{j}}(\mathtt{d}) = \mathtt{d}''$, it is necessary to have $\beta_{\mathtt{j}}(\mathtt{d}') = \beta_{\mathtt{i}}(\mathtt{d}'')$. Since $\mathtt{d}' = f(\mathtt{d})$, defining $\beta_{\mathtt{j}}^{-1}(\mathtt{d}') = \beta_{\mathtt{i}}(\mathtt{d}'')$ is equivalent to define ${\beta''_{\mathtt{j}}}^{-1}(f(\mathtt{d})) = f(\beta'_{\mathtt{j}}(\mathtt{d}))$: this corresponds to item 2 of the i-closure definition (remember that in the algorithm, β_i denotes also β'_i and β''_i).

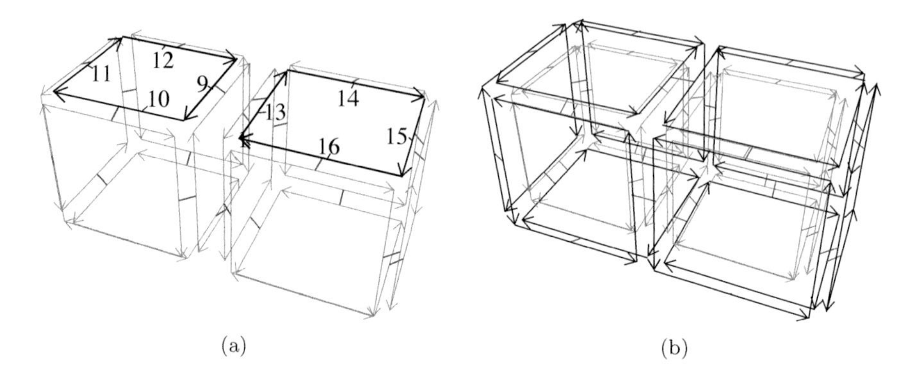

FIGURE 6.10
Closure for 3-map.
(a) The 3-map M' obtained from Fig. 6.9(b) by applying the 2-closure operation. Darts 9 to 16 are new darts.
(b) The 3-map $M'_{3\text{-close}}$. The forty black darts are new darts, describing the external volume enclosing the two initial cubes.

Algorithm 48: `closeNMap(cm,i)`: i-close an n-map

Input: `cm`: an n-map;
`i` $\in \{2, \ldots, n\}$.
Result: `cm` is i-closed.

```
1  if i = 2 then S ← {4,...,n};
2  else S ← {0,1,...,i−2,i+2,...,n};
3  foreach i-free dart d ∈ cm.Darts do
4      d′ ← createDartNMap(cm);
5      d.Betas[i] ← d′; d′.Betas[i] ← d;
6      foreach j ∈ S do
7          d″ ← d.Betas[j];
8          if not isFreeNMap(cm,d,j) and not isFreeNMap(cm,d″,i)
           then
9              d′.Betas[inv(j)] ← d″.Betas[i];
10             d″.Betas[i].Betas[j] ← d′;
11     d″ ← d.Betas[inv(i − 1)];
12     while not isFreeNMap(cm,d″,i) and not
       isFreeNMap(cm,d″.Betas[i],inv(i − 1)) do
13         d″ ← d″.Betas[i].Betas[inv(i − 1)];
14     if not isFreeNMap(cm,d″,i) then
15         d′.Betas[i − 1] ← d″.Betas[i];
16         d″.Betas[i].Betas[inv(i − 1)] ← d′;
```

Note that β_j is defined for d only if d is not j-free and d″ is not i-free. Indeed, if d is j-free, then d′ must be also j-free, and thus there is nothing to do as darts are created free in all dimensions. If d″ is i-free, it is not possible yet to define the β_j links between d′ and β_i(d″), as the new dart associated with β_j(d) is not yet created. This will be done in a future step of the algorithm, when dart β_j(d) will be processed. This is why this definition is done in both directions.

β_{i-1} is initialized by following the path of darts $\beta_{i-1}^{-1} \circ (\beta_i \circ \beta_{i-1}^{-1})^k(\mathtt{d})$ until obtaining either an i-free dart or a dart such that its associated dart by β_i is (inv(i − 1))-free. The first case corresponds to the definition of β'_{i-1} in the i-boundary map. However, it is not possible to define β_{i-1} as the second dart concerned by this definition is not yet created. The definition will be done when the second extremity of this path will be processed. For this second dart, the first dart is no longer i-free, because its corresponding dart has already been created, and linked with it by β_i. However, the new dart is (inv(i − 1))-free as it is not yet (inv(i − 1))-linked. In this second case, β_{i-1}(d′) is defined as β_i(d″), and reciprocally. As d′ $= f$(d) and $\beta_i(\mathtt{d''}) = f(\beta_{i-1}^{-1} \circ (\beta_i \circ \beta_{i-1}^{-1})^k(\mathtt{d}))$, this is equivalent to $\beta''_{i-1}(f(\mathtt{d})) = f(\beta_{i-1}^{-1} \circ (\beta_i \circ \beta_{i-1}^{-1})^k(\mathtt{d}))$, which corresponds to item 2 of the i-closure definition.

Note that β_{i-1}^{-1} is used, because the i-closure definition uses the inverse of the i-boundary map (this is only necessary for i = 2).

The complexity of Algorithm 48 is linear in number of darts of cm. Indeed, each dart is considered exactly once in the main foreach loop. In this loop, each operation is atomic except the while loop (line 12) which is linear in number of darts in the path $\beta_{i-1}^{-1} \circ (\beta_i \circ \beta_{i-1}^{-1})^k(\mathtt{d})$. However these darts are considered exactly twice, the first time when d is processed, and the second time when the second i-free dart is processed. In other words, these paths are disjoint for each different pair of i-free darts. So, the overall complexity of the algorithm which is linear in number of darts of the n-map.

The behaviour of this algorithm can be illustrated on the examples provided in Fig. 6.7(a) and Fig. 6.10(a). Since it is close to the behavior of Algorithm 47, this is left to the reader.

The 1-closure operation for n-maps is implemented by Algorithm 49. It is similar to Algorithm 48, the two main differences are that both β_0 and β_1 links have to be defined, and second that β_{i-1} does not exist.

As for Algorithm 48, a new dart d′ is created and linked by β_1 with an existing 1-free dart. Then, e $= \beta_0^k$(d) is computed, such that k is the smallest integer such that e is 0-free. e and d′ are 1-sewn. Then, during the foreach loop (line 7), the technique of the general algorithm is followed in order to link new dart d′ with its neighbors, for each j $\in \{3, \ldots, n\}$.

The complexity of Algorithm 49 is also linear in number of darts of cm, for the same reasons than for the general i-closure algorithm.

The behaviour of this algorithm can be illustrated by considering the 3-map depicted in Fig. 6.9(a). In this 3-map, there are only two 1-free darts. Let us start with d = 3; a new dart d′ = 7 is created, and darts 3 and 7 are

Algorithm 49: `1closeNMap(cm)`: 1-close an n-map

```
Input: cm: an n-map.
Result: cm is 1-closed.
1  foreach 1-free dart d ∈ cm.Darts do
2      d′ ← createDartNMap(cm);
3      d.Betas[1] ← d′; d′.Betas[0] ← d;
4      e ← d;
5      while not isFreeNMap(cm,e,0) do e ← e.Betas[0];
6      d′.Betas[1] ← e; e.Betas[0] ← d′;
7      foreach j ∈ {3,...,n} do
8          d″ ← d.Betas[j];
9          if not isFreeNMap(cm,d,j) and not isFreeNMap(cm,d″,0)
           then
10             d′.Betas[j] ← d″.Betas[0];
11             d″.Betas[0].Betas[j] ← d′;
```

1-sewn. Then, after the while loop (line 5), dart e $= 1$ is reached, and darts 7 and 1 are 1-sewn. During the foreach loop (line 7), only j $= 3$ is considered, and thus d″ $= 6$. However, as dart 6 is 0-free, β_3 is not updated for dart 7. Then dart d $= 4$ is considered: a new dart d′ $= 8$ is created, and darts 4 and 8 are 1-sewn. Dart e $= 2$, and darts 8 and 2 are 1-sewn. At last, since dart d $= 4$ is not 3-free and dart d″ $= \beta_3(4) = 1$ is not 0-free, darts 8 and 7 are 3-sewn.

6.2 Removal

The *removal* operation consists in removing a given i-cell c while merging the two $(i+1)$-cells incident to c when they exist; when there is only one $(i+1)$-cell incident to c, no cells are merged. The basic operation is the i-removal, which consists in removing an i-cell. Two examples are depicted in Fig. 6.11. Starting from the 2D object in Fig. 6.11(a), first, the two edges e_1 and e_2 are removed, producing the object given in Fig. 6.11(b). The two faces incident to edge e_1 in the initial object are merged into one face (this is similar for the two faces incident to edge e_2). Then, the two vertices v_1 and v_2 are removed, producing the object in Fig. 6.11(c). The two edges incident to vertex v_1 are merged into one edge (as the two edges incident to vertex v_2).

Given an nD object, the i-removal operation is defined for $i \in \{0, \ldots, n-1\}$. Indeed, for $i = n$, there is no $(i+1)$-cell incident to c. Moreover, it is not possible to remove any i-cell: at most two $(i+1)$-cells are incident to c,

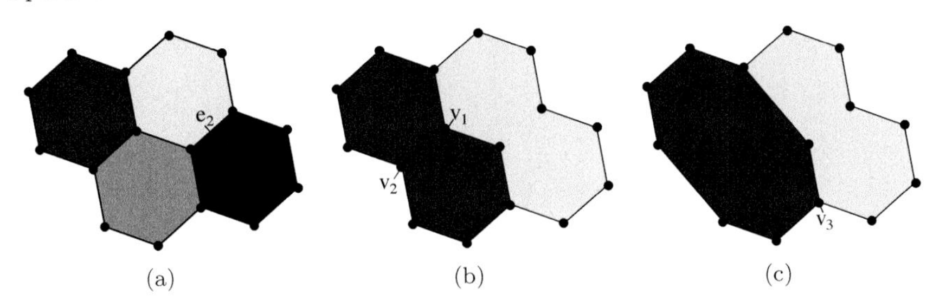

FIGURE 6.11
Examples of removal operation.
(a) A 2D object.
(b) The result of the removal of the two edges e_1 and e_2.
(c) The result of the removal of the two vertices v_1 and v_2.

in order to be able to merge these cells when removing c. It is for example impossible to remove vertex v_3 in Fig. 6.11(c): v_3 is incident to three edges, which cannot be merged into one edge.

6.2.1 For n-Gmaps

Before studying the definitions and algorithms of removal for n-Gmaps [77], let us analyze some examples in 1D and in 2D. For instance, vertex $\{6, 7\}$ is removed from the 1-Gmap G depicted in Fig. 6.12(a): this invalidates α_0 for darts 5 and 8. Indeed, these darts are linked by α_0 to one removed dart. In order to define $\alpha'_0(5)$, i.e. the new value for $\alpha_0(5)$, a path of darts is constructed, starting from $\alpha_0(5)$, by applying $\alpha_0 \circ \alpha_1$. Dart 8 is the first dart of this path which does not belong to the removed vertex, so $\alpha'_0(5) = 8$. For dart 8, a similar construction leads to state $\alpha'_0(8) = 5$ (and thus α'_0 is still an involution). The removal of vertex $\{6, 7\}$ produces the 1-Gmap $G_{R_0}(\{6, 7\})$ (cf. Fig. 6.12(b)); the two edges incident to the removed vertex are merged into one edge.

Look now at the 2-Gmap G in Fig. 6.13(a); let us remove the edge $\{3, 4, 9, 10\}$. Removing these darts invalidates α_1 for darts 2, 5, 11 and 18. As before, a path is constructed for each dart. For instance for dart 2, this path starts with $\alpha_1(2)$, and it is constructed by applying $\alpha_1 \circ \alpha_2$. Dart 11 is the first dart of this path which does not belong to the removed edge, so $\alpha'_1(2) = 11$. The same process is applied for the other darts, so $\alpha'_1(11) = 2$, $\alpha'_1(5) = 18$ and $\alpha'_1(18) = 5$. The 2-Gmap $G_{R_1}(\{3, 4, 9, 10\})$ is constructed: cf. Fig. 6.13(b); it corresponds to the initial 2-Gmap, where the edge $\{3, 4, 9, 10\}$ is removed, and its two incident faces are merged.

Now, vertex $\{2, 3, 15, 20\}$ is removed from the 2-Gmap depicted in Fig. 6.14(a). α_0 is modified for darts 1, 4, 16 and 19: for each of these darts, a path of darts is followed, starting from α_0 and applying $\alpha_0 \circ \alpha_1$. Thus

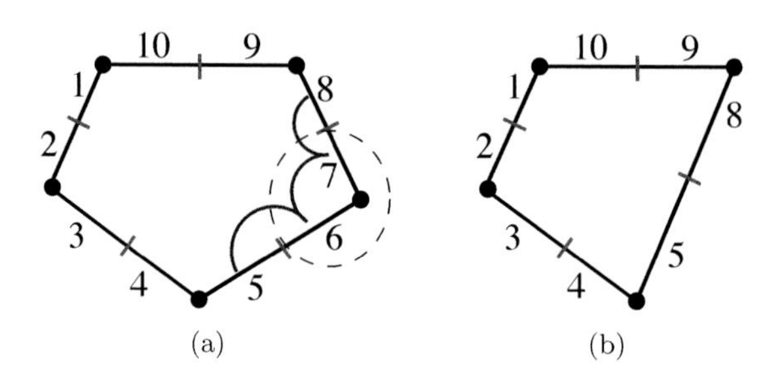

FIGURE 6.12
0-removal for 1-Gmap.
(a) 1-Gmap G.
(b) 1-Gmap $G_{R_0}(\{6,7\})$.

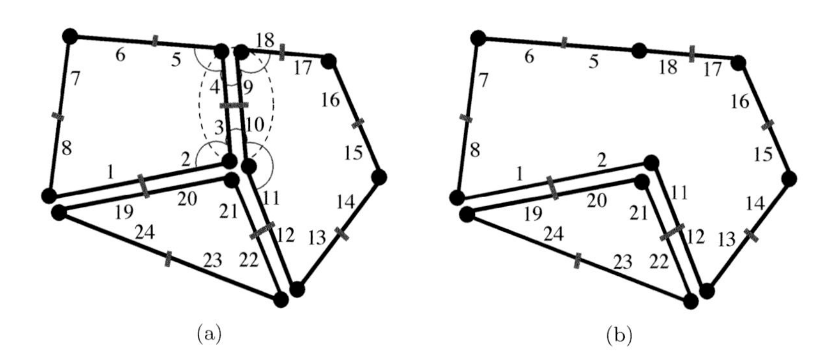

FIGURE 6.13
1-removal for 2-Gmap.
(a) 2-Gmap G.
(b) 2-Gmap $G_{R_1}(\{3,4,9,10\})$.

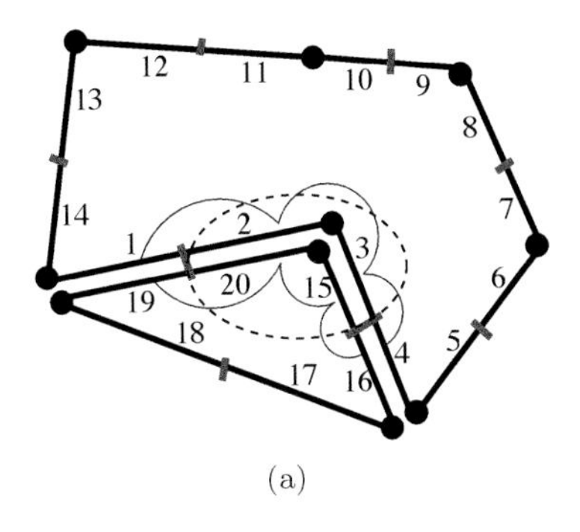

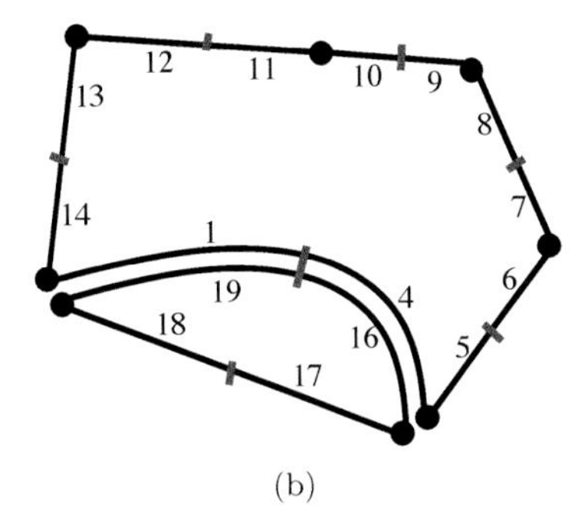

FIGURE 6.14
0-removal for 2-Gmap.
(a) 2-Gmap G.
(b) 2-Gmap $G_{R_0}(\{2, 3, 15, 20\})$.

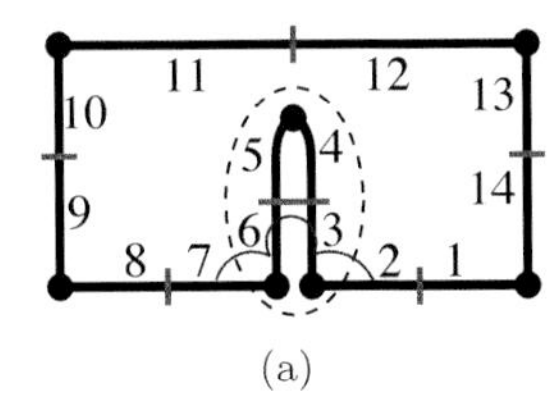

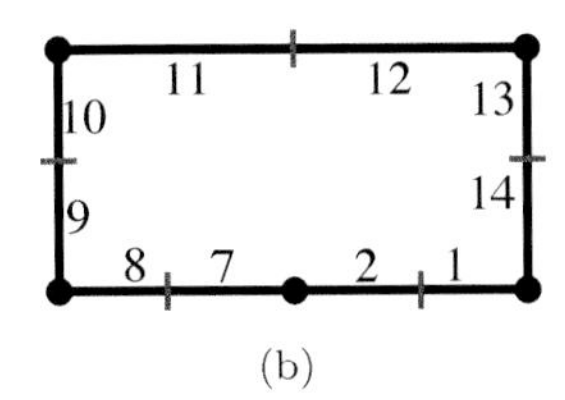

FIGURE 6.15
1-removal of a dangling edge.
(a) 2-Gmap G; edge $\{3, 4, 5, 6\}$ is a dangling edge.
(b) 2-Gmap $G_{R_1}(\{3, 4, 5, 6\})$.

$\alpha'_0(1) = 4$, $\alpha'_0(4) = 1$, $\alpha'_0(16) = 19$ and $\alpha'_0(19) = 16$. This produces the 2-Gmap $G_{R_0}(\{2, 3, 15, 20\})$ in Fig. 6.14(b); it corresponds to the initial 2-Gmap, where the vertex $\{2, 3, 15, 20\}$ is removed, and its two incident edges are merged.

In these three examples, two $(i+1)$-cells are incident to the removed i-cell, and they are merged into one $(i+1)$-cell by the operation. In the example of Fig. 6.15, only one face is incident to edge $\{3, 4, 5, 6\}$, which is thus a dangling edge. This edge is the only edge incident to vertex $\{4, 5\}$, thus its removal involves also the removal of vertex $\{4, 5\}$: cf. Fig. 6.15(b).

Before defining the i-removal operation, the i-removability condition is defined in Def. 58: it characterizes the cells which can be removed. As explained above, an i-cell can be removed if it has at most two $(i + 1)$-cells incident.

Definition 58 (removable cell) *An i-cell C in an n-Gmap $G = (D, \alpha_0, \ldots, \alpha_n)$ is* removable *if:*

- $i = n - 1$;
- *or* $0 \leq i < n - 1$, *and*, $\forall d \in C$, $\alpha_{i+1} \circ \alpha_{i+2}(d) = \alpha_{i+2} \circ \alpha_{i+1}(d)$.

Note that an $(n-1)$-cell is always removable in an n-Gmap. This is a consequence of the quasi-manifold definition: in nD, at most two n-cells are incident to a given $(n-1)$-cell. When $0 \leq i < n-1$, the condition "$\forall d \in C$, $\alpha_{i+1} \circ \alpha_{i+2}(d) = \alpha_{i+2} \circ \alpha_{i+1}(d)$" ensures that at most two $(i+1)$-cells C_1, incident to a given dart d of C, and C_2, incident to $\alpha_{i+1}(d)$, are incident to C (possibly with $C_1 = C_2$).

The i-removal operation is defined in Def. 59. It can be applied to a removable i-cell C; as explained above, removing C invalidates α_i for the darts linked with C. Let D^S denote this set of darts. The resulting n-Gmap is obtained by removing all the darts of C, and by redefining α_i for the darts of D^S. For each dart $d \in D^S$, the path of darts $(\alpha_i \circ \alpha_{i+1})^k \circ \alpha_i(d)$ is followed until obtaining a dart belonging to D^S. All others α_j, $j \neq i$, are not modified, nor α_i for the darts which do not belong to D^S.

Definition 59 (cell removal) *Let* $G = (D, \alpha_0, \ldots, \alpha_n)$ *be an* n*-Gmap and* C *be a removable* i*-cell. Let* $D^S = \alpha_i(C) \setminus C$ *be the set of darts* i*-linked with* C *which do not belong to* C*. The* n*-Gmap resulting from the* i*-removal of* C *in* G *is* $G_{R_i}(C) = (D', \alpha'_0, \ldots, \alpha'_n)$*, defined by:*

- $D' = D \setminus C$;
- $\forall j \in \{0, \ldots, n\}$, $j \neq i$, $\alpha'_j = \alpha_{j|D'}$;
- $\forall d \in D' \setminus D^S$, $\alpha'_i(d) = \alpha_i(d)$;
- $\forall d \in D^S$, $\alpha'_i(d) = (\alpha_i \circ \alpha_{i+1})^k \circ \alpha_i(d)$,
  k *being the smaller positive integer s.t.* $(\alpha_i \circ \alpha_{i+1})^k \circ \alpha_i(d) \in D^S$.

The path of darts followed for modifying α_i corresponds to $(\alpha_i \circ \alpha_{i+1})^k \circ \alpha_i$, k being the smaller positive integer such that a dart which belongs to D^S is reached. For the four previous examples, k is always equal to 1. But k can be greater than 1, for instance for configurations in which multi-incidence occurs, as illustrated in Fig. 6.16. In the initial 2-Gmap in Fig. 6.16(a), an edge is a loop. When this edge is removed, the path of darts associated with dart 5 (resp. 12) is $(4 = \alpha_1(5), 2 = \alpha_1 \circ \alpha_2 \circ \alpha_1(5), 12 = (\alpha_1 \circ \alpha_2)^2 \circ \alpha_1(5))$ (resp. $(3, 1, 5)$).

As illustrated in Fig. 6.17, other configurations can lead to $k > 1$, for instance when the n-Gmap has boundaries. In this example, the initial 2-Gmap in Fig. 6.17(a) has one 1-boundary; edge $\{1, 2, 9, 10\}$ is incident to this boundary. For its removal, the path associated with 8 is $(1 = \alpha_1(8), 9 = \alpha_1 \circ \alpha_2 \circ \alpha_1(8), 8 = (\alpha_1 \circ \alpha_2)^2 \circ \alpha_1(8))$. Thus, after applying the 1-removal, edge $\{7, 8\}$ is incident to the boundary.

It is easy to prove that, given a removable i-cell C of an n-Gmap G,

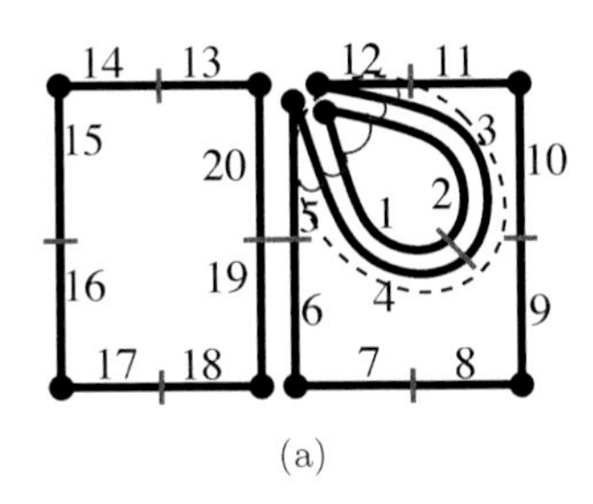

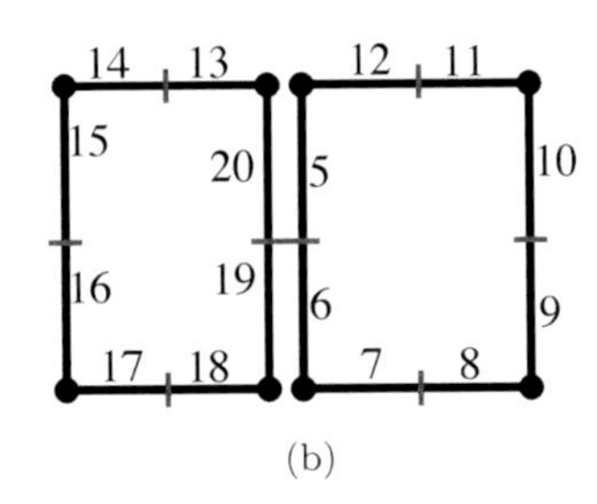

(a) (b)

FIGURE 6.16
1-removal of a loop in a 2-Gmap.
(a) 2-Gmap G; edge $\{1, 2, 3, 4\}$ is a loop.
(b) $G_{R_1}(\{1, 2, 3, 4\})$.

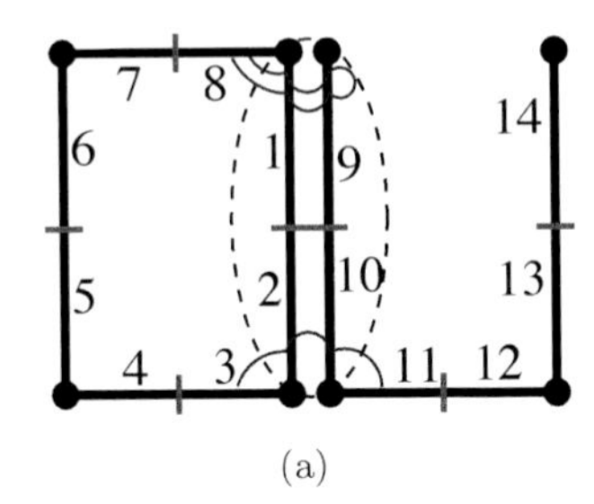

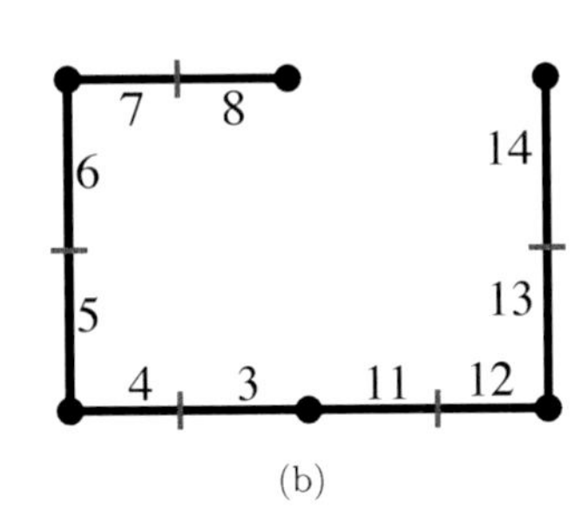

(a) (b)

FIGURE 6.17
1-removal of an edge in a 2-Gmap with boundaries.
(a) 2-Gmap G.
(b) $G_{R_1}(\{1, 2, 9, 10\})$.

$G_{R_i}(C)$, obtained by removing C from G, is an n-Gmap. For that, the conditions of the n-Gmap definition have to be checked. This can be done, thanks to the removable property, which allows to commute α_{i+1} and α_{i+2}.

Note that the removal operation can be extended in order to simultaneously remove several disjoint cells of same or different dimensions (see [77]).

Algorithm 50 allows to check that a given i-cell is removable or not. This is a direct implementation of the corresponding definition.

Algorithm 50: `isRemovableNGMap(gm,d,i)`: test if an `i`-cell is removable for n-Gmaps

---

Input: `gm`: an n-Gmap;
`d` $\in$ `gm.Darts`: a dart;
`i` $\in \{0, \ldots, n\}$.
Output: True iff $c_{\mathtt{i}}(\mathtt{d})$ is removable.

1 **if** `i` $= n$ **then return** *false*;
2 **if** `i` $= n-1$ **then return** *true*;
3 **foreach** *dart* `d'` $\in c_{\mathtt{i}}(\mathtt{d})$ **do**
4   **if** `d'.Alphas[i + 2].Alphas[i + 1]` $\neq$ `d'.Alphas[i + 1].Alphas[i + 2]` **then**
5   **return** *false*;
6 **return** *true*;

The i-removal operation is implemented by Algorithm 51. Instead of building the new n-Gmap $G_{R_i}(C)$, the algorithm directly modifies the given n-Gmap G. This allows to avoid to copy the initial n-Gmap when this is not required. If needed, $G_{R_i}(C)$ is obtained by first copying G (using Algorithm 14 page 117), second by modifying the copy by applying Algorithm 51.

As explained above, α_i is only modified for darts linked with the removed i-cell (denoted by D^S). So, the first step consists in marking all the darts of the removed i-cell, in order to be able to test in constant time if a given dart belongs to this cell or not. α_i is modified by the main loop of the algorithm (line 3). To retrieve all the darts of D^S, we iterate through the darts of the removed i-cell. For each dart d' of C, let $d_1 = \alpha_i(d')$. If d_1 does not belong to D^S, nothing has to be done. If d_1 belongs to D^S, the corresponding path of darts for $(\alpha_i \circ \alpha_{i+1})^k \circ \alpha_i$ is followed until a nonmarked dart d_2 is reached, i.e. d_2 belongs to D^S. Then d_1 and d_2 are linked by α_i (line 9). At the end of the main loop, all darts of D^S have been processed, and the darts of C are removed. Note that there is no dart to unmark, as all the marked darts have been removed.

The complexities of Algorithms 50 and 51 are linear in the number of darts of the removed i-cell. For the i-removal operation, each path of darts is considered twice: a first time for dart d_1 and a second time in reverse direction for the second extremity of the path. The first time allows to define $\alpha_i(d_1) = d_2$ and the second time to define $\alpha_i(d_2) = d_1$. The algorithm can be modified

Algorithm 51: `removeNGMap(gm,d,i)`: remove a given `i`-cell for n-Gmaps

Input: `gm`: an n-Gmap;
`d` $\in$ `gm.Darts`: a dart such that $c_{\mathtt{i}}(\mathtt{d})$ is removable;
`i` $\in \{0, \ldots, n-1\}$.
Result: Remove the `i`-cell $c_{\mathtt{i}}(\mathtt{d})$.

```
1  ma ← reserveMarkNGMap(gm);
2  mark all the darts in c_i(d) for ma;
3  foreach dart d' ∈ c_i(d) do
4      d1 ← d'.Alphas[i];
5      if not isMarkedNGMap(d1,ma) then
6          d2 ← d'.Alphas[i + 1].Alphas[i];
7          while isMarkedNGMap(d2,ma) do
8              d2 ← d2.Alphas[i + 1].Alphas[i];
9          d1.Alphas[i] ← d2;
10 foreach dart d' ∈ c_i(d) do
11     remove d' from gm.Darts;
12 freeMarkNGMap(gm,ma);
```

in order to consider each path exactly once, but it is necessary to conceive a more complex algorithm.

6.2.2 For n-maps

Now the definitions and algorithms of the removal operation for n-maps are studied. They are similar to the ones conceived for n-Gmaps; but as usual for n-maps, special cases occur when β_1 is taken into account. First, several examples are studied, in fact the ones used for n-Gmaps.

An example of 0-removal in 1D is provided in Fig. 6.18. Vertex $\{4\}$ is removed, and $\beta_1'(3)$ is stated to 5, producing the 1-map in Fig. 6.18(b). The two edges incident to vertex $\{4\}$ are merged. Note the difference with the same operation for 1-Gmap. For a 0-removal, β_1 is modified for a 1-map, while α_0' is modified for a 1-Gmap (0 being the dimension of the removed cell). Moreover, the path of darts is now oriented, as β_1 is a partial permutation and not an involution.

An example of 1-removal in a 2-map is depicted in Fig. 6.19. Starting from the initial 2-map in Fig. 6.19(a), edge $\{4, 5\}$ is removed. β_1 is modified for darts 3 and 9. A path of darts is constructed for dart 3, starting from $\beta_1(3)$ and applying $\beta_1 \circ \beta_2$. Dart 6 is obtained, which does not belong to the removed edge; thus, $\beta_1'(3) = 6$. Similarly, $\beta_1'(9) = 1$. At last, the two darts 4 and 5 are removed, producing the 2-map in Fig. 6.19(b); edge $\{4, 5\}$ is removed, and its two incident faces are merged. Here, the principle is exactly the same than for

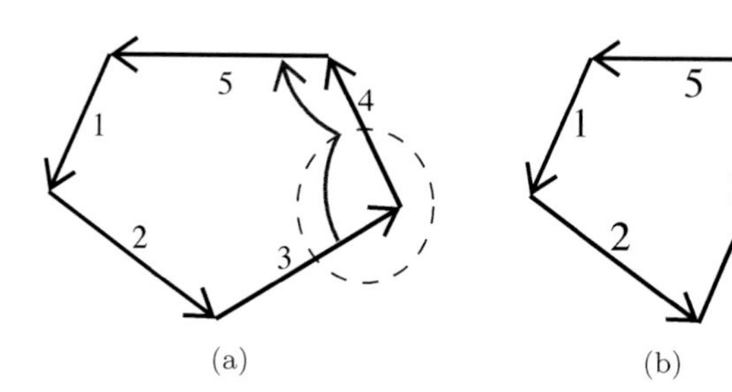

FIGURE 6.18
0-removal for 1-map.
(a) 1-map M.
(b) 1-map $M_{R_0}(\{4\})$.

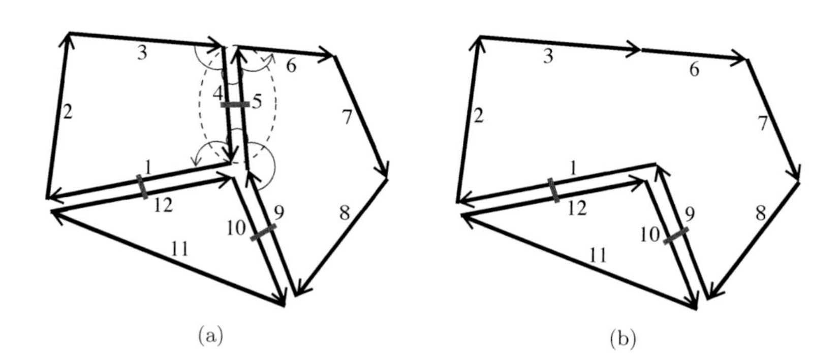

FIGURE 6.19
1-removal for 2-map.
(a) 2-map M.
(b) 2-map $M_{R_1}(\{4, 5\})$.

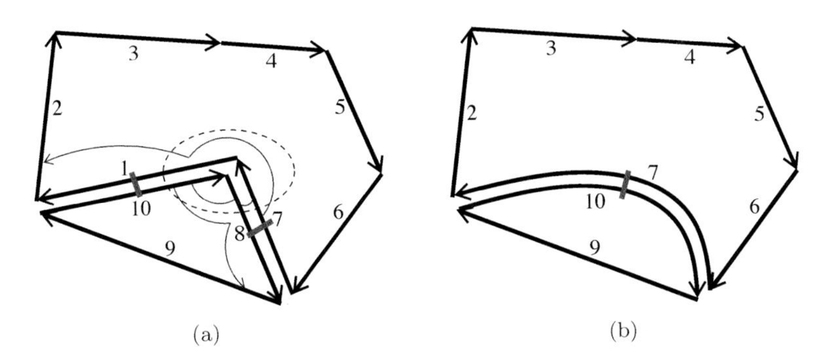

FIGURE 6.20
0-removal for 2-map.
(a) 2-map M.
(b) 2-map $M_{R_0}(\{1,8\})$.

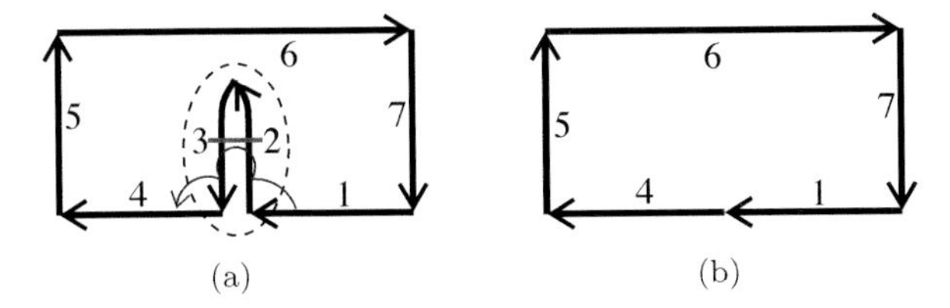

FIGURE 6.21
1-removal of a dangling edge in a 2-map.
(a) 2-map M; edge $\{2,3\}$ is dangling.
(b) 2-map $M_{R_1}(\{2,3\})$.

the 1-removal for 2-Gmaps. The only difference is that β_1 is modified for the two darts 3 and 9 (which are 0-sewn with a dart of the removed edge), but not for the two darts 1 and 6 (which are 1-sewn with a dart of the removed edge), while α_1 is modified for the four corresponding darts in the associated 2-Gmap. Moreover, as in the previous example, the paths of darts are now oriented.

An example of 0-removal in a 2-map is illustrated in Fig. 6.20, showing that the modifications are slightly different for the dimension 0. Vertex $\{1,8\}$ is removed from the initial 2-map in Fig. 6.20(a). β'_1 is computed for the two darts 7 and 10, by following the path of darts $\beta_1 \circ \beta_1$. However, this is not sufficient, since darts 7 and 10 are linked by β_2 with darts 8 and 1. It is thus necessary to define $\beta'_2(7) = 10$ and $\beta'_2(10) = 7$. This example shows that the 0-removal operation is a special case for n-maps: all β have to be modified, not only β_i.

The example in Fig. 6.21 shows the removal of a dangling edge.

The removability condition is defined in Def. 60: it is the direct transposition of the same definition for n-Gmap. The only difference is the use of β_{i+1}^{-1} instead of β_{i+1}; since β_{i+1} is a partial involution for $i > 0$, this difference is important only for $i = 0$. For example in the 2-map of Fig. 6.20(a), the vertex $\{1, 8\}$ is 0-removable, because $\beta_1 \circ \beta_2(1) = \beta_2 \circ \beta_0(1) = 8$ and $\beta_1 \circ \beta_2(8) = \beta_2 \circ \beta_0(8) = 1$.

Definition 60 (removable cell) *An i-cell C in an n-map $M = (D, \beta_1, \ldots, \beta_n)$ is* i-removable *if:*

- $i = n - 1$;
- *or* $0 \leq i < n - 1$, *and* $\forall d \in C$, $\beta_{i+1} \circ \beta_{i+2}(d) = \beta_{i+2} \circ \beta_{i+1}^{-1}(d)$.

The i-removal operation, for $i \geq 1$, is defined in Def. 61. It is also a direct transposition of the same definition for n-Gmap. Let D^S be the set of darts i-linked with the removed i-cell C. β_i is modified for all darts of D^S; in fact, β_i^{-1} is used in order to take correctly into account the case when $i = 1$. For each dart $d \in D^S$, the path of darts $(\beta_i \circ \beta_{i+1})^k \circ \beta_i(d)$ is followed, until either a dart in D^S is reached, which is the second extremity of the path, or $\varnothing$ is obtained, when the n-map has boundaries. These two cases correspond to the same condition: k is the smaller positive integer such that the dart of the path does not belong to C.

Definition 61 (cell removal) *Let $M = (D, \beta_1, \ldots, \beta_n)$ be an n-map and C be a removable i-cell, $i \geq 1$. Let $D^S = \beta_i^{-1}(C) \setminus C$. The n-map resulting from the i-removal of C in M is $M_{R_i}(C) = (D', \beta'_1, \ldots, \beta'_n)$, defined by:*

- $D' = D \setminus C$;
- $\forall j \in \{1, \ldots, n\}$, $j \neq i$, $\beta'_j = \beta_{j|D'}$;
- $\forall d \in D' \setminus D^S$, $\beta'_i(d) = \beta_i(d)$;
- $\forall d \in D^S$, $\beta'_i(d) = (\beta_i \circ \beta_{i+1})^k \circ \beta_i(d)$,
  k being the smaller positive integer s.t. $(\beta_i \circ \beta_{i+1})^k \circ \beta_i(d) \notin C$.

A loop is removed in the example depicted in Fig. 6.22. The path associated with dart 6 begins with $\beta_1(6) = 2$; then dart 1, which belongs to the removed edge, is obtained by applying $\beta_1 \circ \beta_2$. Thus $\beta_1 \circ \beta_2$ is applied once more, and dart 3 is reached, which does not belong to the removed edge.

A 2-map with a 1-boundary is depicted in Fig. 6.23. The removed edge $\{1, 7\}$ is incident to this boundary. The path of darts associated with dart 4 corresponds to $\beta_1 \circ \beta_2 \circ \beta_1$, producing $\varnothing$ which does not belong to C; thus, $k = 1$ and $\beta'_1(4) = \varnothing$.

The 0-removal operation is defined in Def. 62. As seen above, all β_j have to be modified. This is due to the fact that the darts of a vertex are linked with other darts by β_j (while this is not true for other i-cells, $i > 0$). β_j is modified,

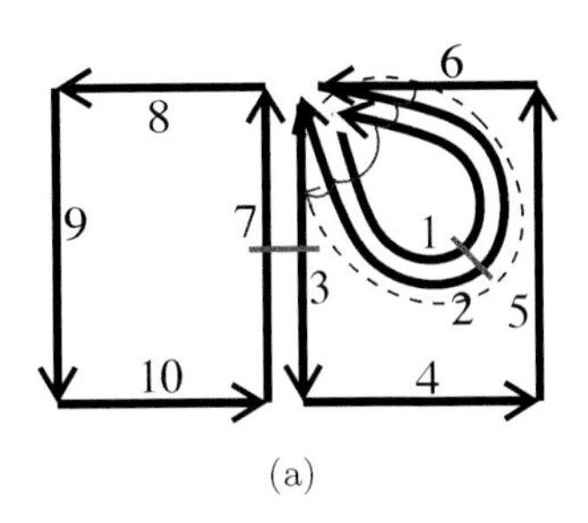

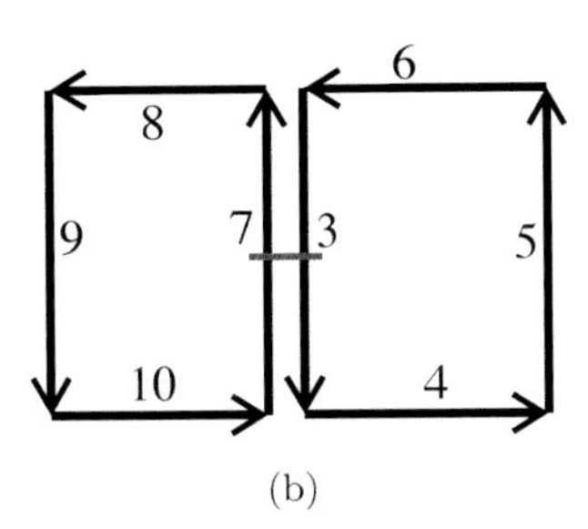

FIGURE 6.22
Removal of a loop in a 2-map.
(a) 2-map M; edge $\{1, 2\}$ is a loop.
(b) 2-map $M_{R_1}(\{1, 2\})$.

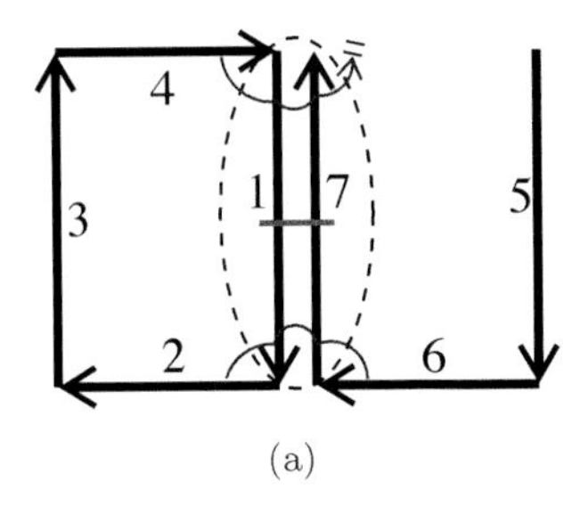

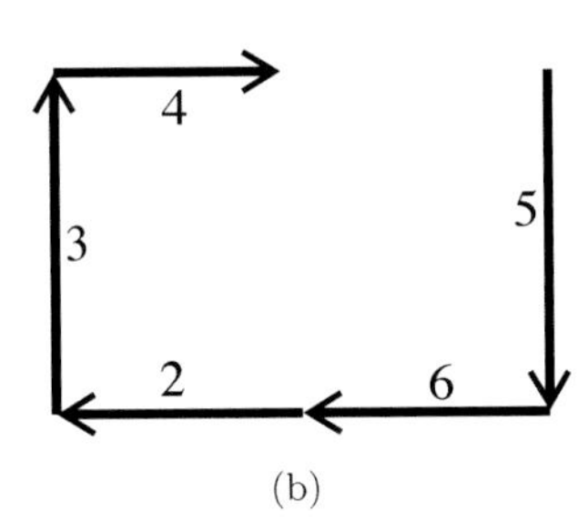

FIGURE 6.23
Removal of an edge in a 2-map with boundaries.
(a) 2-map M.
(b) 2-map $M_{R_1}(\{1, 7\})$.

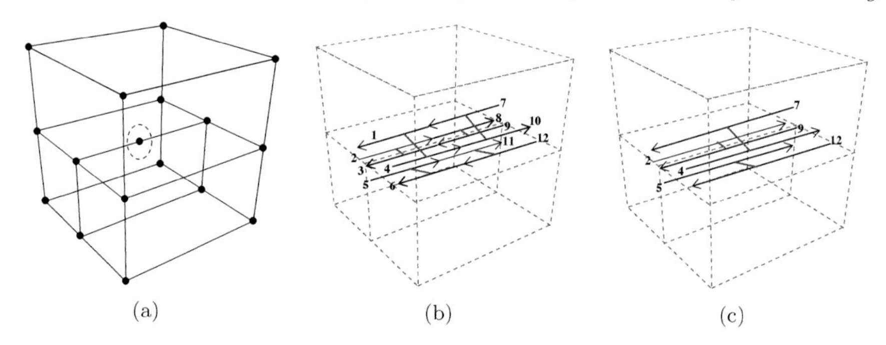

FIGURE 6.24
Removal of a vertex in a 3-map.
(a) A 3D subdivided object.
(b) Corresponding 3-map M (partial representation).
(c) 3-map $M_{R_0}(\{1,3,6,8,10,11\})$.

by following the path of darts $\beta_j \circ (\beta_1)^k$, for all the darts in $\beta_j^{-1}(C) \setminus C$. First, β_1 is applied as many times as necessary to reach a dart which does not belong to C, then β_j is applied.

Definition 62 (0-cell removal) *Let $M = (D, \beta_1, \dots, \beta_n)$ be an n-map and C be a removable 0-cell. The n-map resulting from the 0-removal of C in M is $M_{R_0}(C) = (D', \beta'_1, \dots, \beta'_n)$, defined by:*

- $D' = D \setminus C$;
- $\forall j \in \{1, \dots, n\}$, $\forall d \in D'$:
  - *if* $d \notin \beta_j^{-1}(C) \setminus C$, $\beta'_j(d) = \beta_j(d)$;
  - *otherwise,* $\beta'_j(d) = \beta_j \circ (\beta_1)^k(d)$,
    k being the smaller positive integer s.t. $\beta_j \circ (\beta_1)^k(d) \notin C$.

Look again at Fig. 6.20; $C = \{1,8\}$. For $j = 1$, $\beta_1^{-1}(C) \setminus C = \{7,10\}$; thus, $\beta'_1(7) = 2$ and $\beta'_1(10) = 8$. For $j = 2$, $\beta_2^{-1}(C) \setminus C = \{7,10\}$, thus $\beta'_2(7) = 10$ and $\beta'_2(10) = 7$. $k = 1$ for all cases.

An example of 0-removal in 3D is presented in Fig. 6.24. Vertex $\{1,3,6,8,10,11\}$ is removed from the 3-map depicted in Fig. 6.24(b) (only darts around the removed vertex are drawn). The vertex is 0-removable, because $\beta_1 \circ \beta_2 = \beta_2 \circ \beta_0$ for each of its darts (for example $\beta_1 \circ \beta_2(1) = \beta_2 \circ \beta_0(1) = 10$). For $j = 1$, $\beta_1^{-1}(C) \setminus C = \{2,4,5,7,9,12\}$, and, for example, $\beta'_1(2) = \beta_1(8)$. For $j = 2$, $\beta_2^{-1}(C) \setminus C = \{2,4,5,7,9,12\}$, and, for example, $\beta'_2(2) = \beta_2 \circ \beta_1(2) = 9$. For $j = 3$, $\beta_3^{-1}(C) \setminus C = \{2,4,5,7,9,12\}$, and, for example, $\beta'_3(2) = \beta_3 \circ \beta_1(2) = 7$.

Algorithm 52 allows to check that a given i-cell is removable or not. This is a direct implementation of the corresponding definition.

Algorithm 52: `isRemovableNMap(cm,d,i)`: test if an `i`-cell is removable for n-maps

Input: `cm`: an n-map;
 `d` $\in$ `cm.Darts`: a dart;
 `i` $\in \{0, \ldots, n\}$.
Output: True iff $c_{\mathtt{i}}(\mathtt{d})$ is removable.

```
1 if i = n then return false;
2 if i = n − 1 then return true;
3 foreach dart d′ ∈ c_i(d) do
4     if d′.Betas[i + 2].Betas[i + 1] ≠ d′.Betas[inv(i + 1)].Betas[i + 2]
      then
5         return false;
6 return true;
```

The i-removal operation, for $1 \leq i \leq n-1$, is implemented by Algorithm 53. This algorithm is very similar to the i-removal operation for n-Gmaps. The differences are related to the facts that β_1 is a partial permutation (and not a partial involution like other β_i), and that an i-free dart is linked with $\varnothing$. First, β_i^{-1} is applied in order to initialize the first extremity d_1. Second, after the computation of the two extremities of the path, four possible configurations are taken into account, depending if these extremities are equal to or different from $\varnothing$. At last, if $i = 1$, $\beta_0(d_2)$ is also defined (line 12), in order to consider only once the path, which is oriented in this case.

The 0-removal operation is implemented by Algorithm 54. The definition of β_1 is similar to the general case, by following the path of darts corresponding to $(\beta_1)^k$. Only darts in $\beta_0(C)\backslash C$ are processed (cf. test in line 5). d_2, the second extremity of the path, is computed, and the four possible configurations are taken into account. The main difference with the general case is the additional loop (line 14), allowing to define β_j, $\forall j \in \{2, \ldots, n\}$. According to Def. 62, $\beta'_j(d_1)$ has to be set to $\beta_j \circ (\beta_1)^k(d_1)$: since $d_2 = \beta_1 \circ (\beta_1)^k(d_1)$, $\beta_1^k(d_1) = \beta_0(d_2)$ and thus $\beta'_j(d_1)$ is set to $\beta_j \circ \beta_0(d_2)$. Note that if d_1 is j-free or equal to $\varnothing$, nothing has to be done since $d_1 \notin \beta_j^{-1}(C) \setminus C$. Note also that if $d_2 = \varnothing$, $\beta'_j(d_1)$ is set to $\varnothing$ (cf. line 16), since $\varnothing$ is linked with itself for all β. At last, if $d_1 = \varnothing$ and $d_2 \neq \varnothing$, $\beta'_0(d_2)$ is set to $\varnothing$ (line 19), since no dart is linked with d_2 by β'_1 (all the darts of the path belong to the removed vertex).

The main loop of Algorithm 54 iterates through the darts of $c_0(\mathtt{d})$. However, since the β_j links of darts in $\beta_j^{-1}(C) \setminus C$ are modified, the use of the vertex iterator defined in Algorithm 30 page 163 could lead to errors. So, the darts belonging to $c_0(\mathtt{d})$ are stored in a data structure (for example a stack) before the loop; then, they are processed by using this data structure, and not the vertex iterator.

As for the removal operation for n-Gmaps, the complexities of Algorithm 52,

Algorithm 53: `removeNMap(cm,d,i)`: remove an `i`-cell for n-maps

```
   Input: cm: an n-map;
          d ∈ cm.Darts: a dart such that c_i(d) is removable;
          i ∈ {1,...,n−1}.
   Result: Remove the i-cell c_i(d).
1  ma ← reserveMarkNMap(cm);
2  mark all the darts in c_i(d) for ma;
3  foreach dart d' ∈ c_i(d) do
4      d1 ← d'.Betas[inv(i)];
5      if not isMarkedNMap(cm,d1,ma) then
6          d2 ← d'.Betas[i + 1].Betas[i];
7          while isMarkedNMap(cm,d2,ma) do
8              d2 ← d2.Betas[i + 1].Betas[i];
9          if d1 ≠ cm.null_dart then
10             if d2 ≠ cm.null_dart then
11                 d1.Betas[i] ← d2;
12                 if i = 1 then d2.Betas[0] ← d1;
13             else
14                 d1.Betas[i] ← cm.null_dart;
15         else
16             if d2 ≠ cm.null_dart then
17                 d2.Betas[inv(i)] ← cm.null_dart;
18 foreach dart d' ∈ c_i(d) do
19     remove d' from cm.Darts;
20 freeMarkNMap(cm,ma);
```

Algorithm 54: `0removeNMap(cm,d)`: remove a 0-cell for n-maps

```
   Input: cm: an n-map;
          d ∈ cm.Darts: a dart such that c0(d) is removable.
   Result: Remove the 0-cell c0(d).
 1 ma ← reserveMarkNMap(cm);
 2 mark all the darts in c0(d) for ma;
 3 foreach dart d' ∈ c0(d) do
 4     d1 ← d'.Betas[0];
 5     if not isMarkedNMap(cm,d1,ma) then
 6         d2 ← d'.Betas[1];
 7         while isMarkedNMap(cm,d2,ma) do
 8             d2 ← d2.Betas[1];
 9         if d1 ≠ cm.null_dart then
10             if d2 ≠ cm.null_dart then
11                 d1.Betas[1] ← d2; d2.Betas[0] ← d1;
12             else
13                 d1.Betas[1] ← cm.null_dart;
14             for j ← 2 to n do
15                 if not isFreeNMap(cm,d1,j) then
16                     d1.Betas[j] ← d2.Betas[0].Betas[j];
17         else
18             if d2 ≠ cm.null_dart then
19                 d2.Betas[0] ← cm.null_dart;
20 foreach dart d' ∈ c0(d) do
21     remove d' from cm.Darts;
22 freeMarkNMap(cm,ma);
```

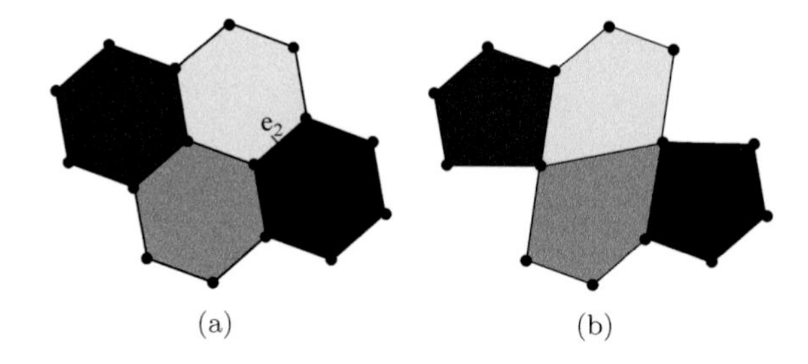

FIGURE 6.25
Contraction operation.
(a) A 2D object.
(b) The result of the contraction of edges e_1 and e_2.

Algorithm 53 and Algorithm 54 are linear in number of darts of the removed cell.

6.3 Contraction

The *contraction* operation is the dual of the removal operation. It consists in removing a given i-cell c while merging the two $(i-1)$-cells[1] incident to c when they exist. When only one $(i-1)$-cell is incident to c, no cells are merged.

The contraction operation is illustrated in Fig. 6.25. Starting from the 2D object in Fig. 6.25(a), the two edges e_1 and e_2 are contracted, producing the object in Fig. 6.25(b). The two vertices incident to edge e_1 in the initial object are merged into one vertex (this is similar for the two vertices incident to edge e_2).

6.3.1 For n-Gmaps

In 1D, as illustrated in Fig. 6.26, the edge contraction is the dual of the vertex removal. In this example, edge $\{7,8\}$ is contracted: this corresponds to modify α_1 for darts 6 and 9. So, a path of darts is computed, starting from $\alpha_1(6)$ and applying $\alpha_1 \circ \alpha_0$. Dart 9 is the first dart of this path which does not belong to the contracted edge: thus, $\alpha'_1(6) = 9$. Similarly, dart 6 is obtained for dart 9, thus $\alpha'_1(9) = 6$. The resulting 1-Gmap $G_{C_1}(\{7,8\})$ is depicted in Fig. 6.26(b): note that the two vertices incident to edge $\{7,8\}$ are merged.

[1] In nD, duality associates an i-cell c and an $(n-i)$-cell c^*; thus, an $(i+1)$-cell incident to c corresponds to a $(n-i-1)$-cell incident to c^*. Let $j = n-i$: contracting a j-cell merges its incident $(j-1)$-cells.

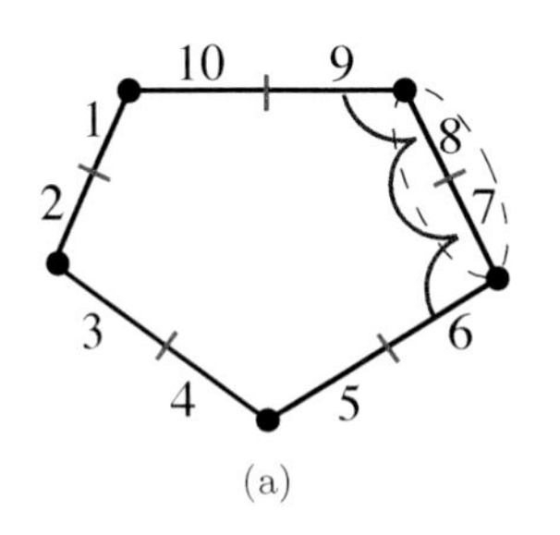

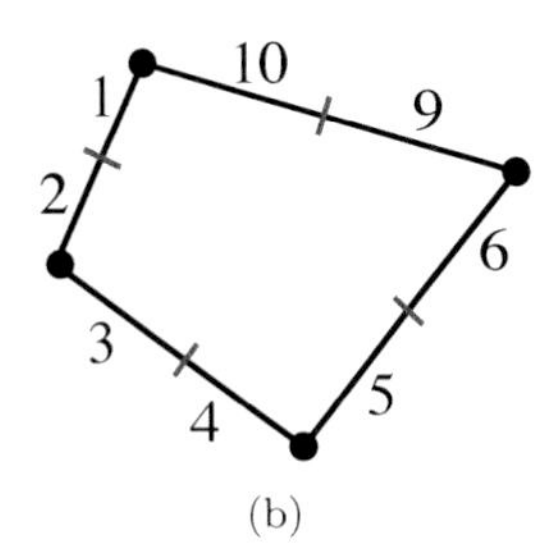

FIGURE 6.26
1-contraction for 1-Gmap.
(a) 1-Gmap G.
(b) 1-Gmap $G_{C_1}(\{7, 8\})$.

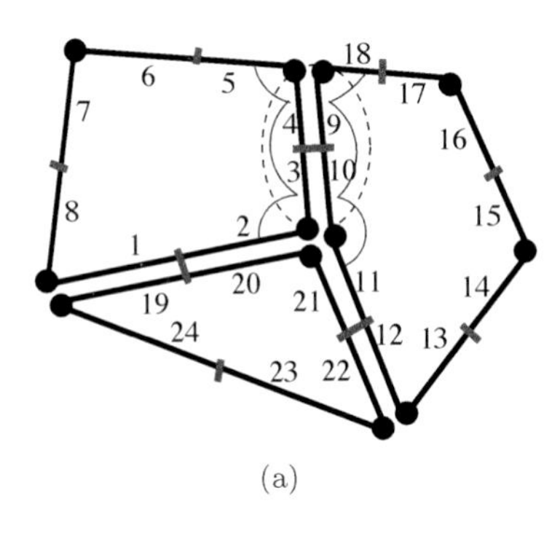

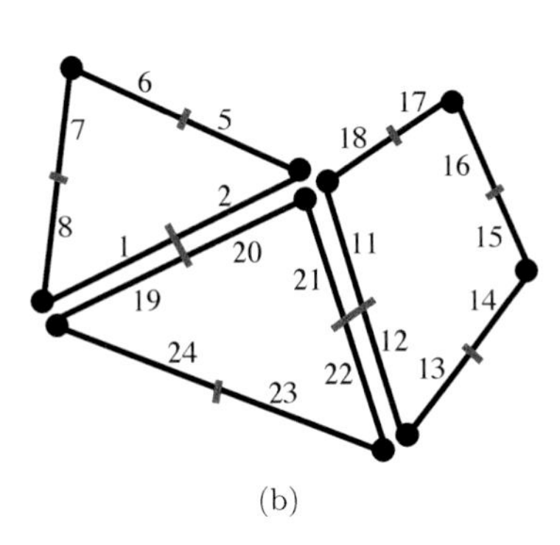

FIGURE 6.27
1-contraction for 2-Gmap.
(a) 2-Gmap G.
(b) 2-Gmap $G_{C_1}(\{3, 4, 9, 10\})$.

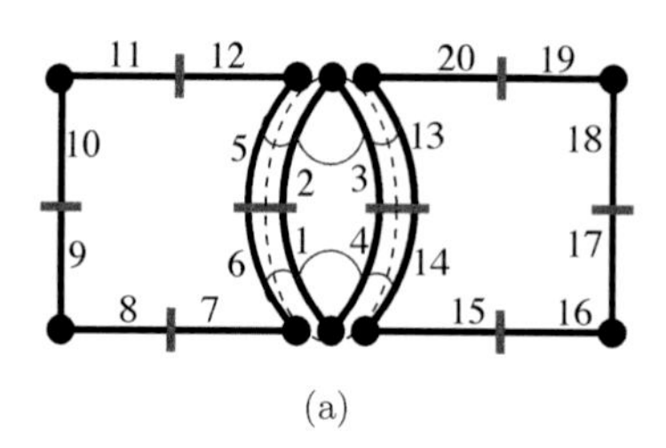

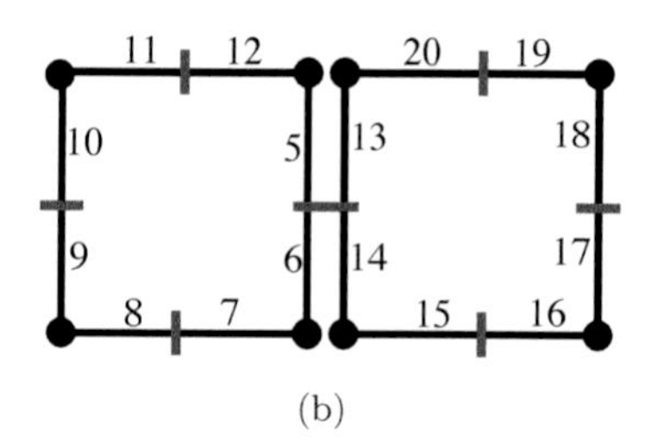

FIGURE 6.28
2-contraction for 2-Gmap.
(a) 2-Gmap G.
(b) 2-Gmap $G_{C_2}(\{1,2,3,4\})$.

In 2D, the edge contraction is the dual of the edge removal: cf. Fig. 6.27. In this example, edge $\{3,4,9,10\}$ is contracted, by modifying α_1 for the four darts 2, 5, 11 and 18. For each dart, a path of darts is followed, starting with α_1 and applying $\alpha_1 \circ \alpha_0$. In the resulting 2-Gmap $G_{C_1}(\{3,4,9,10\})$ (cf. Fig. 6.27(b)), edge $\{3,4,9,10\}$ is contracted and its two incident vertices are merged.

In 2D, the face contraction is the dual of the vertex removal: cf. Fig. 6.28. Face $\{1,2,3,4\}$ is contracted by modifying α_2 for the four darts 5, 6, 13 and 14. For each dart, a path of darts is computed, starting with α_2 and applying $\alpha_2 \circ \alpha_1$. In the resulting 2-Gmap $G_{C_2}(\{1,2,3,4\})$ (cf. Fig. 6.28(b)), face $\{1,2,3,4\}$ is contracted, and its two incident edges are merged.

The contractibility condition characterizes the i-cells which can be contracted: cf. Def. 63. This condition can be deduced from the i-removability condition, by replacing $i+1$ (resp. $i+2$) by $i-1$ (resp. $i-2$), according to duality.

Definition 63 (contractible cell) *An i-cell C in an n-Gmap $G = (D, \alpha_0, \ldots, \alpha_n)$ is* contractible *if:*

- $i = 1$*;*
- *or* $1 < i \le n$*, and,* $\forall d \in C$*,* $\alpha_{i-1} \circ \alpha_{i-2}(d) = \alpha_{i-2} \circ \alpha_{i-1}(d)$*.*

The i-contraction operation can be deduced from the i-removal operation, by duality. Let C be a contractible i-cell. The n-Gmap resulting from the i-contraction of C is obtained by removing the darts of C, and by modifying α_i for the darts linked with C, by following the paths of darts corresponding to $(\alpha_i \circ \alpha_{i-1})^k \circ \alpha_i$.

Definition 64 (cell contraction) *Let $G = (D, \alpha_0, \ldots, \alpha_n)$ be an n-Gmap and C be a contractible i-cell. Let $D^S = \alpha_i(C) \setminus C$ be the set of darts i-linked*

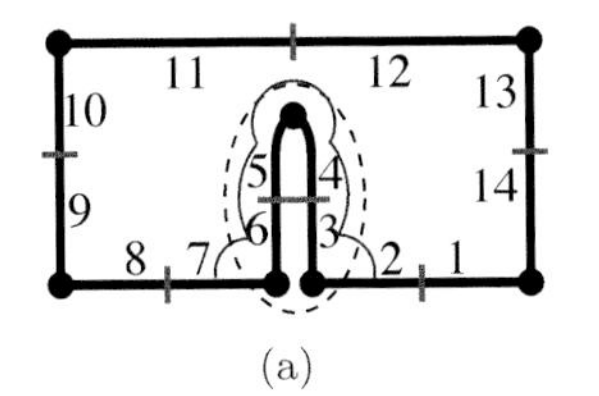

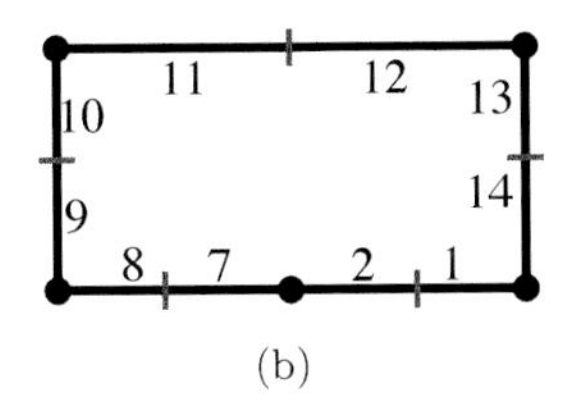

FIGURE 6.29
1-contraction of a dangling edge.
(a) 2-Gmap G; edge $\{3,4,5,6\}$ is a dangling edge.
(b) 2-Gmap $G_{C_1}(\{3,4,5,6\})$.

with C which do not belong to C. The n-Gmap resulting from the i-contraction of C in G is $G_{C_i} = (D', \alpha'_0, \ldots, \alpha'_n)$ defined by:

- $D' = D \setminus C$;
- $\forall j \in \{0,\ldots,n\}$, $j \neq i$, $\alpha'_j = \alpha_{j|D'}$;
- $\forall d \in D' \setminus D^S$, $\alpha'_i(d) = \alpha_i(d)$;
- $\forall d \in D^S$, $\alpha'_i(d) = (\alpha_i \circ \alpha_{i-1})^k \circ \alpha_i(d)$,
  k being the smaller positive integer such that $(\alpha_i \circ \alpha_{i-1})^k \circ \alpha_i(d) \in D^S$.

As for the removal operation, k can be greater than 1. For instance, a dangling edge is contracted in the example presented in Fig. 6.29. Starting from dart 2, the path of darts corresponding to $(\alpha_1 \circ \alpha_0)^2 \circ \alpha_1$ leads to dart 7, which does not belong to the contracted edge. Note that removing or contracting edge $\{3,4,5,6\}$ produce, in this configuration, the same 2-Gmap.

It is easy to prove that given an n-Gmap G and a contractible i-cell C, $G_{C_i}(C)$ obtained by the contraction of C in G is an n-Gmap. Note that the contraction operation can be extended in order to simultaneously contract several disjoint cells of same or different dimensions. Moreover, a general operation can be defined in order to simultaneously contract and remove several disjoint cells of same or different dimensions (see [77]).

The algorithms implementing the test of i-contractibility condition and the contraction operation, given in Algorithms 55 and 56, correspond to the algorithms described for the removal operation, in which $i+1$ (resp. $i+2$) is replaced by $i-1$ (resp. $i-2$).

The complexities of these two algorithms are the same than the complexities for the removal operation, i.e. linear in number of darts of the contracted cell.

Algorithm 55: `isContractibleNGMap(gm,d,i)`: test if an i-cell is contractible for n-Gmaps

Input: gm: an n-Gmap;
d ∈ gm.Darts: a dart;
i ∈ $\{0, \ldots, n\}$.
Output: True iff c_i(d) is contractible.

```
1 if i = 0 then return false;
2 if i = 1 then return true;
3 foreach dart d' ∈ c_i(d) do
4 |   if d'.Alphas[i − 2].Alphas[i − 1] ≠ d'.Alphas[i − 1].Alphas[i − 2]
    |   then
5 |   |   return false;
6 return true;
```

Algorithm 56: `contractNGMap(gm,d,i)`: contract an i-cell for n-Gmaps

Input: gm: an n-Gmap;
d ∈ gm.Darts: a dart such that c_i(d) is contractible;
i ∈ $\{1, \ldots, n\}$.
Result: Contract the i-cell c_i(d).

```
1 ma ← reserveMarkNGMap(gm);
2 mark all the darts in c_i(d) for ma;
3 foreach dart d' ∈ c_i(d) do
4 |   d_1 ← d'.Alphas[i];
5 |   if not isMarkedNGMap(d_1,ma) then
6 |   |   d_2 ← d'.Alphas[i − 1].Alphas[i];
7 |   |   while isMarkedNGMap(d_2,ma) do
8 |   |   |   d_2 ← d_2.Alphas[i − 1].Alphas[i];
9 |   |   d_1.Alphas[i] ← d_2;
10 foreach dart d' ∈ c_i(d) do
11 |   remove d' from gm.Darts;
12 freeMarkNGMap(gm,ma);
```

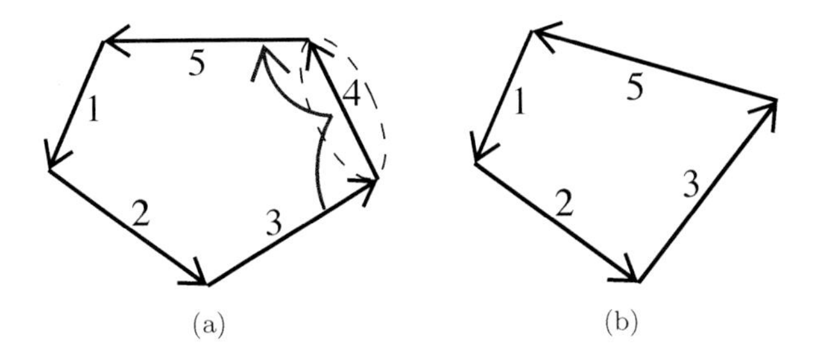

FIGURE 6.30
1-contraction for 1-map.
(a) 1-map M.
(b) 1-map $M_{C_1}(\{4\})$.

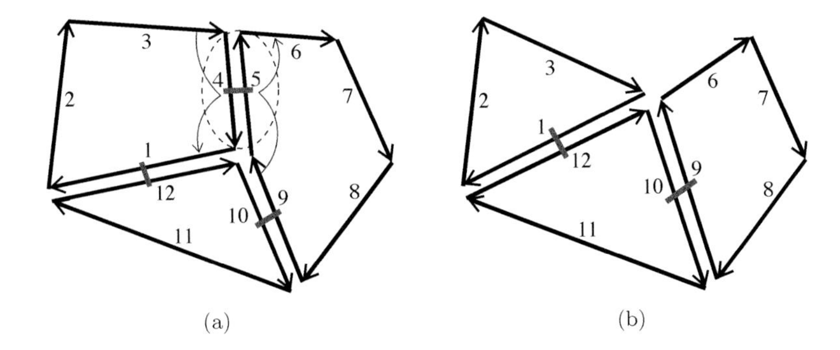

FIGURE 6.31
1-contraction for 2-Gmap.
(a) 2-map M.
(b) 2-map $M_{C_1}(\{4,5\})$.

6.3.2 For n-maps

As explained above, the i-contraction operation is the dual of the $(n-i)$-removal operation. In 1D, as illustrated in the example given in Fig. 6.30, the 1-contraction is the dual of the 0-removal. In this example, edge $\{4\}$ is contracted, by setting $\beta_1'(3) = 5$. Note that this produces the same 1-map than the removal of vertex $\{4\}$.

Figure 6.31 represents an example of 1-contraction in a 2-map. Edge $\{4,5\}$ is contracted by modifying β_1 for darts 3 and 9, by following paths of darts related to $\beta_1 \circ \beta_1$. So, $\beta_1'(3) = 1$ and $\beta_1'(9) = 6$. Edge $\{4,5\}$ is contracted and its two incident vertices are merged.

For the example represented in Fig. 6.32, face $\{1,2\}$ is contracted, by modifying β_2 for darts 3 and 7, by following paths of darts related to $\beta_2 \circ \beta_1 \circ \beta_2$.

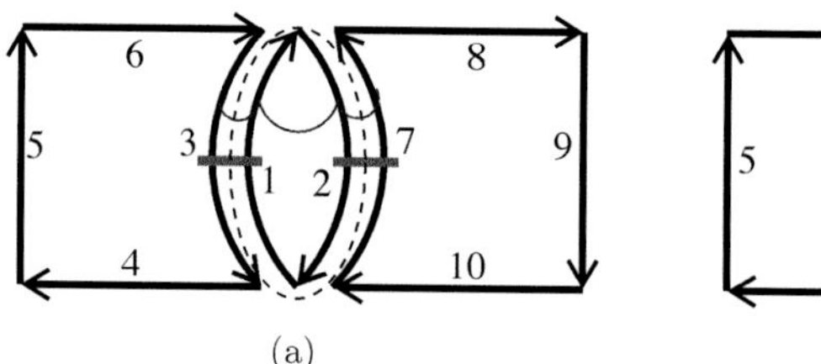

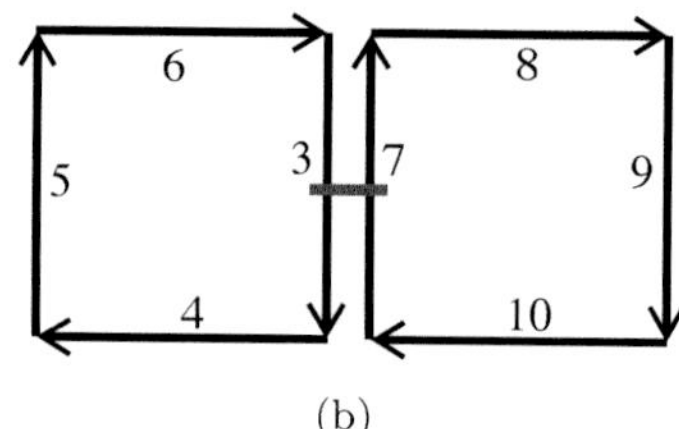

FIGURE 6.32
2-contraction for 2-map.
(a) 2-map M.
(b) 2-map $M_{C_2}(\{1,2\})$.

So, $\beta_2'(3) = 7$ and $\beta_2'(7) = 3$. Face $\{1,2\}$ is contracted and its two incident edges are merged.

The contractibility condition is defined in Def. 65. This definition is the direct transposition of the same definition for n-Gmaps. The only difference is the use of β_{i-2}^{-1} instead of α_{i-2}; this is important for $i = 2$ and $i = 3$, since β_1 is a partial permutation.

Definition 65 (contractible cell) *An i-cell C in an n-map $M = (D, \beta_1, \ldots, \beta_n)$ is* contractible *if:*

- $i = 1$;
- *or* $1 < i \leq n$, *and* $\forall d \in C$, $\beta_{i-1} \circ \beta_{i-2}(d) = \beta_{i-2}^{-1} \circ \beta_{i-1}(d)$.

Note that when $i = 2$, this condition becomes $\beta_1 \circ \beta_0(d) = \beta_1 \circ \beta_1(d)$. Thus a face, i.e. a 2-cell, is contractible if each of its darts is either a loop, i.e. it is linked with itself by β_1, or belongs to a cycle of two darts linked by β_1. Indeed, all other configurations are prohibited by this condition.

The i-contraction operation, for $i \geq 2$, is defined in Def. 66.

Definition 66 (cell contraction) *Let $M = (D, \beta_1, \ldots, \beta_n)$ be an n-map, and C be a contractible i-cell, $i \geq 2$. Let $D^S = \beta_i(C) \setminus C$. The n-map resulting from the i-contraction of C in M is $M_{C_i} = (D', \beta_1', \ldots, \beta_n')$, defined by:*

- $D' = D \setminus C$;
- $\forall j \in \{1, \ldots, n\}$, $j \neq i$, $\beta_j' = \beta_{j|D'}$;
- $\forall d \in D' \setminus D^S$, $\beta_i'(d) = \beta_i(d)$;
- $\forall d \in D^S$, $\beta_i'(d) = (\beta_i \circ \beta_{i-1})^k \circ \beta_i(d)$,
  k being the smaller positive integer s.t. $(\beta_i \circ \beta_{i-1})^k \circ \beta_i(d) \notin C$.

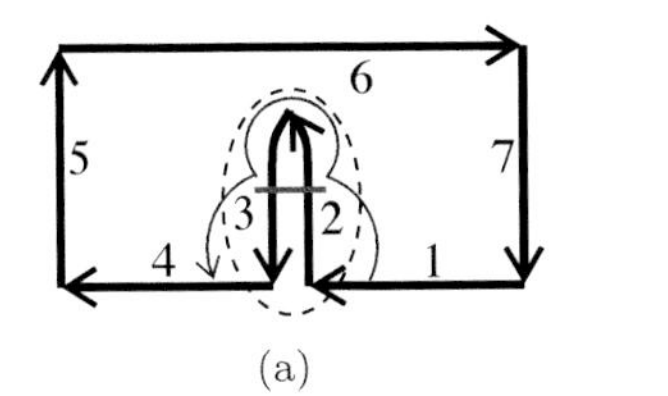

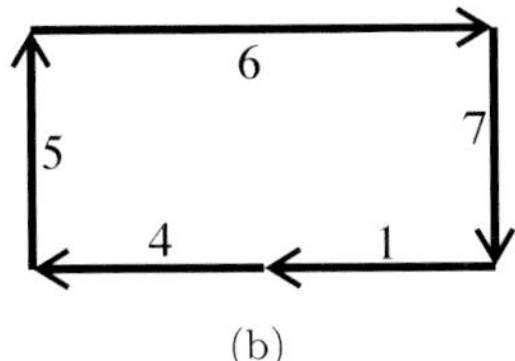

FIGURE 6.33
1-contraction of a dangling edge.
(a) 2-map M; edge $\{2,3\}$ is dangling.
(b) 2-map $M_{C_1}(\{2,3\})$.

The 1-contraction operation is defined in Definition 67 (cf. Figs. 6.30 and 6.31). There are two differences with the general case. First, the set of darts for which β_1 has to be modified is $D^S = \beta_0(C) \setminus C$: β_0 is applied, instead of β_1, since β_1 is a partial permutation. Second, a path of darts corresponding to $(\beta_1)^k(d)$ is followed until obtaining a dart which does not belong to C.

Definition 67 (1-cell contraction) *Let $M = (D, \beta_1, \ldots, \beta_n)$ be an n-map, and C be a contractible 1-cell. Let $D^S = \beta_0(C) \setminus C$. The n-map resulting from the 1-contraction of C in M is $M_{C_1} = (D', \beta'_1, \ldots, \beta'_n)$, defined by:*

- $D' = D \setminus C$;
- $\forall j \in \{2, \ldots, n\}$, $\beta'_j = \beta_{j|D'}$;
- $\forall d \in D' \setminus D^S$, $\beta'_1(d) = \beta_1(d)$;
- $\forall d \in D^S$, $\beta'_1(d) = (\beta_1)^k(d)$,
  k being the smaller positive integer s.t. $(\beta_1)^k(d) \notin C$.

An example of 1-contraction is presented in Fig. 6.33. Edge $\{2,3\}$ is dangling: cf. Fig. 6.33(a). $D^S = \{1\}$, and $\beta'_1(1) = 4$ (with $k = 2$). Note that the same 2-map is produced by the removal of edge $\{2,3\}$.

Algorithm 57 implements the test of the contractibility condition. This is a direct implementation of the corresponding definition.

The i-contraction operation is implemented by Algorithm 58, for $i \geq 2$, and Algorithm 59 implements the 1-contraction operation.

As for the contraction operation for n-Gmaps, the complexities of these algorithms are linear in number of darts of the contracted cell.

Algorithm 57: `isContractibleNMap(cm,d,i)`: test if an i-cell is contractible for n-maps

Input: cm: an n-map;
d ∈ cm.Darts: a dart;
i ∈ $\{0, \dots, n\}$.
Output: True iff c_i(d) is contractible.

```
1 if i = 0 then return false;
2 if i = 1 then return true;
3 foreach dart d' ∈ c_i(d) do
4     if d'.Betas[i − 2].Betas[i − 1] ≠ d'.Betas[i − 1].Betas[inv(i − 2)]
      then
5         return false;
6 return true;
```

Algorithm 58: `contractNMap(cm,d,i)`: contract an i-cell for n-maps

Input: cm: an n-map;
d ∈ cm.Darts: a dart such that c_i(d) is contractible;
i ∈ $\{2, \dots, n\}$.
Result: Contract the i-cell c_i(d).

```
1 ma ← reserveMarkNMap(cm);
2 mark all the darts in c_i(d) for ma;
3 foreach dart d' ∈ c_i(d) do
4     d1 ← d'.Betas[i];
5     if not isMarkedNMap(cm,d1,ma) then
6         d2 ← d'.Betas[i − 1].Betas[i];
7         while isMarkedNMap(cm,d2,ma) do
8             d2 ← d2.Betas[i − 1].Betas[i];
9         if d1 ≠ cm.null_dart then
10            if d2 ≠ cm.null_dart then
11                d1.Betas[i] ← d2;
12            else
13                d1.Betas[i] ← cm.null_dart;
14        else
15            if d2 ≠ cm.null_dart then
16                d2.Betas[i] ← cm.null_dart;
17 foreach dart d' ∈ c_i(d) do
18     remove d' from cm.Darts;
19 freeMarkNMap(cm,ma);
```

Algorithm 59: `1contractNMap(cm,d)`: contract a 1-cell for n-maps

```
Input: cm: an n-map;
       d ∈ cm.Darts: a dart such that c1(d) is contractible.
Result: Contract the 1-cell c1(d).
1  ma ← reserveMarkNMap(cm);
2  mark all the darts in c1(d) for ma;
3  foreach dart d' ∈ c1(d) do
4      d1 ← d'.Betas[0];
5      if not isMarkedNMap(cm,d1,ma) then
6          d2 ← d'.Betas[1];
7          while isMarkedNMap(cm,d2,ma) do
8              d2 ← d2.Betas[1];
9          if d1 ≠ cm.null_dart then
10             if d2 ≠ cm.null_dart then
11                 d1.Betas[1] ← d2; d2.Betas[0] ← d1;
12             else
13                 d1.Betas[1] ← cm.null_dart;
14         else
15             if d2 ≠ cm.null_dart then
16                 d2.Betas[0] ← cm.null_dart;
17 foreach dart d' ∈ c1(d) do
18     remove d' from cm.Darts;
19 freeMarkNMap(cm,ma);
```

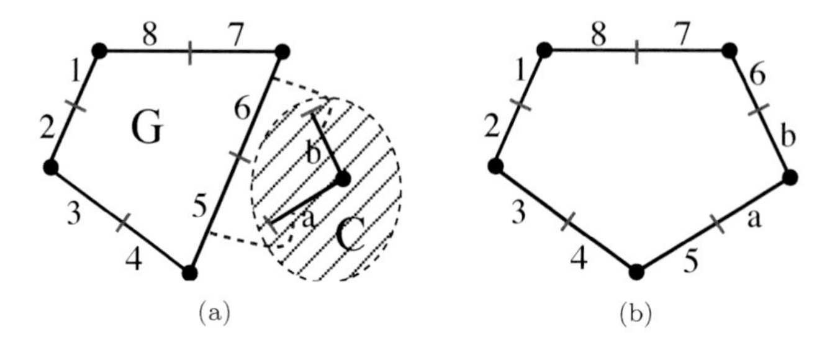

FIGURE 6.34
0-insertion for 1-Gmap.
(a) Two 1-Gmaps G and C, linked by bijection γ.
(b) 1-Gmap $G_{I_0}(C, \gamma)$.

6.4 Insertion

The *insertion* operation is the inverse of the removal operation. It consists in adding a new i-cell inside an existing $(i+1)$-cell, possibly cutting this $(i+1)$-cell in two. The insertion operation is defined in this chapter only for n-Gmaps; the conversion for n-maps can be done without particular difficulty. Since insertion is the inverse of removal, the examples presented in Section 6.2.1 illustrate also the insertion operation, but starting here from the results.

Look at 1-Gmaps G and C in Fig. 6.34(a). C describes a 0-cell, which is inserted into a 1-cell of G, producing 1-Gmap $G_{I_0}(C, \gamma)$ presented in Fig. 6.34(b). γ is a bijection (represented by gray dashed curves in the figures), which maps C onto the darts of the 1-cell; more precisely, $\gamma(a) = 5$ and $\gamma(b) = 6$. The set of darts of the resulting 1-Gmap is the union of the two sets of darts of G and C; its involutions are either defined according to G (when two darts of G are concerned), or to C (when two darts of C are concerned), or to γ otherwise (i.e. when a dart of G and a dart of C are concerned).

Two 2-Gmaps G and C are represented in Fig. 6.35(a). C describes an edge, which is inserted into a face of G, producing 2-Gmap $G_{I_1}(C, \gamma)$, represented in Fig. 6.35(b). Bijection γ describes the way the 1-cell is inserted into the face: more precisely, $\gamma(a) = 2$, $\gamma(b) = 11$, $\gamma(c) = 10$ and $\gamma(d) = 3$. The construction of the resulting 2-Gmap is similar to the 1D case.

Note that if the insertion is defined according to γ', defined by $\gamma'(a) = 3, \gamma'(b) = 11, \gamma'(c) = 10, \gamma'(d) = 2$, the face in which the edge is inserted is not split. The resulting 2-Gmap is not orientable; more precisely, its topological characteristics are that of a Möbius strip. This illustrates the interest of describing by bijection γ the way the cell is inserted.

Examples of 0-insertion and 1-insertion in a 2-Gmap are presented in Figs. 6.36 and 6.37. In Fig. 6.37, C describes a dangling edge, and γ is defined

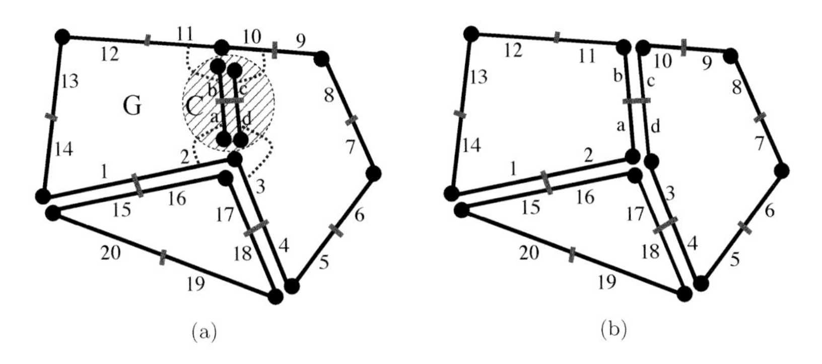

FIGURE 6.35
1-insertion for 2-Gmap.
(a) Two 2-Gmaps G and C, linked by bijection γ.
(b) 2-Gmap $G_{I_1}(C, \gamma)$.

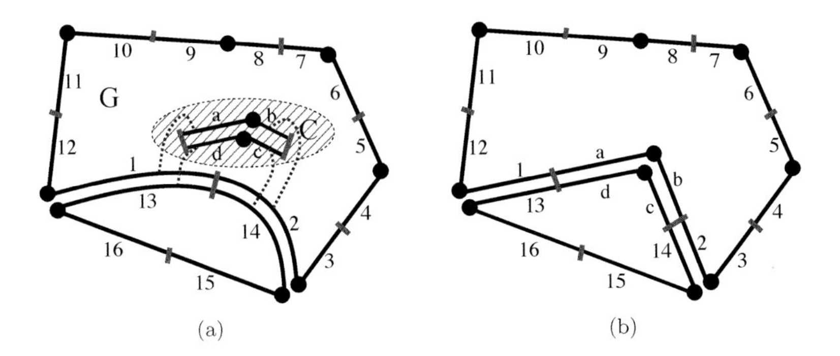

FIGURE 6.36
0-insertion for 2-Gmap.
(a) Two 2-Gmaps G and C, linked by bijection γ.
(b) 2-Gmap $G_{I_0}(C, \gamma)$.

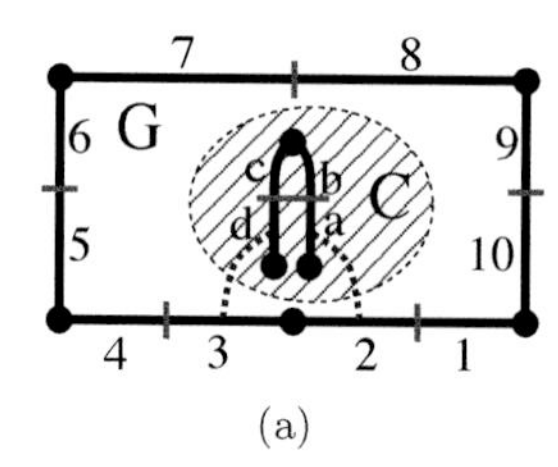

(a)

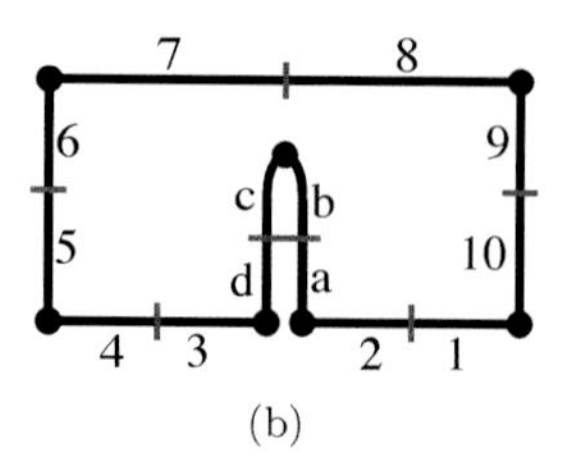

(b)

FIGURE 6.37
1-insertion of a dangling edge.
(a) Two 2-Gmaps G and C, linked by bijection γ. The edge in C is dangling.
(b) 2-Gmap $G_{I_1}(C, \gamma)$.

only for a subset of the darts of C, i.e. $\alpha_1(b) = c, \alpha_1(c) = b, \gamma(a) = 2, \gamma(d) = 3$. Note that since darts b and c are linked by α_1 in C, they are still linked by α_1 in $G_{I_1}(C, \gamma)$. This illustrates the interest of describing the cell to insert by an n-Gmap.

The general i-insertion operation takes three parameters as input:

- n-Gmap G, in which an i-cell is inserted;
- n-Gmap C, describing the i-cell to insert;
- γ, the function describing the way the i-cell is inserted. More precisely, γ is a bijection between a subset BV' of the darts of C, and a subset BV of the darts of G.

The insertion can be achieved if C is i-insertable into G according to γ (cf. Def. 68). More precisely, four conditions have to be satisfied:

- C contains exactly one removable i-cell. This condition is required for ensuring the invertibility of insertion and removal operations;
- the darts of BV' are i-free in C, since these darts will be i-linked with the darts of BV;
- γ commutes with involution α_j in G and C for all darts in $BV \cup BV'$, for all j such that $|i - j| \geq 2$. This conditions ensures that the resulting α_j is an involution in the resulting n-Gmap;
- the fourth condition is also required to ensure the invertibility of insertion and removal operations: the paths of darts followed if C is removed in the resulting n-Gmap lead to the darts linked by α_i in the initial n-Gmap.

Definition 68 (insertable cell) *Let* $G = (D, \alpha_0, \ldots, \alpha_n)$ *and* $C = (D', \alpha'_0, \ldots, \alpha'_n)$ *be two* n*-Gmaps,* $i \in \{0, \ldots, n\}$*, and* γ *be a bijection from* $BV' \subseteq D'$ *to* $BV \subseteq D$*.* C *is* i*-insertable into* G *according to* γ *if:*

- *C contains exactly one removable i-cell;*
- *$\forall d' \in BV'$, d' is i-free;*
- *$\forall j \in \{0, \ldots, n\}$ s.t. $|i - j| \geq 2$:*
  - *$\forall d \in BV$ s.t. $\alpha_j(d) \in BV$, $\gamma^{-1} \circ \alpha_j(d) = \alpha'_j \circ \gamma^{-1}(d)$;*
  - *$\forall d' \in BV'$ s.t. $\alpha'_j(d') \in BV'$, $\gamma \circ \alpha'_j(d') = \alpha_j \circ \gamma(d')$;*
- *$\forall d \in BV$, $\alpha_i(d) = \gamma \circ (\alpha'_{i+1} \circ \alpha'_i)^k \circ \alpha'_{i+1} \circ \gamma^{-1}(d)$,
  k being the smaller positive integer s.t. $(\alpha'_{i+1} \circ \alpha'_i)^k \circ \alpha'_{i+1} \circ \gamma^{-1}(d) \in BV'$.*

For instance in Fig. 6.35, $BV = \{2, 3, 10, 11\}$ and $BV' = \{a, b, c, d\}$. The two first conditions are obviously satisfied (C is removable, and $\forall d \in BV'$, d is 1-free). The third condition is trivially satisfied (there is no $j \in \{0, 1, 2\}$ such that $|1 - j| \geq 2$). The fourth condition is also satisfied: for example, $\alpha_1(2) = 3$ and $\gamma \circ \alpha'_2 \circ \gamma^{-1}(2) = 3$ ($k = 0$); this is similar for the other darts of BV.

For Fig. 6.36, $BV = \{1, 2, 13, 14\}$ and $BV' = \{a, b, c, d\}$. C is 0-removable. The darts of BV' are 0-free. The third condition is satisfied: for example, $\gamma^{-1} \circ \alpha_2(1) = d = \alpha'_2 \circ \gamma^{-1}(1)$ (this is similar for the other darts). The fourth condition is also satisfied: for example, $\alpha_0(1) = 2$ and $\gamma \circ \alpha'_1 \circ \gamma^{-1}(1) = 2$ ($k = 0$); this is similar for the other darts in BV.

The i-insertion operation is defined in Def. 69. The n-Gmap resulting from the i-insertion of C in G is $G_{I_i}(C, \gamma)$: its set of darts is the union of the two sets of darts of G and C; its involutions are defined according to the involutions of G (when two darts of G are concerned), or that of C (when two darts of C are concerned), or according to γ otherwise (when one dart of G and one dart of C are concerned).

Definition 69 (insertion) *Let $G = (D, \alpha_0, \ldots, \alpha_n)$ and $C = (D', \alpha'_0, \ldots, \alpha'_n)$ be two n-Gmaps, $BV \subseteq D$, $BV' \subseteq D'$, γ be a bijection from BV' onto BV, such that C is an insertable i-cell according to γ. The n-Gmap resulting from the i-insertion of C in G is $G_{I_i}(C, \gamma) = (D'', \alpha''_0, \ldots, \alpha''_n)$ defined by:*

- *$D'' = D \cup D'$;*
- *$\forall j \in \{0, \ldots, n\}$, $j \neq i$:*

$$\forall d \in D, \alpha''_j(d) = \alpha_j(d);\ \ \forall d' \in D', \alpha''_j(d') = \alpha'_j(d');$$

- *$\forall d \in D \setminus BV$, $\alpha''_i(d) = \alpha_i(d)$; $\forall d' \in D' \setminus BV'$, $\alpha''_i(d') = \alpha'_i(d')$;*
- *$\forall d \in BV$, $\alpha''_i(d) = \gamma^{-1}(d)$; $\forall d' \in BV'$, $\alpha''_i(d') = \gamma(d')$.*

The definition of the resulting n-Gmap $G_{I_i}(C, \gamma)$ is straightforward, the complex part of the definition being the i-insertability condition, which guaranty the validity of the operation.

Algorithm 60 is a direct implementation of the `i-insertability` condition.

Algorithm 60: `isInsertableNGMap(gm,c,i,assoc)`: test if an `i`-cell is insertable for n-Gmaps

Input: `gm`, `c`: two n-Gmaps;
`i` $\in \{0, \dots, n\}$;
`assoc`: a mapping from `c.Darts` to `gm.Darts`.
Output: True iff `c` can be inserted in `gm` according to `assoc`.

```
1  if c contains more than one i-cell then return false;
2  if the i-cell in c is not removable then return false;
3  assoc^-1 ← an associative array between darts, inverse of assoc;
4  foreach dart d ∈ assoc do
5      if not isFreeNGMap(d,i) then return false;
6      d' ← assoc[d];
7      foreach j ∈ {0,...,i−2,i+2,...,n} do
8          if d.Alphas[j] ∈ assoc and
           assoc[d.Alphas[j]] ≠ assoc[d].Alphas[j] then
9              return false;
10         if d'.Alphas[j] ∈ assoc^-1 and
           assoc^-1[d'.Alphas[j]] ≠ assoc^-1[d'].Alphas[j] then
11             return false;
12     d2 ← d.Alphas[i+1];
13     while d2 ∉ assoc do
14         d2.Alphas[i].Alphas[i+1];
15     if assoc[d2] ≠ d'.Alphas[i] then return false;
16 return true;
```

The mapping `assoc`, corresponding to γ in Def. 68, is implemented as an associative array (for example a hash table or a binary search tree) from `c.Darts` to `gm.Darts`. Testing if a dart `d` belongs to `assoc` is equivalent to test if `d` belongs to BV' in the insertability condition definition. $\texttt{assoc}^{-1}$ is computed from `assoc` ($\texttt{assoc}^{-1}$ corresponds to γ^{-1}), and it is stored into a second associative array. It allows to test if a given dart `d` belongs to BV. Note that, by definition of `assoc` and $\texttt{assoc}^{-1}$, iterating through the darts of `assoc` (thus BV') and considering $\texttt{d}' \leftarrow \texttt{assoc[d]}$ is equivalent to iterate through the darts of $\texttt{assoc}^{-1}$ (thus BV).

The i-insertion operation is implemented by Algorithm 61. Instead of building a new n-Gmap $G_{I_i}(C, \gamma)$, n-Gmap `gm` is modified in order to obtain an n-Gmap which is isomorphic to $G_{I_i}(C, \gamma)$. So, the `Alphas[i]` links have to be modified only for the darts of `assoc` and of $\texttt{assoc}^{-1}$.

Algorithm 61: `insertNGMap(gm,c,i,assoc)`: insert an `i`-cell in an n-Gmap

Input: `gm`: an n-Gmap;
  `c`: an n-Gmap, which is an insertable `i`-cell according to `assoc`;
  $\texttt{i} \in \{0, \ldots, n\}$;
  `assoc`: a mapping from `c.Darts` to `gm.Darts`.
Result: Insert the `i`-cell `c` in `gm` along `assoc`.

```
1 foreach dart d' ∈ c.Darts do
2     add d' in gm.Darts;
3     remove d' from c.Darts;
4     if d' ∈ assoc then
5         d'.Alphas[i] ← assoc[d'];
6         assoc[d'].Alphas[i] ← d';
```

The complexity of Algorithm 60 is linear in the number of darts $\#d$ of the n-Gmap `c` times the complexity for accessing an element in the associative array $\log \#d$. Indeed, it is possible to iterate through the `i`-cell `c` in linear time, in order to test if `c` contains more than one `i`-cell. Then, it is possible again to test in linear time if this `i`-cell is removable. The number of elements in `assoc` is bounded by the number of darts in `c`, thus each access in `assoc` or $\texttt{assoc}^{-1}$ is done in $\log \#d$. At last, the complexity of the loop over the path of darts $(\alpha'_{\texttt{i+1}} \circ \alpha'_{\texttt{i}})^k \circ \alpha'_{\texttt{i+1}} \circ \gamma^{-1}(d)$ is bounded by the number of darts in `c`, because the considered paths are disjoint. This gives the global complexity of the algorithm.

The complexity of Algorithm 61 is also linear in number darts $\#d$ of the n-Gmap `c` times $\log \#d$. This is straightforward, since all darts in `c` are processed, using the links described by `assoc`.

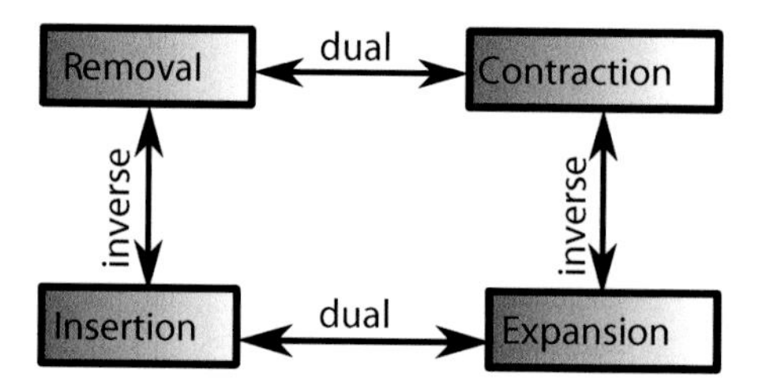

FIGURE 6.38
Links between the four operations: removal, contraction, insertion and expansion.

6.5 Expansion

The *expansion* operation is the inverse of the contraction operation. It consists in adding a new i-cell inside an existing given $(i-1)$-cell, possibly cutting this $(i-1)$-cell in two. As before, the expansion operation is defined here only for n-Gmaps; the conversion for n-maps can be done without particular difficulty.

Figure 6.38 summarizes the links between the four operations: removal, contraction, insertion and expansion.

The expansibility condition and the expansion operation, defined in Defs. 70 and 71, are directly deduced from the insertability condition and the insertion operation, by duality (i.e. $i+1$ is replaced by $i-1$).

Definition 70 (expansible cell) *Let $G = (D, \alpha_0, \ldots, \alpha_n)$ and $C = (D', \alpha'_0, \ldots, \alpha'_n)$ be two n-Gmaps, $i \in \{0, \ldots, n\}$, and γ be a bijection from $BV' \subseteq D'$ to $BV \subseteq D$. C is i-expansible into G according to γ if:*

- *C contains exactly one contractible i-cell;*
- *$\forall d' \in BV', d'$ is i-free;*
- *$\forall j \in \{0, \ldots, n\}$ s.t. $|i-j| \geq 2$:*
  - *$\forall d \in BV$ s.t. $\alpha_j(d) \in BV$, $\gamma^{-1} \circ \alpha_j(d) = \alpha'_j \circ \gamma^{-1}(d)$;*
  - *$\forall d' \in BV'$ s.t. $\alpha'_j(d') \in BV'$, $\gamma \circ \alpha'_j(d') = \alpha_j \circ \gamma(d')$;*
- *$\forall d \in BV$, $\alpha_i(d) = \gamma \circ (\alpha'_{i-1} \circ \alpha'_i)^k \circ \alpha'_{i-1} \circ \gamma^{-1}(d)$,*
  k being the smaller positive integer s.t. $(\alpha'_{i-1} \circ \alpha'_i)^k \circ \alpha'_{i-1} \circ \gamma^{-1}(d) \in BV'$.

Definition 71 (expansion) *Let $G = (D, \alpha_0, \ldots, \alpha_n)$ and $C = (D', \alpha'_0, \ldots, \alpha'_n)$ be two n-Gmaps, $BV \subseteq D$, $BV' \subseteq D'$, γ be a bijection from BV' onto BV, such that C is an expansible i-cell according to γ. The n-Gmap resulting from the i-expansion of C in G is $G_{E_i}(C, \gamma) = (D'', \alpha''_0, \ldots, \alpha''_n)$, defined by:*

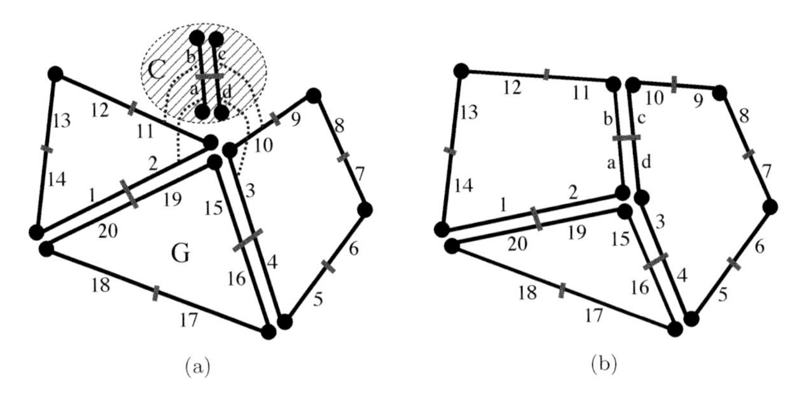

FIGURE 6.39
1-expansion for 2-Gmap.
(a) Two 2-Gmaps G and C, linked by bijection γ.
(b) 2-Gmap $G_{E_1}(C, \gamma)$.

- $D'' = D \cup D'$;
- $\forall j \in \{0, \ldots, n\}$, $j \neq i$:

$$\forall d \in D, \alpha''_j(d) = \alpha_j(d); \;\; \forall d' \in D', \alpha''_j(d') = \alpha'_j(d');$$

- $\forall d \in D \setminus BV$, $\alpha''_i(d) = \alpha_i(d)$; $\;\; \forall d' \in D' \setminus BV'$, $\alpha''_i(d') = \alpha'_i(d')$;
- $\forall d \in BV$, $\alpha''_i(d) = \gamma^{-1}(d)$; $\;\; \forall d' \in BV'$, $\alpha''_i(d') = \gamma(d')$.

Note that the definition of the i-expansion operation is the same than the definition of the i-insertion operation, the only difference being the fact that an expansible i-cell is taken into account instead of an insertable i-cell.

Since the expansion operation is the inverse of the contraction operation, the examples presented in Section 6.3.1 illustrate also the expansion operation, but starting here from the results. Fig. 6.39 represents a 1-expansion in a 2-Gmap. Starting from Fig. 6.39(a), the edge described by 2-Gmap C is expanded into 2-Gmap G according to γ, producing $G_{E_1}(C, \gamma)$ represented in Fig. 6.39(b).

Figure 6.40 illustrates a 2-expansion in a 2-Gmap. Starting from Fig. 6.40(a), the face described by 2-Gmap C is expanded into 2-Gmap G, according to γ. This produces $G_{E_2}(C, \gamma)$, represented in Fig. 6.40(b).

Since expansion and insertion are very similar, the corresponding algorithms are also very similar: cf. Algorithms 62 and 63. Thus, the complexities of these two algorithms are also linear in the number of darts $\#d$ of the n-Gmap c times $\log \#d$.

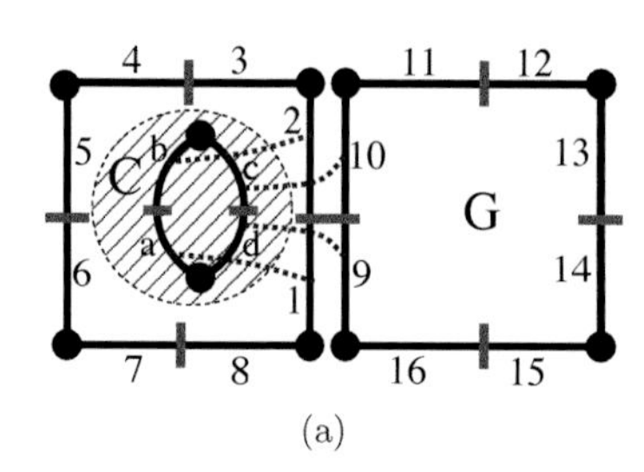

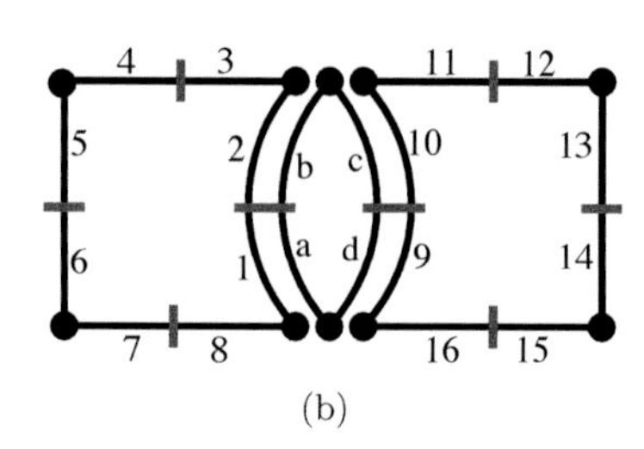

FIGURE 6.40
2-expansion for 2-Gmap.
(a) Two 2-Gmaps G and C, linked by bijection γ.
(b) 2-Gmap $G_{E_2}(C, \gamma)$.

Algorithm 62: `isExpansibleNGMap(gm,c,i,assoc)`: test if an `i`-cell is expansible for n-Gmaps

---

Input: `gm`, `c`: two n-Gmaps;
`i` $\in \{0, \ldots, n\}$;
`assoc`: a mapping from `c.Darts` to `gm.Darts`.
Output: True iff `c` can be expanded in `gm` according to `assoc`.

1 **if** `c` *contains more than one* `i`*-cell* **then return** *false*;
2 **if** *the* `i`*-cell in* `c` *is not contractible* **then return** *false*;
3 $\texttt{assoc}^{-1} \leftarrow$ an associative array between darts, inverse of `assoc`;
4 **foreach** *dart* `d` $\in$ `assoc` **do**
5   **if not** `isFreeNGMap(d,i)` **then return** *false*;
6   $\texttt{d}' \leftarrow \texttt{assoc[d]}$;
7   **foreach** $\texttt{j} \in \{0, \ldots, \texttt{i}-2, \texttt{i}+2, \ldots, n\}$ **do**
8   **if** `d.Alphas[j]` $\in$ `assoc` **and** `assoc[d.Alphas[j]]` $\neq$ `assoc[d].Alphas[j]` **then**
9   **return** *false*;
10   **if** $\texttt{d}'\texttt{.Alphas[j]} \in \texttt{assoc}^{-1}$ **and** $\texttt{assoc}^{-1}\texttt{[d}'\texttt{.Alphas[j]]} \neq \texttt{assoc}^{-1}\texttt{[d}'\texttt{].Alphas[j]}$ **then**
11   **return** *false*;
12   $\texttt{d}_2 \leftarrow \texttt{d.Alphas[i}-1\texttt{]}$;
13   **while** $\texttt{d}_2 \notin \texttt{assoc}$ **do**
14     $\texttt{d}_2\texttt{.Alphas[i].Alphas[i}-1\texttt{]}$;
15   **if** $\texttt{assoc[d}_2\texttt{]} \neq \texttt{d}'\texttt{.Alphas[i]}$ **then return** *false*;
16 **return** *true*;

Algorithm 63: `expandNGMap(gm,c,i,assoc)`: expand an `i`-cell in an n-Gmap

Input: `gm`: an n-Gmap;
`c`: an expansible `i`-cell according to `assoc`;
`i` $\in \{0, \ldots, n\}$;
`assoc`: a mapping from `c.Darts` to `gm.Darts`.
Result: Expand the `i`-cell `c` in `gm` along `assoc`.

```
1 foreach dart d' ∈ c.Darts do
2     add d' in gm.Darts;
3     remove d' from c.Darts;
4     if d' ∈ assoc then
5         d'.Alphas[i] ← assoc[d'];
6         assoc[d'].Alphas[i] ← d';
```

6.6 Chamfering

The *chamfering* operation takes an i-cell as parameter, and replaces it by an n-cell. It allows to smooth an object by rounding its sharp cells. In 2D, vertices or edges can be chamfered (cf. examples in Fig. 6.41). In both cases, when no multi-incidence occurs, the chamfered cell is replaced by a new face which has as many edges in its boundary than the number of faces incident to the chamfered cell. For instance in Fig. 6.41:

- when the vertex is chamfered, the new face is incident to three edges;
- when the edge is chamfered, the new face is incident to two edges.

Note that only the given cell is chamfered: for instance, when chamfering an edge, its incident vertices are not chamfered.

In $3D$, vertices (cf. Fig. 6.42), edges (cf. Fig. 6.43) or faces can be chamfered. When no multi-incidence occurs, the chamfered cell is replaced by a volume, which has as many faces in its boundary than the number of volumes incident to the chamfered cell. For instance in Fig. 6.42, the new volume is incident to eight faces; in Fig. 6.43, the new volume is incident to four faces. Note again that only the given cell is chamfered, not its incident cells.

The chamfering operation presented here is a basic operation. As illustrated in Fig. 6.44, it is a basis for a *generalized chamfering* operation, which consists in applying several times the basic chamfering operation. In this example, starting from the 3D object represented in Fig. 6.44(a), edge e is first chamfered, then the two vertices incident to edge e before its chamfering are chamfered. This produces the object represented in Fig. 6.44(b), after a cleanup step which merges several new cells.

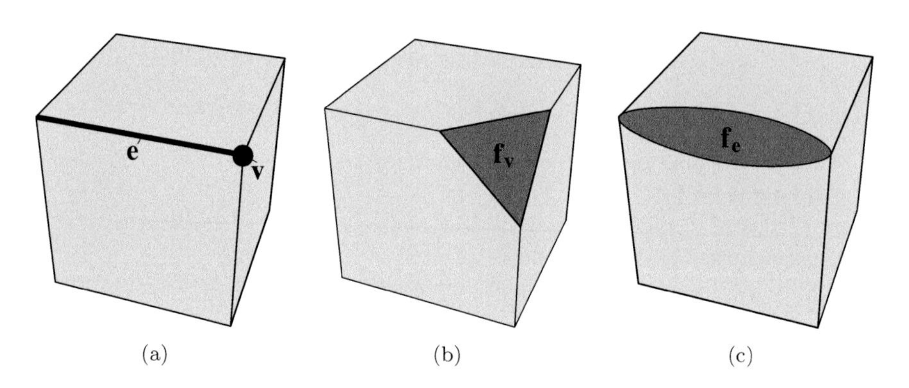

FIGURE 6.41
(a) A 2D subdivided object describing the surface of a cube.
(b) Chamfering vertex v.
(c) Chamfering edge e.

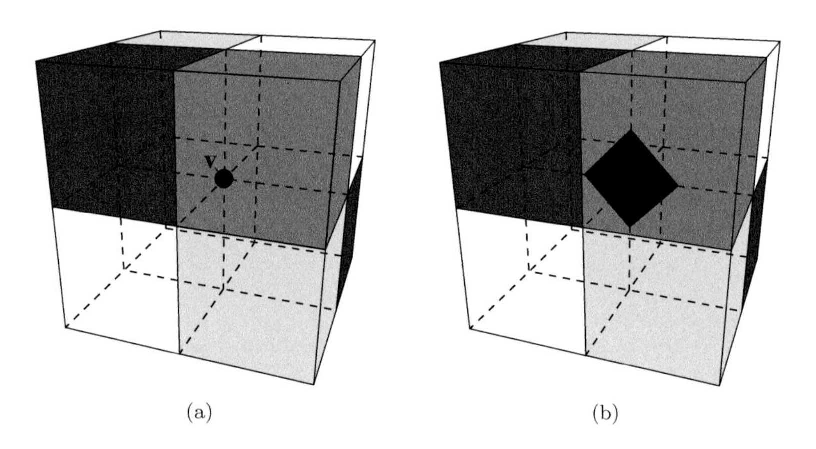

FIGURE 6.42
(a) A 3D subdivided object made of eight cubes.
(b) Chamfering vertex v.

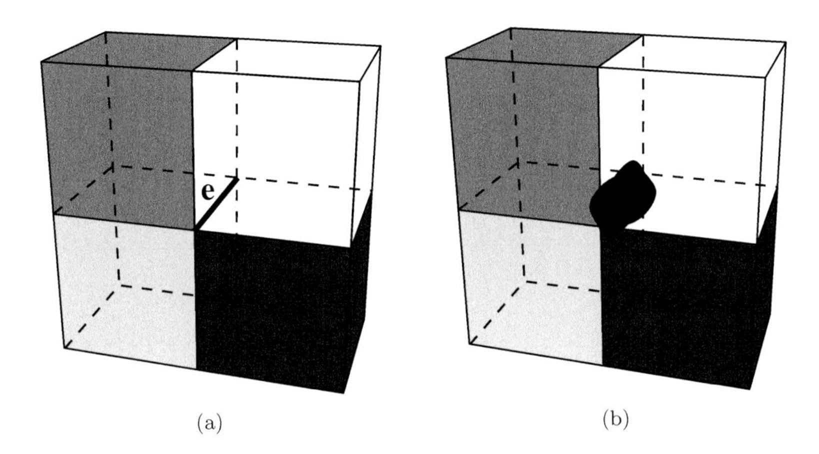

FIGURE 6.43
(a) A 3D subdivided object made of four cubes.
(b) Chamfering edge e.

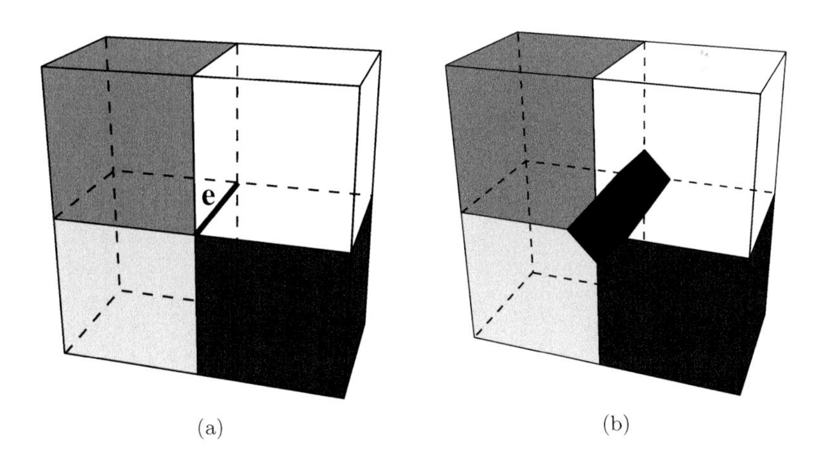

FIGURE 6.44
(a) A 3D subdivided object made of four cubes.
(b) The result of the generalized chamfering operation applied to edge e.

The i-chamfering operation is defined in Def. 72. This operations takes as input an n-Gmap G and one of its i-cell C. It mainly consists in creating new sets of darts D^j, for each $i+1 \leq j \leq n$. Intuitively, each D^j corresponds to the j-cells required to chamfer the initial i-cell.

Definition 72 (chamfering) *Let $G = (D, \alpha_0, \ldots, \alpha_n)$ be an n-Gmap, C be one of its i-cells, with $i \in \{0, \ldots, n-1\}$. The n-Gmap resulting from the i-chamfering of C in G is $G_{i\text{-}chf}(C) = (D', \alpha'_0, \ldots, \alpha'_n)$ defined by:*

- *$\forall j \in \{i+1, \ldots, n\}$, let D^j be a set of new darts, such that a bijection φ^j maps C onto D^j. Let φ^i be the identity on C;*
- *$D' = D \cup_{j=i+1}^{n} D^j$;*
- 1. *$\forall j \in \{0, \ldots, n\}$, $j \neq i+1$, $\forall d \in D$, $\alpha'_j(d) = \alpha_j(d)$;*
  2. *$\forall d \in D \setminus C$, $\alpha'_{i+1}(d) = \alpha_{i+1}(d)$;*
  3. *$\forall d \in C$, $\alpha'_{i+1}(d) = \varphi^{i+1}(d)$.*
- *$\forall j \in \{i+1, \ldots, n\}$, $\forall d \in C$:*

  4. *$\forall k \in \{0, \ldots, i-1\}$, $\alpha'_k(\varphi^j(d)) = \varphi^j(\alpha_k(d))$;*
  5. *$\forall k \in \{i, \ldots, j-1\}$, $\alpha'_k(\varphi^j(d)) = \varphi^j(\alpha_{k+1}(d))$;*
  6. *$\alpha'_j(\varphi^j(d)) = \varphi^{j-1}(d)$;*
  7. *if $j < n$, $\alpha'_{j+1}(\varphi^j(d)) = \varphi^{j+1}(d)$;*
  8. *$\forall k \in \{j+2, \ldots, n\}$, $\alpha'_k(\varphi^j(d)) = \varphi^j(\alpha_k(d))$.*

The chamfering operation is illustrated in Fig. 6.45; it is applied to vertex $C = \{1, 2, 3, 4, 5, 6\}$ of a 2-Gmap describing the surface of a cube. $D^1 = \{1^1, 2^1, 3^1, 4^1, 5^1, 6^1\}$ and $D^2 = \{1^2, 2^2, 3^2, 4^2, 5^2, 6^2\}$ are the sets of new darts, φ^1 (resp. φ^2) associates darts d and d^1 (resp. d^2).

α' involutions are defined on D by items $1-3$ of the definition. For the example, nothing is changed, except for the darts of C, which are 1-linked with the darts of D^1 according to φ^1 (item 3): so, each dart incident to a corner of a face in the initial object is now linked with a new dart (for instance, $\alpha'_1(1) = 1^1$); this is the first part of the insertion of a new edge into each corner of each face.

α' involutions are defined on the new darts by items $4-8$ of the definition:

- for D^1:
  - each edge inserted into an initial face is constructed, by defining α'_0, according to α_1 (item 5); this is the second part of the insertion of a new edge into each corner of each face (for instance, $\alpha'_0(1^1) = 2^1$);
  - α'_1 is defined, according to the definition of α'_1 for the darts of C (item 6): so α'_1 is an involution (for instance, $\alpha'_1(1^1) = 1$); this is the last part of the insertion of a new edge into each corner of each face;

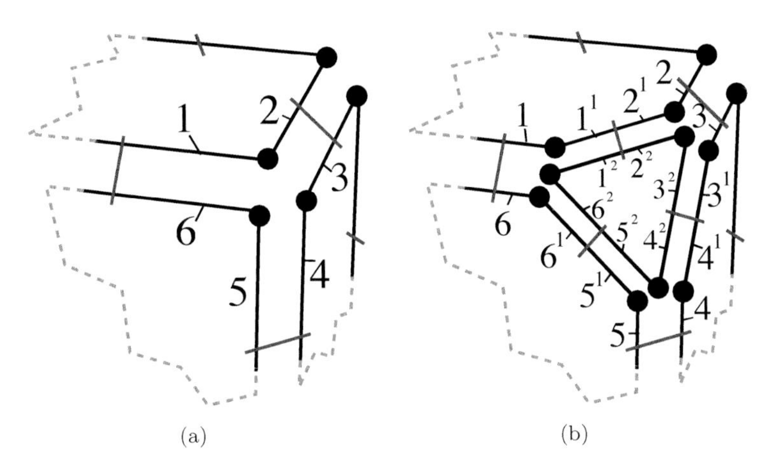

FIGURE 6.45
0-chamfering for 2-Gmap.
(a) 2-Gmap G describing the surface of a cube (partially drawn).
(b) 2-Gmap $G_{0\text{-chf}}(\{1,2,3,4,5,6\})$.

  – each dart of D^1 is 2-linked with its corresponding dart of D^2 (item 7): this is the first part of the insertion of the new face (for instance, $\alpha'_2(1^1) = 1^2$);

- for D^2:

  – the face is constructed, by defining α'_0 and α'_1, according to α_1 and α_2 (item 5); this is the second part of the insertion of the new face (for instance, $\alpha'_0(1^2) = 2^2$ and $\alpha'_1(1^2) = 6^2$); this is the second part of the insertion of the new face;

  – each dart of D^2 is 2-linked with its corresponding dart of D^1 (item 6): so, α'_2 is an involution (for instance, $\alpha'_2(1^2) = 1^1$); this is the last part of the insertion of the new face.

The application of the chamfering operation to an edge $C = \{1,2,3,4\}$ in a 2-Gmap is illustrated in Fig. 6.46. $D^2 = \{1^2, 2^2, 3^2, 4^2\}$ is the set of new darts. Nothing is changed for the darts of D, except that the darts of C are linked with the corresponding new darts by α'_2 (items $1-3$): this is the first part of the insertion of the new face (for instance, $\alpha'_2(1^1) = 1^2$). For D^2:

- α'_0 is defined, according to α_0 (item 4); this is the first part of the construction of the new face (for instance, $\alpha'_0(1^2) = 2^2$);

- α'_1 is defined, according to α_2 (item 5); this is the second part of the construction of the new face (for instance, $\alpha'_1(1^2) = 3^2$);

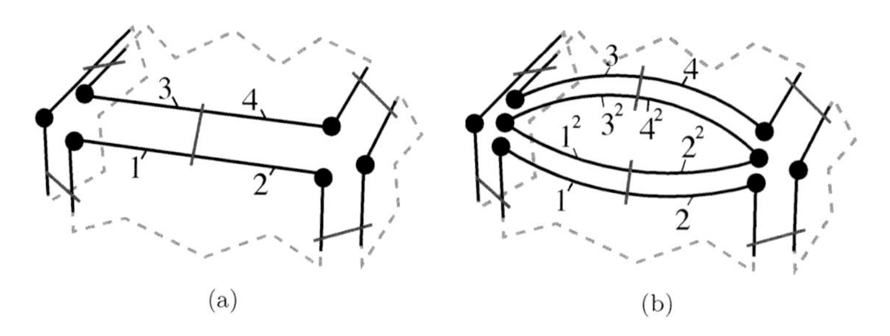

FIGURE 6.46
1-chamfering for 2-Gmap.
(a) 2-Gmap G describing the surface of a cube (partially drawn).
(b) 2-Gmap $G_{1\text{-chf}}(\{1, 2, 3, 4\})$.

- α'_2 is defined, according to the definition of α'_2 for the darts of C (item 6); so, α'_2 is an involution (for instance, $\alpha'_2(1^2) = 1^1$). This is the last part of the insertion of the new face.

The i-chamfering operation is implemented by Algorithm 64. As above, the given n-Gmap is modified, instead of building a new n-Gmap as result of the operation. The algorithm is a direct translation of the definition. First, new darts are created, and stored into `phi` associative arrays (there is an associative array `phi[j]` for each `j` $\in \{$`i`$, \dots, n\}$). Then, the α links are defined for these darts according to the definition.

Remember that α_{i+1} is modified for the darts of $c_{\mathtt{i}}(\mathtt{d})$ (line 18). In order to avoid side effects, this is done in a second loop, after having defined all the α links for the new darts. Thus, the definition of the links for these new darts is based on the original α links of the n-Gmap. Moreover, and due to the same reason, it is also necessary to be careful for the second loop, in order to correctly iterate through the darts of $c_{\mathtt{i}}(\mathtt{d})$. For that, the darts of $c_{\mathtt{i}}(\mathtt{d})$ are stored in a data structure (for example a stack) during the first loop (line 2); the iteration of the second loop is done through this data structure, and not through a cell iterator (line 17).

The complexity of Algorithm 64 is linear in the number of darts $\#d$ of the chamfered i-cell times $\log \#d$. Indeed, n being a constant, the loops iterating through `j` and `k` (lines 7, 8, 10 and 15) are bounded and thus can be considered as constants, and the accesses to the associative array are done in log time.

Algorithm 64: `chamferingNGMap(gm,d,i)`: chamfer an i-cell for n-Gmaps

```
  Input: gm: an n-Gmap;
         d ∈ gm.Darts: a dart;
         i ∈ {0,...,n−1}.
  Result: Chamfer the i-cell c_i(d).
1  phi ← an array of n − i + 1 empty associative arrays between darts;
2  foreach dart d' ∈ c_i(d) do
3      phi[i][d'] ← d';
4      for j ← i + 1 to n do
5          phi[j][d'] ← createDartNGMap(gm);
6  foreach dart d' ∈ c_i(d) do
7      for j ← i + 1 to n do
8          for k ← 0 to i − 1 do
9              phi[j][d'].Alphas[k] ← phi[j][d'.Alphas[k]];
10         for k ← i to j − 1 do
11             phi[j][d'].Alphas[k] ← phi[j][d'.Alphas[k + 1]];
12         phi[j][d'].Alphas[j] ← phi[j − 1][d'];
13         if j < n then
14             phi[j][d'].Alphas[j + 1] ← phi[j + 1][d'];
15         for k ← j + 2 to n do
16             phi[j][d'].Alphas[k] ← phi[j][d'.Alphas[k]];
17 foreach dart d' ∈ c_i(d) do
18     d'.Alphas[i + 1] ← phi[i + 1][d'];
```

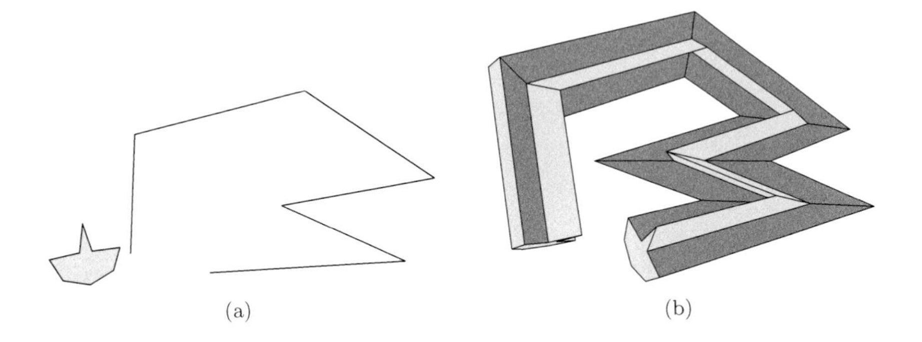

(a) (b)

FIGURE 6.47
Example of extrusion operation.
(a) A 2D object (a face) and a path.
(b) The result of the extrusion of the face along the path.

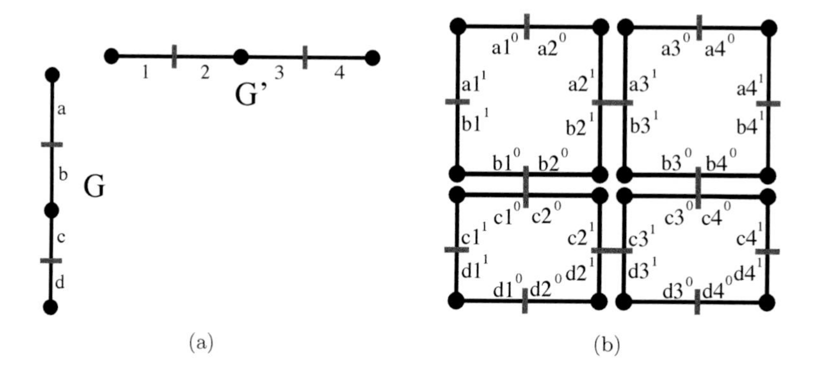

FIGURE 6.48
Extrusion for 1-Gmap.
(a) 1-Gmap G describing the object, and 1-Gmap G' describing the path.
(b) 2-Gmap $G_{\text{extr}}(G')$.

6.7 Extrusion

The *extrusion* operation takes as parameters an nD object and a path (i.e. a curve); it builds an $(n+1)$D object by "moving" the nD object along the path. In the example presented in Fig. 6.47, a face is extruded along a path, producing the volume corresponding to the trace of the face moved along the path.

More formally, the operation takes as input an n-Gmap G, which describes the object, and a 1-Gmap G', which describes the path. An $(n+1)$-Gmap $G_{\text{extr}}(G')$ is built, corresponding to the extrusion of G along G'.

Each n-cell of G is extruded as many times as the number of edges in the path. Thus the number of $(n+1)$-cells produced by the extrusion operation is the number of n-cells of G times the number of edges of G'. Similarly, $n+1$ darts of $G_{\text{extr}}(G')$ are issued from one dart of G and one dart of G'. The extrusion operation mainly consists in linking these darts according to α involutions of G and G'.

Look at Fig. 6.48: two darts dd'^0 and dd'^1 are associated with each pair of darts (d, d'), such that d (resp. d') belongs to G (resp. G'). These darts are linked together according to the following rules:

- $\alpha''_0(dd'^0) = d(\alpha'_0(d'))^0$. For example, $\alpha''_0(a1^0) = a2^0$;
- $\alpha''_0(dd'^1) = (\alpha_0(d))d'^1$. For example, $\alpha''_0(a1^1) = b1^1$;
- $\alpha''_1(dd'^0) = dd'^1$ (and reciprocally). For example, $\alpha''_1(a1^0) = a1^1$;
- $\alpha''_2(dd'^0) = (\alpha_1(d))d'^0$. For example, $\alpha''_2(b1^0) = c1^0$;

- $\alpha_2''(dd'^1) = d(\alpha_1'(d'))^1$. For example, $\alpha_2''(b2^1) = b3^1$.

In Fig. 6.49, a 2-Gmap is extruded along a path. Three darts in $G_{\text{extr}}(G')$, dd'^0, dd'^1 and dd'^2, correspond to each pair (d, d'), such that d (resp. d') is a dart of G (resp. G'). The rules for linking the darts are similar to the 1D case:

- $\alpha_0''(dd'^0) = d(\alpha_0'(d'))^0$. For example, $\alpha_0''(a1^0) = a2^0$;
- $\alpha_0''(dd'^1) = (\alpha_0(d))d'^1$. For example, $\alpha_0''(a1^1) = b1^1$;
- $\alpha_0''(dd'^2) = (\alpha_0(d))d'^2$. For example, $\alpha_0''(a1^2) = b1^2$;
- $\alpha_1''(dd'^0) = dd'^1$ (and reciprocally). For example, $\alpha_1''(a1^0) = a1^1$;
- $\alpha_1''(dd'^2) = (\alpha_1(d))d'^2$. For example, $\alpha_1''(a1^2) = h1^2$;
- $\alpha_2''(dd'^0) = (\alpha_1(d))d'^0$. For example, $\alpha_2''(a1^0) = h1^0$;
- $\alpha_2''(dd'^1) = dd'^2$ (and reciprocally). For example, $\alpha_2''(a1^1) = a1^2$;
- $\alpha_3''(dd'^0) = (\alpha_2(d))d'^0$. For example, $\alpha_3''(h1^0) = i1^0$;
- $\alpha_3''(dd'^1) = (\alpha_2(d))d'^1$. For example, $\alpha_3''(h1^1) = i1^1$;
- $\alpha_3''(dd'^2) = d(\alpha_1'(d'))^2$. For example, $\alpha_3''(a2^2) = a3^2$.

These rules can be generalized for dimension n, and the extrusion operation is defined in Def. 73. $n+1$ bijections φ^i are defined in order to associate the different darts issued from the same pair of original darts (i.e. $\varphi^i(dd') = dd'^i$). The first three items in the definition define α'' for the darts of D^0, the three next items define α'' for the darts of D^n, and the last four items define α'' for the darts of $D^1, \ldots, D^{n-1}$ (note that the definitions for D^0 and D^n are simply particular cases of the general definitions).

Definition 73 (extrusion) *Let $G = (D, \alpha_0, \ldots, \alpha_n)$ be an n-Gmap and $G' = (D', \alpha_0', \alpha_1')$ be a 1-Gmap. The $(n+1)$-Gmap resulting from the extrusion of G along G' is $G_{extr}(G') = (D'', \alpha_0'', \ldots, \alpha_{n+1}'')$, defined by:*

- *$\forall j \in \{0, \ldots, n\}$, let D^j be a set of new darts, such that a bijection φ^j maps $D \times D'$ onto D^j;*
- *$D'' = \cup_{j=0}^{n} D^j$;*
- *$\forall d \in D$, $\forall d' \in D'$:*
    1. *$\alpha_0''(\varphi^0(d, d')) = \varphi^0(d, \alpha_0'(d'))$;*
    2. *$\alpha_1''(\varphi^0(d, d')) = \varphi^1(d, d')$;*
    3. *$\forall i \in \{2, \ldots, n+1\}$, $\alpha_i''(\varphi^0(d, d')) = \varphi^0(\alpha_{i-1}(d), d')$;*
    4. *$\forall i \in \{0, \ldots, n-1\}$, $\alpha_i''(\varphi^n(d, d')) = \varphi^n(\alpha_i(d), d')$;*

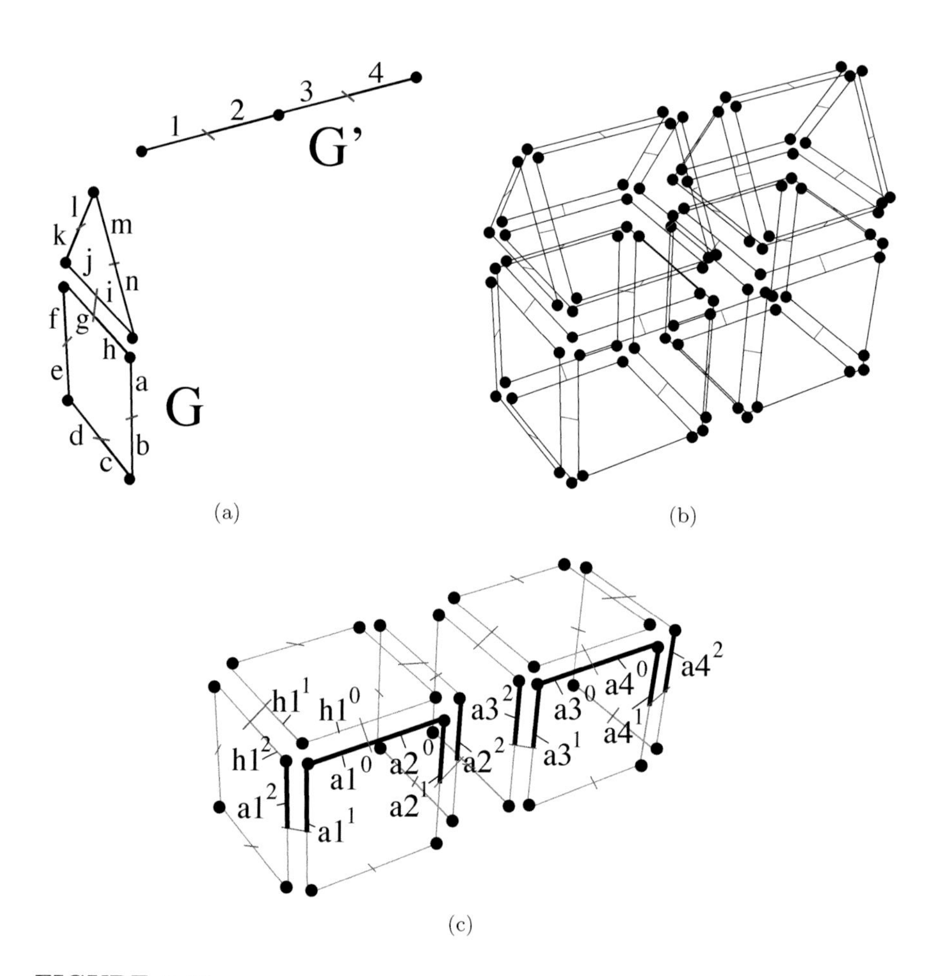

FIGURE 6.49
Extrusion for 2-Gmap.
(a) 2-Gmap G describing the object to extrude along the path G'.
(b) 3-Gmap $G_{\text{extr}}(G')$.
(c) Zoom on the resulting 3-Gmap, showing the darts issued from dart a.

5. $\alpha''_n(\varphi^n(d,d')) = \varphi^{n-1}(d,d')$;
6. $\alpha''_{n+1}(\varphi^n(d,d')) = \varphi^n(d,\alpha'_1(d'))$.

- $\forall j \in \{1,\ldots,n-1\}$, $\forall d \in D$, $\forall d' \in D'$:

7. $\forall i \in \{0,\ldots,j-1\}$, $\alpha''_i(\varphi^j(d,d')) = \varphi^j(\alpha_i(d),d')$;
8. $\alpha''_j(\varphi^j(d,d')) = \varphi^{j-1}(d,d')$;
9. $\alpha''_{j+1}(\varphi^j(d,d')) = \varphi^{j+1}(d,d')$;
10. $\forall i \in \{j+2,\ldots,n+1\}$, $\alpha''_i(\varphi^j(d,d')) = \varphi^j(\alpha_{i-1}(d),d')$.

The extrusion operation is implemented by Algorithm 65. Contrary to the previous operations, the n-Gmap `gm` is not modified. A new $(n+1)$-Gmap is built, which is the result of the extrusion of `gm` along the path defined by `gm`$'$. Indeed, it is not possible to modify `gm`, since the dimension of the resulting Gmap is $n+1$. The algorithm follows directly the definition. First, the darts of `gm`$'$ are created, and the bijections φ^i are stored in $n+1$ associative arrays. Second, all pairs of darts (d,d') are processed, in order to define all α involutions.

Note that contrary to the chamfering algorithm, no side effects can occur, since the given n-Gmap is not modified.

The complexity of Algorithm 65 is linear in the number of darts $\#d_1$ of the n-Gmap `gm` times the number of darts $\#d_2$ of the 1-Gmap `gm`$'$ times $\log(\#d_1 + \#d_2)$.

Note that several operations can be deduced from the extrusion operation. For instance, let G be a connected n-Gmap and Q be the associated quasi-manifold. The *cone* operation, applied to G, creates a $(n+1)$-Gmap, denoted G_{cone}, such that its associated quasi-manifold is the cone of Q (cf. Section 8.1).

Definition 74 (cone) *Let $G = (D,\alpha_0,\ldots,\alpha_n)$ be an n-Gmap. The $(n+1)$-Gmap resulting from the cone of G is $G_{cone} = (D'',\alpha''_0,\ldots,\alpha''_{n+1})$, defined by:*

- *$\forall j \in \{0,\ldots,n\}$, let D^j be a set of new darts, such that a bijection φ^j maps D onto D^j; let D' be a set of new darts, such that a bijection φ' maps D onto D';*

- *$D'' = D' \cup_{j=0}^{n} D^j$;*

- *$\forall d \in D$:*

1. $\alpha''_0(\varphi^0(d)) = \varphi'(d)$;
2. $\alpha''_1(\varphi^0(d)) = \varphi^1(d)$;
3. $\forall i \in \{2,\ldots,n+1\}$, $\alpha''_i(\varphi^0(d)) = \varphi^0(\alpha_{i-1}(d))$;
4. $\forall i \in \{0,\ldots,n-1\}$, $\alpha''_i(\varphi^n(d)) = \varphi^n(\alpha_i(d))$;

Algorithm 65: `extrudeNGMap(gm,gm')`: extrude an n-Gmap along a path

```
   Input: gm: an n-Gmap;
          gm': a 1-Gmap.
   Output: The (n + 1)-Gmap result of the extrusion of gm along gm'.
1  phi ← an array of n + 1 empty associative arrays from pairs of darts to
   darts;
2  gm'' ← createNGMap(n + 1);
3  foreach dart d ∈ gm.Darts do
4      foreach dart d' ∈ gm'.Darts do
5          for i ← 0 to n do
6              phi[i][(d, d')] ← createDartNGMap(gm'');
7  foreach dart d ∈ gm.Darts do
8      foreach dart d' ∈ gm'.Darts do
9          phi[0][(d, d')].Alphas[0] ← phi[0][(d, d'.Alphas[0])];
10         phi[0][(d, d')].Alphas[1] ← phi[1][(d, d')];
11         for i ← 2 to n + 1 do
12             phi[0][(d, d')].Alphas[i] ← phi[0][(d.Alphas[i − 1], d')];
13         for i ← 0 to n − 1 do
14             phi[n][(d, d')].Alphas[i] ← phi[n][(d.Alphas[i], d')];
15         phi[n][(d, d')].Alphas[n] ← phi[n − 1][(d, d')];
16         phi[n][(d, d')].Alphas[n + 1] ← phi[n][(d, d'.Alphas[1])];
17         for j ← 1 to n − 1 do
18             for i ← 0 to j − 1 do
19                 phi[j][(d, d')].Alphas[i] ← phi[j][(d.Alphas[i], d')];
20             phi[j][(d, d')].Alphas[j] ← phi[j − 1][(d, d')];
21             phi[j][(d, d')].Alphas[j + 1] ← phi[j + 1][(d, d')];
22             for i ← j + 2 to n + 1 do
23                 phi[j][(d, d')].Alphas[i] ← phi[j][(d.Alphas[i − 1], d')];
```

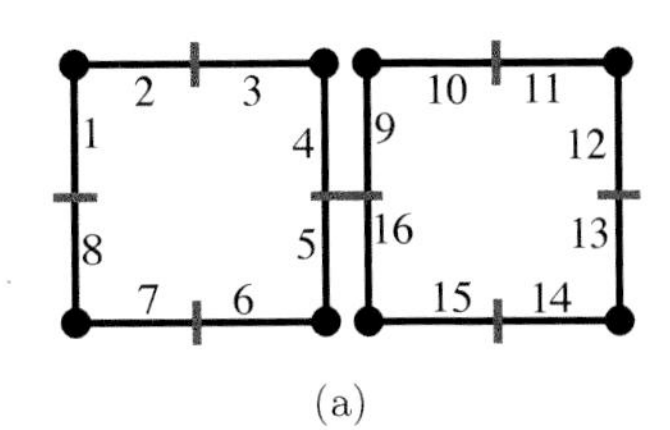

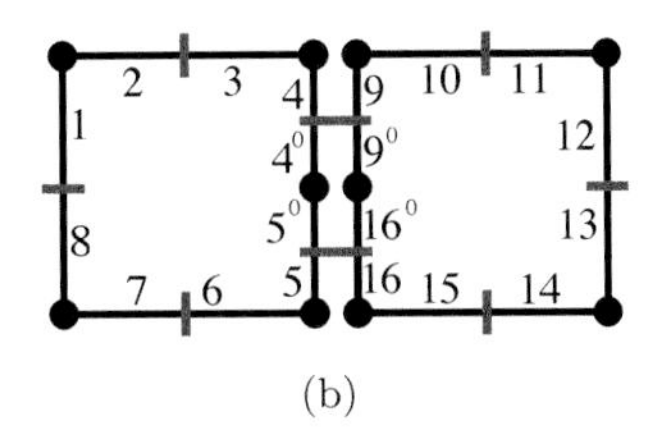

FIGURE 6.50
1-triangulation for 2-Gmap.
(a) 2-Gmap G.
(b) 2-Gmap $G_{\text{1-triangulation}}(\{4,5,9,16\})$.

5. $\alpha''_n(\varphi^n(d)) = \varphi^{n-1}(d)$;
6. $\alpha''_{n+1}(\varphi^n(d)) = \varphi^n(d)$.

- $\forall j \in \{1,\ldots,n-1\}$, $\forall d \in D$:

7. $\forall i \in \{0,\ldots,j-1\}$, $\alpha''_i(\varphi^j(d)) = \varphi^j(\alpha_i(d))$;
8. $\alpha''_j(\varphi^j(d)) = \varphi^{j-1}(d)$;
9. $\alpha''_{j+1}(\varphi^j(d)) = \varphi^{j+1}(d)$;
10. $\forall i \in \{j+2,\ldots,n+1\}$, $\alpha''_i(\varphi^j(d)) = \varphi^j(\alpha_{i-1}(d))$;

- $\forall d \in D$:

11. $\alpha''_0(\varphi'(d)) = \varphi^0(d)$;
12. $\forall i \in \{1,\ldots,n+1\}$, $\alpha''_i(\varphi'(d)) = \varphi'(\alpha_{i-1}(d))$.

The cone operation can be generalized in order to create cones on the cells of an n-Gmap: this corresponds to the *triangulation*, described in the next section.

6.8 Triangulation

The last operation presented in this chapter is the *triangulation*. This operation consists in splitting a given i-cell by inserting a vertex inside it.

As illustrated in Fig. 6.50, the 1-triangulation operation consists in adding a vertex inside an edge. For each dart d of the edge, a new dart d^0 is created: $\alpha'_0(d) = d^0$, $\alpha'_1(d^0) = (\alpha_0(d))^0$, and $\alpha'_2(d^0) = (\alpha_2(d))^0$ (and reciprocally). Other α links are not modified.

The 2-triangulation operation consists in adding a vertex inside a face. The

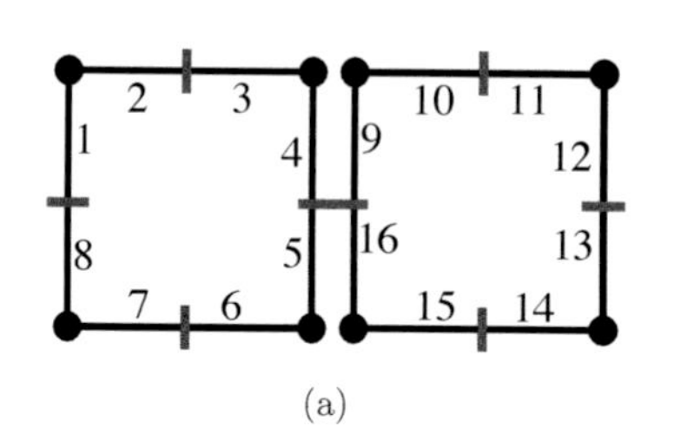

(a)

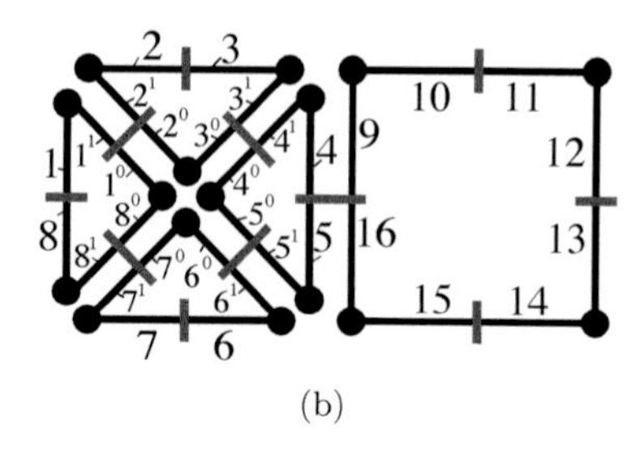

(b)

FIGURE 6.51
2-triangulation for 2-Gmap.
(a) 2-Gmap G.
(b) 2-Gmap $G_{2\text{-triangulation}}(\{1, 2, 3, 4, 5, 6, 7, 8\})$.

initial face is split into k faces, one for each of its k incident edges: cf. Fig. 6.51. Two darts d^0 and d^1 are created for each dart d of the face: $\alpha'_0(d^0) = d^1$, $\alpha'_1(d) = d^1$, $\alpha'_1(d^0) = (\alpha_0(d))^0$, $\alpha'_2(d^0) = (\alpha_1(d))^0$, and $\alpha'_2(d^1) = (\alpha_1(d))^1$ (and reciprocally). Other α links are not modified.

The operation is similar in higher dimension. Fig. 6.52(a) represents the result of the 3-triangulation of the 3-cell of a 3-Gmap representing a cube. A zoom on one volume is represented in Fig. 6.52(b): three new darts d^0, d^1 and d^2 are created for each dart d of the volume: $\alpha'_0(d^0) = d^1$, $\alpha'_0(d^2) = (\alpha_0(d))^2$, $\alpha'_1(d^0) = (\alpha_0(d))^0$, $\alpha'_1(d^1) = d^2$, $\alpha'_2(d) = d^2$, $\alpha'_2(d^0) = (\alpha_1(d))^0$, $\alpha'_2(d^1) = (\alpha_1(d))^1$, $\alpha'_3(d^0) = (\alpha_2(d))^0$, $\alpha'_3(d^1) = (\alpha_2(d))^1$, and $\alpha'_3(d^2) = (\alpha_2(d))^2$ (and reciprocally). Other α links are not modified.

This process can be generalized for any dimension. The general i-triangulation operation for n-Gmaps is defined in Def. 75. This operation takes as input an n-Gmap G and one of its i-cell C. As for the two previous operations, new sets of darts D^j are created for each $j \in \{0, \dots, i-1\}$, associated with the initial darts by bijections φ^j. Item 1 states that the α' links are not modified for the darts which do not belong to C. Item 2, 3 and 4 define the α' links for the darts of D^0. The next 2 items define the α' links for the darts in C. They are not modified, except for α'_{i-1}. The last five items give the definitions of α' links for the darts of D^k, $\forall k \in \{1, \dots, i-1\}$.

Definition 75 (triangulation) *Let $G = (D, \alpha_0, \dots, \alpha_n)$ be an n-Gmap, and C be one of its i-cells, with $1 \leq i \leq n$. The n-Gmap resulting from the i-triangulation of C in G is $G_{i\text{-triangulation}}(C) = (D', \alpha'_0, \dots, \alpha'_n)$, defined by:*

- *$\forall j \in \{0, \dots, i-1\}$, let D^j be a set of new darts, such that a bijection φ^j maps C onto D^j. Let φ^i be the identity on C;*
- *$D' = D \cup_{j=0}^{i-1} D^j$;*
- *1. $\forall d \in D \setminus C$, $\forall j \in \{0, \dots, n\}$, $\alpha'_j(d) = \alpha_i(d)$.*

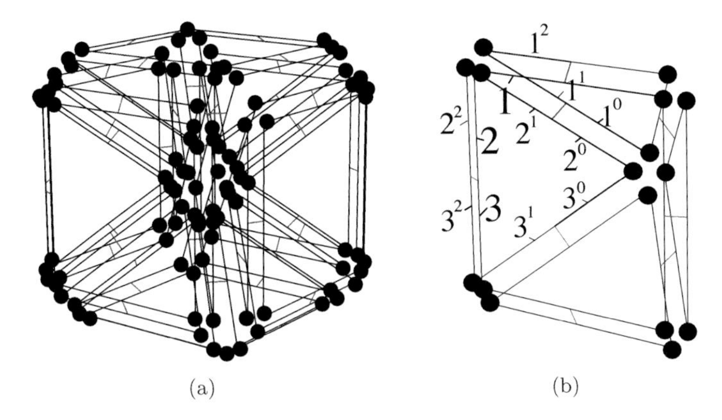

FIGURE 6.52
3-triangulation for 3-Gmap.
(a) The result of the 3-triangulation of the 3-cell in a 3-Gmap representing a cube.
(b) Zoom on one volume issued from the 3-triangulation.

- $\forall d \in C$:

  2. $\alpha'_0(\varphi^0(d)) = \varphi^1(d)$;
  3. $\forall j \in \{1, \ldots, i\}$, $\alpha'_j(\varphi^0(d)) = \varphi^0(\alpha_{j-1}(d))$;
  4. $\forall j \in \{i+1, \ldots, n\}$, $\alpha'_j(\varphi^0(d)) = \varphi^0(\alpha_j(d))$;
  5. $\forall j \in \{0, \ldots, n\}$, $j \neq i-1$, $\alpha'_j(d) = \alpha_j(d)$;
  6. $\alpha'_{i-1}(d) = \varphi^{i-1}(d)$;

- $\forall k \in \{1, \ldots, i-1\}$, $\forall d \in C$:

  7. $\forall j \in \{0, \ldots, k-2\}$, $\alpha'_j(\varphi^k(d)) = \varphi^k(\alpha_j(d))$;
  8. $\alpha'_{k-1}(\varphi^k(d)) = \varphi^{k-1}(d)$;
  9. $\alpha'_k(\varphi^k(d)) = \varphi^{k+1}(d)$;
  10. $\forall j \in \{k+1, \ldots, i\}$, $\alpha'_j(\varphi^k(d)) = \varphi^k(\alpha_{j-1}(d))$;
  11. $\forall j \in \{i+1, \ldots, n\}$, $\alpha'_j(\varphi^k(d)) = \varphi^k(\alpha_j(d))$.

This operation is implemented by Algorithm 66. The n-Gmap `gm` is modified in order to obtain the result of the i-triangulation operation. The operation is similar to the two previous operations: new darts are created and stored in `phi` associative arrays (there is one associative array `phi[j]` for each `j` $\in \{0, \ldots,$ `i`$\}$). α involutions are modified for these darts, according to the definition. The algorithm returns a dart belonging to the new vertex, since this dart could be used after the operation.

Algorithm 66: `triangulationNGMap(gm,d,i)`: triangulation of an `i`-cell for n-Gmaps

```
   Input: gm: an n-Gmap;
          d ∈ gm.Darts: a dart;
          i ∈ {1, ..., n}.
   Result: Make a triangulation of the i-cell c_i(d).
   Output: A dart of the new vertex.
 1 phi ← an array of i + 1 empty associative arrays between darts;
 2 foreach dart d' ∈ c_i(d) do
 3     for j ← 0 to i − 1 do
 4         phi[j][d'] ← createDartNGMap(gm);
 5     phi[i][d'] ← d';
 6 foreach dart d' ∈ c_i(d) do
 7     phi[0][d'].Alphas[0] ← phi[1][d'];
 8     for j ← 1 to i do
 9         phi[0][d'].Alphas[j] ← phi[0][d'.Alphas[j − 1]];
10     for j ← i + 1 to n do
11         phi[0][d'].Alphas[j] ← phi[0][d'.Alphas[j]];
12     for k ← 1 to i − 1 do
13         for j ← 0 to k − 2 do
14             phi[k][d'].Alphas[j] ← phi[k][d'.Alphas[j]];
15         phi[k][d'].Alphas[k − 1] ← phi[k − 1][d'];
16         phi[k][d'].Alphas[k] ← phi[k + 1][d'];
17         for j ← k + 1 to i do
18             phi[k][d'].Alphas[j] ← phi[k][d'.Alphas[j − 1]];
19         for j ← i + 1 to n do
20             phi[k][d'].Alphas[j] ← phi[k][d'.Alphas[j]];
21 foreach dart d' ∈ c_i(d) do
22     d'.Alphas[i − 1] ← phi[i − 1][d'];
23 return phi[0][d];
```

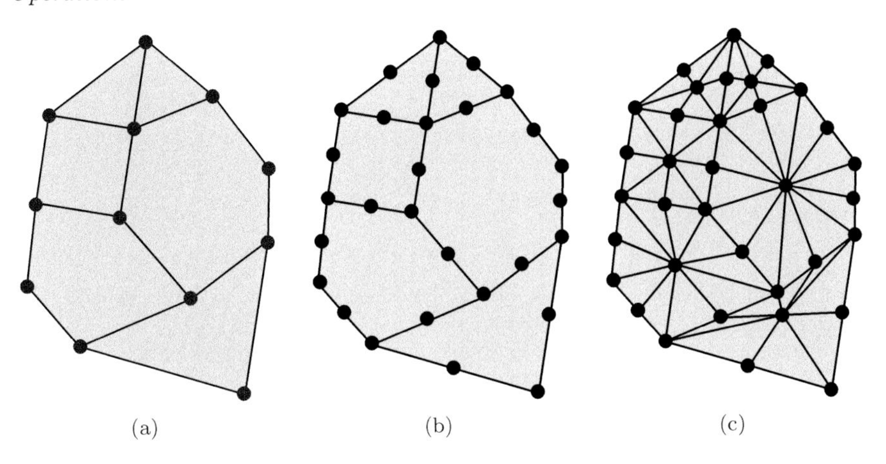

FIGURE 6.53
(a) A 2D object.
(b) The result of the 1-triangulation operation applied to the edges of (a).
(c) The result of the 2-triangulation operation applied to the faces of (b). This is the barycentric triangulation of the initial object.

As for the chamfering operation, the α_{i-1} links are modified for the darts in $c_{\mathtt{i}}(\mathtt{d})$ (line 22). To avoid side effects, this modification is done in a second loop, after having defined the α links for the new darts. Moreover, the darts of $c_{\mathtt{i}}(\mathtt{d})$ are stored in a data structure during the first loop (line 2); the iteration of the second loop is done through the data structure, and not through a cell iterator (line 21). At last, the algorithm returns one dart belonging to the new vertex, as it can be used in other algorithms (as we will see in Algorithm 74 page 265).

The complexity of Algorithm 66 is linear in the number of darts $\#d$ of the given i-cell times $\log \#d$.

The *barycentric triangulation* of a given n-Gmap can be defined by applying the i-triangulation operation to the i-cells of the n-Gmap, from $i = 1$ to $i = n$. For example, look at the 2D object represented in Fig. 6.53(a). It contains 12 vertices, 17 edges and 6 faces. The 1-triangulation operation is applied to all its edges, producing the 2D object represented in Fig. 6.53(b). This object contains 29 vertices, 34 edges and 6 faces. Then, the 2-triangulation operation is applied to all its faces, producing the 2D object represented in Fig. 6.53(c). This object contains 35 vertices, 84 edges and 50 faces. This is the barycentric triangulation of the initial object: a vertex corresponds to each cell of the initial object.

7

Embedding for Geometric Modeling and Image Processing

In the previous chapters, all the main notions about n-Gmaps and n-maps have been defined, and also related data structures and algorithms. However, only the combinatorial aspects have been studied corresponding to the structure (or topology) of the associated subdivisions. But the combinatorial description is not sufficient for many applications. It is often needed to associate different information to cells. For example a material (resp. a color, a length) can be associated with each volume (resp. each face, each edge) of a 3D object.

Most applications require to describe the geometry of the subdivided object. This is done through a specific type of information which is called an embedding[1]. For instance a point in $\mathbb{R}^3$ can be associated with each vertex of a 2D subdivided object, or a B-spline curve with each edge... Different types of embeddings are introduced in Section 7.1. We show how they can be associated with n-Gmaps and n-maps, and discuss the corresponding modifications for the operations. Then an example of conception of a 3D geometrical modeler based on 3-Gmaps is detailled in Section 7.2 and an example of conception of a 2D image processing framework is given in Section 7.3. These two applications illustrate the different interests of using combinatorial maps: they can be used in different applications, with different dimensions and different types of embeddings.

7.1 Embedding

Generally speaking, an *embedding* of an n-map M into a space E is an application f from the sets of cells of M onto E which "preserves the structure" of M: that means that each i-cell of M is associated with an i-dimensional part of E (this part of E is called the embedding of the i-cell) so that any two incident cells of M are associated with incident parts of E. Intuitively, the

[1]This denomination comes from the fact that this information makes it possible to embed the structure of the subdivision into a geometric space, providing a shape to this structure.

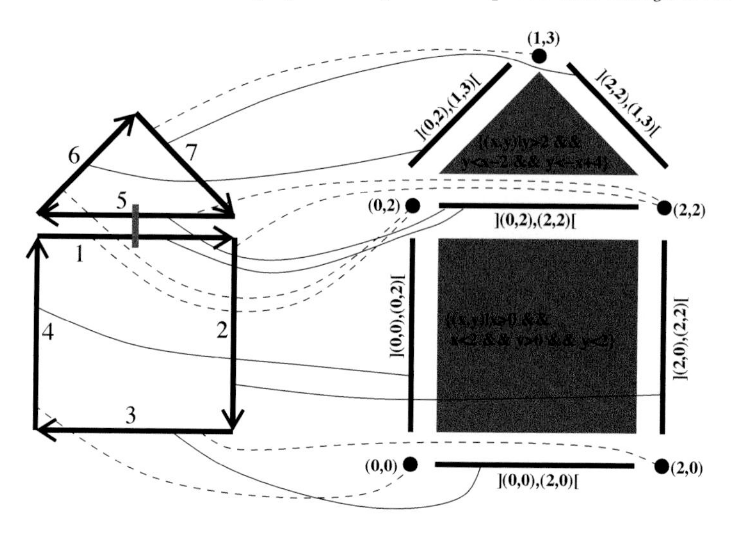

FIGURE 7.1
Example of embedding of a 2-map in $\mathbb{R}^2$. Embedding of vertices is represented by dashed lines, embedding of edges is represented by lines and embedding of faces is not represented.

embedding of M has the same structure than M. Note that the mechanisms explained here are valid for both n-maps and n-Gmaps.

A classical way consists in embedding an n-map in the Euclidean space $\mathbb{R}^d$. Note that the dimension n of the n-map is not necessarily equal to d, the dimension of the Euclidean space. n will be referred to as the *combinatorial dimension*, and d will be referred to as the *ambient dimension*.

A 2-map embedded into $\mathbb{R}^2$ is represented in Fig. 7.1. Each vertex of the 2-map is associated with a point in $\mathbb{R}^2$ (for example vertex $\{1, 6\}$ is associated with point $(0, 2)$ and vertex $\{7\}$ is associated with point $(1, 3)$); each edge is associated with an open segment (for example edge $\{6\}$ is associated with segment $](0, 2), (1, 3)[$ and edge $\{1, 5\}$ is associated with segment $](0, 2), (2, 2)[$); each face is associated with an open surface area (face $\{5, 6, 7\}$ is associated with the surface area defined by $\{(x, y) \in \mathbb{R}^2 | y > 2$ and $y < x - 2$ and $y < -x + 4\}$ and face $\{1, 2, 3, 4\}$ is associated with the surface area defined by $\{(x, y) \in \mathbb{R}^2 | x > 0$ and $x < 2$ and $y > 0$ and $y < 2\}$).

We can verify that this example is a correct embedding by checking that two incident cells in the 2-map are associated with two incident elements of the plane and reciprocally. For example vertex $\{1, 6\}$ associated with point $(0, 2)$ is incident to edge $\{1, 5\}$ associated with segment $](0, 2), (2, 2)[$ and point $(0, 2)$ belongs to the boundary of the segment.

This 2-map is "linearly embedded" into $\mathbb{R}^2$: vertices are associated with points, edges are associated with segments and faces are associated with poly-

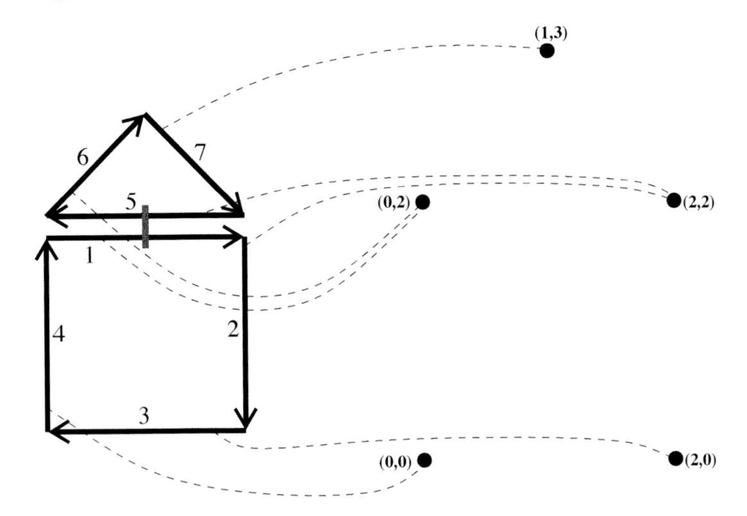

FIGURE 7.2
Embedding a 2-map in $\mathbb{R}^2$: edges and faces are implicitely embedded.

gons. More generally, a linear embedding consists in associating a part of an iD hyperplane with each i-cell. This type of embedding is often used because it is simple to handle. However many other types of embedding exist. For example splines can be used to embed edges or nurbs surfaces to embed faces. However even in these cases, the embedding still consists in associating iD geometrical elements to i-cells, and constraints related to the incidence between the cells still have to be satisfied. The choice of a given embedding will depend on the specific needs of the applications.

Note that it can be avoided to explicitly describe all the geometrical objects, when some of them can be computed using the embedding of other cells and the incidence relations in n-maps. This is for example the case for the linear embedding presented above, explained here for $n = 2$ in order to simplify. In this case, the embedding of any edge is the open segment which extremities are the two points associated with the two vertices incident to the edge. Thus it is not necessary to explicitly associate open segments with edges as they can be retrieved from the 0-embedding. Similarly, the embedding of any face is a part of a plane which boundary contains all segments associated with the incident edges. Thus it is also not necessary to explicitly describe this 2-embedding as it can be retrieved thanks to the 1-embeddings. So it is possible to only associate points with vertices in order to linearly embed a 2-map (cf. Fig. 7.2). This is usually the way linear embeddings are defined, for any n. Note that consistency constraints still have to be satisfied: for instance, the points associated with two vertices incident to the same edge have to define a line segment, i.e. they must be different; the points associated with the vertices incident to a face have to be coplanar, otherwise the shape of the face is not defined; etc.

Conversely, a *dual* approach can be followed, explained here for linear embedding and $n = 2$ in order to simplify; but in fact, it is usually applied for nonlinear embedding (e.g. *trimmed* surfaces: cf [2]). Associate a plane of $\mathbb{R}^3$ with any face. The plane is the *support space* of the face. The surfaces associated with two adjacent faces which share an edge intersect along a line: this line is the support space of the common edge. All lines associated with the edges which share a vertex intersect on a point, which is the embedding of the vertex. The line segment associated with an edge is defined by its support line, delimited by the points associated with its incident vertices. Similarly, the part of plane associated with a face can be retrieved from its support plane, and the lines associated with its incident edges. In fact, all embedding information can be retrieved from the support spaces associated with the faces, since the embedding of other cells can be deduced. Here also, consistency constraints still have to be satisfied: for instance, the planes associated with two edges which share an edge have to be different. This approach can be generalized for higher dimensions, and also for nonlinear embedding. For instance, support spaces associated with faces are often free-form surfaces (e.g. nurbs).

The main interests of implicit embedding are:

1. decrease the memory size: there are less geometrical elements to store and less links to describe;
2. simplify the operations: only one type of embedding is handled;
3. increase the guaranty of validity: some validity constraints of some cell embedding can be implied by the use of the embedding of incident cells.

As a main drawback, there is an additional cost since it can be necessary to explicitly compute (information about) the different implicit embeddings; but generally this cost is small.

There are several possibilities for associating an embedding with an i-cell. However as cells are set of darts, this association is always done through darts. For example a pointer can be added to each dart, that points to the embedding of the i-cell containing the dart. This solution allows a direct access to all embeddings. However it requires to update the pointers of all darts of the i-cell to initialize its embedding. Moreover, there is one pointer for each dart and for each dimension i having an embedding. It is also possible to create an associative array that links each dart of an i-cell with its corresponding embedding. This allows to add or remove a certain embedding without modifying the dart data structure. Moreover if a hash-table is used, the access to the embedding of a dart can be done in constant time in average. As a last example, it is possible to link only one dart per i-cell with the corresponding i-embedding. Some memory space is saved since describing less relations are described; moreover this can speed up some operations modifying the links between darts and embedding, since there is only one dart to modify. However, this slows down the access to an embedding for a given dart, as it is necessary to iterate through all the darts of its i-cell to find the dart linked with the embedding.

7.2 Geometric Modeling

n-Gmaps and n-maps have been used in geometric modeling to construct and modify geometric objects. This section focuses on the conception of MOKA [211], a 3D geometrical modeler based on 3-Gmaps. The choice of 3-Gmaps was taken in order to simplify the implementation of data structures and operations, thanks to their homogeneity. Many implementation choices are dimension independant. First, we explain the linear embedding of n-Gmaps in $\mathbb{R}^d$ in Section 7.2.1. Then geometric operations are described in Section 7.2.2. At last, an example of object construction is provided in Section 7.2.3.

7.2.1 Embedding of n-Gmaps in $\mathbb{R}^d$

As explained above, an n-Gmap is linearly embedded into $\mathbb{R}^d$ by associating a point in $\mathbb{R}^d$ with each vertex of the n-Gmap. The other cells are implicitly embedded. For representing this association between a vertex and a point, we chose to link *one* dart of each vertex with its corresponding point. This choice was done in order to favor the modification operations, since only one association between a dart and its embedding need to be modified in order to modify the embedding. This solution is made to the detriment of the access operation to the embedding, since all the darts belonging to the vertex need to be traversed to find the one having the link. Note that the choice of the dart linked with the point has no importance: any dart of the same vertex can be chosen.

Such an embedding for a 3-Gmap describing a cube adjacent to a pyramid is illustrated in Fig. 7.3(a). Nine points are associated with the nine vertices of the 3-Gmap, and one dart of each vertex is linked with its associated point. Figure 7.3(b) shows a snapshot of this object represented in MOKA.

In order to represent this embedding, the `Dart` data structure (introduced in Section 4.4 page 108) is modified by adding a pointer (called `Point`) to a structure describing the associated point. A point is represented by an array of *d double* corresponding to the coordinates (see structure `GeometricalPoint`). The `Point` pointer is set to `NULL` for darts not linked with a point (some advanced programming technique make it possible to avoid to represent this pointer for darts not linked with a point, by using two different classes of darts).

Listing 7.1

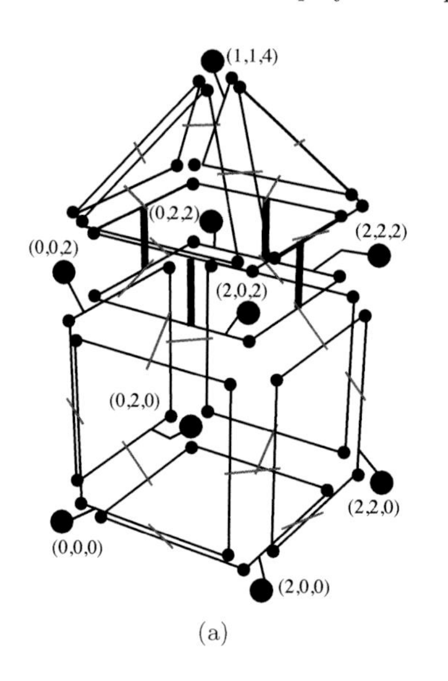

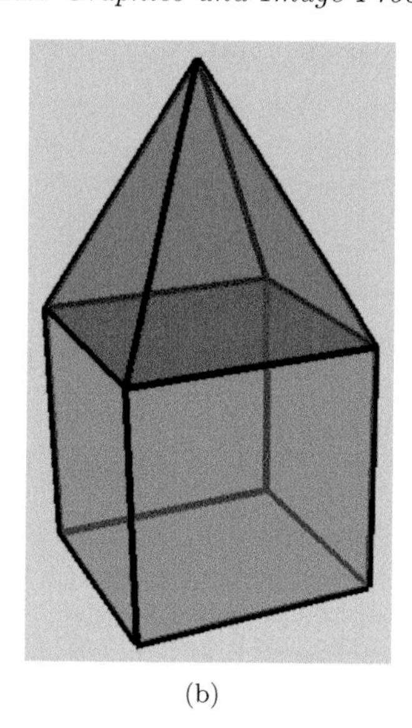

FIGURE 7.3
(a) Embedding of a 3-Gmap in $\mathbb{R}^3$.
(b) Snapshot of this object described in MOKA.

dD point data structure

structure *GeometricalPoint*
{ *double* coord[d] ; } ;

Basic methods allow to manage geometrical points associated with n-Gmaps. Retrieve the point associated with a given vertex is described in Algorithm 67. The darts of the vertex are checked: if the current dart has a non `NULL` `Point` pointer, the associated point is returned. If all the darts are linked with `NULL`, the vertex has no associated point. This case is not supposed to occur when the n-Gmap is correctly embedded, but it can happen during the construction of new objects.

The complexity of Algorithm 67 is linear in the number of darts in c_0(`d`). Note that half of the darts are checked in average, because the loop ends as soon as a dart associated with a `Point` is found.

Algorithm 68 implements the method that sets the point associated with a given vertex to a given value `p`. First it is checked if a point is already associated with the vertex (line 1). When this is not the case, a new point is created and associated with dart `d` (lines 3 and 4). Lastly the value of the point is set to `p` (line 5).

The complexity of Algorithm 68 is also linear in the number of darts in c_0(`d`), due to the first search of an existing point.

Algorithm 69 describes the method for deleting a point associated with a

---

Algorithm 67: `getPoint(d)`: get the geometrical point associated with a vertex

---

Input: `gm`: an embedded n-Gmap;
`d` $\in$ `gm.Darts`: a dart.
Output: The geometrical point associated with c_0(`d`).

```
1 foreach dart e ∈ c0(d) do
2 |   if e.Point ≠ NULL then
3 |   |   return e.Point;
4 return NULL;
```

Algorithm 68: `setPoint(d,p)`: set the geometrical point associated with a vertex

---

Input: `gm`: an embedded n-Gmap;
`d` $\in$ `gm.Darts`: a dart;
`p`: a geometrical point.
Result: Set the geometrical point associated with c_0(`d`) to `p`.

```
1 p' ← getPoint(d);
2 if p' = NULL then
3 |   p' ← create new GeometricalPoint;
4 |   d.Point ←p';
5 set the point in p' to p;
```

---

given vertex. The complexity of this algorithm is also linear in the number of darts in $c_0(\mathtt{d})$.

Algorithm 69: `deletePoint(d)`: delete the geometrical point associated with a vertex

```
  Input: gm: an embedded n-Gmap;
         d ∈ gm.Darts: a dart.
  Result: Delete the geometrical point associated with c0(d).
1 foreach dart e ∈ c0(d) do
2   if e.Point ≠ NULL then
3     delete e'.Point;
4     e'.Point ← NULL;
5     return;
```

7.2.2 Geometric Operations

Several combinatorial operations for handling n-Gmaps have been presented in Section 4.4 and chapter 6 (pages 108 and 185). These operations must be modified for handling embedded n-Gmaps in order to update accordingly the embeddings. Hopefully, this can be done easily, due to the distinction between the topology of the object and its embedding. Indeed, darts and α links correspond to the topology of the object, while the set of geometrical points correspond to the embedding. As illustrated in the following, operations can deal only with the topology, only with the embedding or simultaneously with both.

Generally when a vertex is modified, the associated geometrical point must also be modified. Operations modifying vertices in an n-Gmap are: the creation of a new vertex; the destruction of an existing vertex; the merging of several vertices; the split of one vertex. The corresponding embedding operations are: the creation of a new geometrical point; the deletion of the associated geometrical point; the removing of all the geometrical points except one; the duplication of the geometrical point. Note that some combinatorial operations do not involve embedding modification. This is for example the case of the i-close operation, for $i > 0$ (see Section 6.1 page 185). Indeed, this operation does not create new vertex and there is no merge nor split of existing vertices. The new darts created by the operation are linked with existing darts, so they will belong to existing vertices. All resulting vertices corresponds to initial vertices and their embeddings are not modified.

7.2.2.1 Cube Creation

Algorithm 70 describes the topological creation of a cube in an n-Gmap ($n \geq 2$). Six square faces are created, and 2-sewn (cf. Fig. 7.4(a)). Creating a square face is straightforward: it consists in creating eight darts, two by two sewn by α_0 and α_1 (algorithm not given here). Since this operation is purely combinatorial, all the darts play the same role.

Algorithm 70: `createTopoCube(gm)`: create a topological cube in an n-Gmap

Input: gm: an n-Gmap, $n \geq 2$.
Output: A dart of the new topological cube created in gm.

```
1  for i ← 0 to 5 do
2      d[i] ← one dart of a new square face created in gm;
3  sewNGMap(gm,d[0],d[4].Alphas[0].Alphas[1].Alphas[0],2);
4  sewNGMap(gm,d[0].Alphas[0].Alphas[1],d[3].Alphas[0].Alphas[1],2);
5  sewNGMap(gm,d[0].Alphas[0].Alphas[1].Alphas[0].Alphas[1],
          d[2].Alphas[0].Alphas[1],2);
6  sewNGMap(gm,d[0].Alphas[1],d[1].Alphas[0].Alphas[1],2);
7  sewNGMap(gm,d[5],d[2].Alphas[1],2);
8  sewNGMap(gm,d[5].Alphas[0].Alphas[1],d[1].Alphas[1].Alphas[0],2);
9  sewNGMap(gm,d[5].Alphas[1].Alphas[0].Alphas[1],d[4].Alphas[1],2);
10 sewNGMap(gm,d[5].Alphas[1],d[3].Alphas[1].Alphas[0],2);
11 sewNGMap(gm,d[1].Alphas[1].Alphas[0].Alphas[1],
          d[2].Alphas[1].Alphas[0].Alphas[1],2);
12 sewNGMap(gm,d[1],d[4].Alphas[1].Alphas[0].Alphas[1],2);
13 sewNGMap(gm,d[2],d[3].Alphas[1].Alphas[0].Alphas[1],2);
14 sewNGMap(gm,d[3],d[4],2);
15 return d[0];
```

A geometrical layer can easily be added in order to create a cube embedded in $\mathbb{R}^d$. Since the combinatorial operation creates new vertices, the geometrical layer consists in creating new dD points. Once the topological cube created, dD points are associated with the corresponding darts. Algorithm 71 describes the geometrical creation of a cube in $\mathbb{R}^3$ (cf. Fig. 7.4(b)).

Note that this algorithm can be improved, since no point is associated with any dart when the topological cube is created. It is thus possible to avoid the tests done in the `setPoint` function to search for a dart in the vertex that is already associated with a geometrical point.

7.2.2.2 Geometrical Sew

A second example of geometrical operation is described in Algorithm 72: the geometrical i-sew operation. Several vertices can be merged when applying the combinatorial operation; thus, the additional geometrical layer consists in removing the geometrical points associated with these merged vertices. This removing is a preliminary step, added to Algorithm 20 page 122 which implements the combinatorial i-sew operation. Note that no vertices are identified

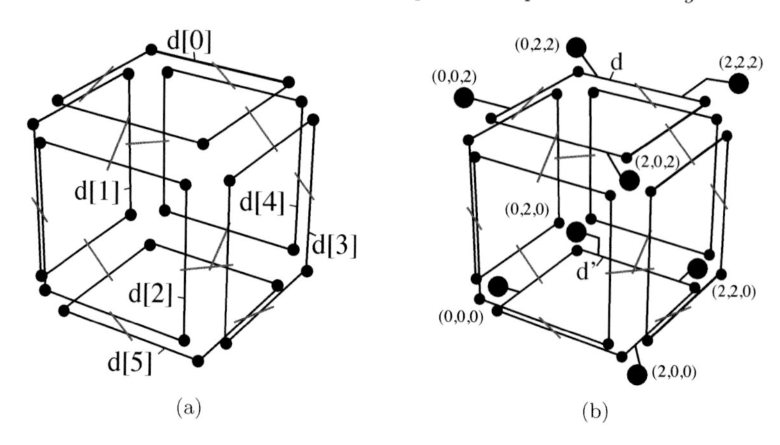

FIGURE 7.4
(a) Creation of a topological cube in an n-Gmap, $n \geq 2$.
(b) Embedding of this topological cube in $\mathbb{R}^3$.

by the 0-sew operation, and thus there is no modification of the embedding in this case.

When two vertices are merged, only one of the associated points has to be kept. The two i-cells are traversed by the iterator `it` and `it'`. If `it` and `it'` do not belong to the same vertex, the second geometrical point is removed. After all removings, the n-Gmap is not correctly embedded, because some vertices are not associated with geometrical points. However, this is corrected by the i-sew operation. Since only one dart of a vertex is linked with a geometrical point, no other modification is necessary.

The complexity of this geometrical i-sew operation is the same than the one of the topological sew, i.e. it is linear in the number of darts in the orbit $\langle \alpha_0, \dots, \alpha_{i-2}, \alpha_{i+2}, \dots, \alpha_n \rangle(\mathtt{d})$. Indeed, a mark can be used in order to test each vertex exactly once in line 5.

An example of geometrical 3-sew between two isolated volumes is described in Fig. 7.5. The two volumes are glued after the operation, thanks to the identification of the two faces containing darts d and d'. It can be verified in Fig. 7.5(b) that the embedding is correctly updated: all the 3D points linked with the vertices incident to the face containing dart d' before the sew are removed, and they are merged with the corresponding vertices of the face containing dart d. Note that the geometry of the pyramid is deformed, since the four 3D points associated to the vertices incident to the face containing dart d' are modified.

Note that in the proposed version, the operation is not symmetric: the geometrical result of 3-sew(d, d') is different from the result of 3-sew(d', d). Instead of keeping the first points, other solutions can be used to update the geometry, such as for example compute the average of the two points.

Algorithm 71: `createGeomCube(gm)`: create a cube in an n-Gmap embedded in $\mathbb{R}^3$

Input: gm: an embedded n-Gmap, $n \geq 2$.
Output: A dart of the new cube created in gm.

```
1  d ← createTopoCube(gm);
2  setPoint(d, (0, 2, 2));
3  setPoint(d.Alphas[0], (2, 2, 2));
4  setPoint(d.Alphas[1].Alphas[0], (0, 0, 2));
5  setPoint(d.Alphas[1].Alphas[0].Alphas[1].Alphas[0], (2, 0, 2));
6  d' ← d.Alphas[2].Alphas[1].Alphas[0].Alphas[1].Alphas[2];
7  setPoint(d', (0, 2, 0));
8  setPoint(d'.Alphas[0], (2, 2, 0));
9  setPoint(d'.Alphas[1].Alphas[0], (0, 0, 0));
10 setPoint(d'.Alphas[0].Alphas[1].Alphas[0], (2, 0, 0));
11 return d;
```

7.2.2.3 Geometrical Removal

Algorithm 73 implements the modification of the removal operation (introduced in Section 6.2.1 page 199) for embedded n-Gmaps. When a cell is removed, two types of modifications can occur: first, some darts of some vertices are removed: so it is necessary to ensure that the darts linked with geometrical points still exist after the operation; second, some vertices could disappear: so the associated points have to be removed.

In the algorithm, we iterate through all the darts in $c_{\mathtt{i}}(\mathtt{d})$ not already considered (line 2). Indeed in the second loop (line 5) all darts in the vertex $c_0(\mathtt{d}')$ have to be processed only once. For that, the darts already considered are marked during the second loop (line 5).

For each vertex incident to the removed i-cell, we test if all its darts are removed or not. If they are all removed, the associated geometrical point is also removed (line 12); otherwise (line 13), we test if the dart linked with the geometrical point belongs to $c_{\mathtt{i}}(\mathtt{d})$. In this case, the dart will be deleted by the operation, and the link between the vertex and its associated point will be lost. To solve this problem, the geometrical point is associated with a dart not removed by the operation (line 14). Here, a second mark is used to test in constant time if a dart belongs to $c_{\mathtt{i}}(\mathtt{d})$ or not.

Note that there is a special case for the 0-removal operation (line 15): in this case the only geometrical modification is the removal of the geometrical point associated with the removed vertex.

At last, the combinatorial i-removal operation is applied since the embedding is now correct (line 16).

The complexity of Algorithm 73 is the same than the one of the combinatorial operation: it is linear in the number of darts of the removed i-cell.

Algorithm 72: `geomSewNGMap(gm,d,d',i)`: i-sew two darts for embedded n-Gmaps

Input: gm: an embedded n-Gmap;
d, d′ ∈ gm.Darts: two i-sewable darts;
i ∈ $\{0, \ldots, n\}$.
Result: i-sew darts d and d′ by updating the embedding.

```
1 if i ≠ 0 then
2   it ← generic iterator(gm,d,(0,...,i − 2,i + 2,...,n));
3   it' ← generic iterator(gm,d',(0,...,i − 2,i + 2,...,n));
4   while it is not to its end do
5     if it' ∉ c0(it) then
6       deletePoint(it');
7     advance it to its next position;
8     advance it' to its next position;
9 sewNGMap(gm,i,d,d');
```

Thanks to the two marks, all operations are in constant time and no dart is processed twice.

This operation is illustrated in Fig. 7.6 by the 1-removal in an embedded 2-Gmap. The initial embedded 2-Gmap is given in Fig. 7.6(a) where the edge $\{3, 4, 5, 6\}$ will be removed. Let us suppose that dart 3 is first processed. Then we iterate through all the darts in $c_0(3)$. Dart 6 is found which is linked with a geometrical point, thus e′ is set to 6. Next dart 2 is found which does not belong to the removed edge thus todelete is set to false and d_1 is set to 2. The vertex is not totally removed, and thus we only test if the dart linked with the geometrical point will be removed or not. Here dart 6 belongs to the removed edge, thus the embedding is updated by linking the geometrical point with dart 2. All darts in $c_0(3)$ are already considered thus they are no longer considered again in the first loop. Then dart 4 is found (for example); it belongs to the second vertex incident to the removed edge. We iterate through all the darts in this vertex $\{4, 5\}$. Dart 4 is found. Since it is linked with a geometrical point, e′ is set to 4, but the two darts belong to the removed edge thus todelete remains true. This means that the vertex will be removed during the combinatorial operation, and thus the associated geometrical point is deleted. The resulting embedded 2-Gmap is given in Fig. 7.6(b).

7.2.2.4 Geometrical Triangulation

A last example of operation for embedded n-Gmaps is the triangulation operation given in Algorithm 74. A new vertex is created, while there is no modification for existing vertices. Thus the additional geometrical layer consists only to create a new geometrical point and to associate it with the new vertex, denoted $c_0(\mathtt{d}')$. Since no point is associated with this new vertex when it is

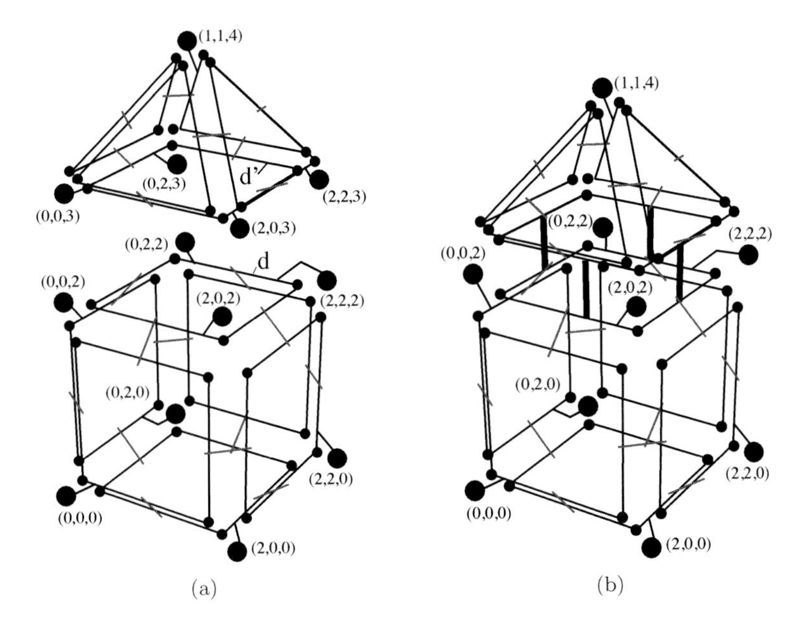

FIGURE 7.5
(a) An embedded 3-Gmap describing two isolated volumes.
(b) The 3-Gmap obtained after the geometrical 3-sew of darts d and d'.

created, we can avoid to search if a dart is already linked with a geometrical point before to set the point of $c_0(\mathtt{d}')$.

We can define the point as the barycenter of a given cell. To compute the barycenter of an i-cell, we iterate through all the vertices incident to this i-cell, and sum up the points associated with all these vertices. As usual, darts belonging to vertices already considered are marked in order to avoid to sum up several times the same geometrical point.

7.2.3 Example of Modeling of an Object

Now the use of Moka for constructing geometrical objects with embedded 3-Gmaps is illustrated. The first available operations are the creation of some basic objects: polylines, polygons, cubes, pyramids, spheres, torus (see examples in Fig. 7.7). Of course all these operations create the combinatorial and the geometrical parts of the objects thanks to darts, α links and 3D points.

Operations can be applied on the basic objects, for instance translation, rotation, scaling, closure, removal, contraction, insertion, expansion, extrusion,

Algorithm 73: `geomRemoveNGMap(gm,d,i)`: remove an i-cell for embedded n-Gmaps

Input: gm: an embedded n-Gmap;
d $\in$ gm.Darts: a dart such that c_i(d) is i-removable;
i $\in \{0, \ldots, n-1\}$.
Result: Remove the i-cell c_i(d) and update the embedding.

```
1  if i ≠ 0 then
2      foreach dart d' ∈ c_i(d) not already considered do
3          todelete ← true;
4          e' ← NULL;
5          foreach dart e ∈ c_0(d') do
6              if e ∉ c_i(d) then
7                  todelete ← false;
8                  d_1 ← e;
9              if e.Point ≠ NULL then
10                 e' ← e;
11         if todelete then
12             delete e'.Point; e'.Point ← NULL;
13         else if e' ∈ c_i(d) then
14             d_1.Point ← e'.Point; e'.Point ← NULL;
15 else deletePoint(d) ;
16 removeNGMap(gm,d,i);
```

chamfering, Boolean operation... An example of construction of a lantern[2] is illustrated in Fig. 7.8.

A square and a path of ten edges are created (Fig. 7.8(a)). Then the four vertices of the square are chamfered in order to smooth the four corners (Fig. 7.8(b)). The face is extruded along the path giving the volume shown in Fig. 7.8(c). In the next step (Fig. 7.8(d)), the geometry of the points are modified, to make the shape of the lantern: All the points belonging to a same level (i.e. having the same z coordinate) are selected, and are translated along the z axis to position the points correctly; then a homothety is applied in order to reduce the size of the selected section. Then, the four main faces of the lantern are subdivided (Fig. 7.8(e)): given a face and a number of subdivisions k, the corresponding operation subdivides the face into k^2 faces. This operation applied the insertion operation several times: vertices are inserted along the boundary of the face and into the inserted edges, and edges are inserted between these new vertices. In the last step, the four central faces resulting

[2]This construction is inspired from the 3D Studio Max tutorial "Modeling a Lantern Using Splines and Loft".

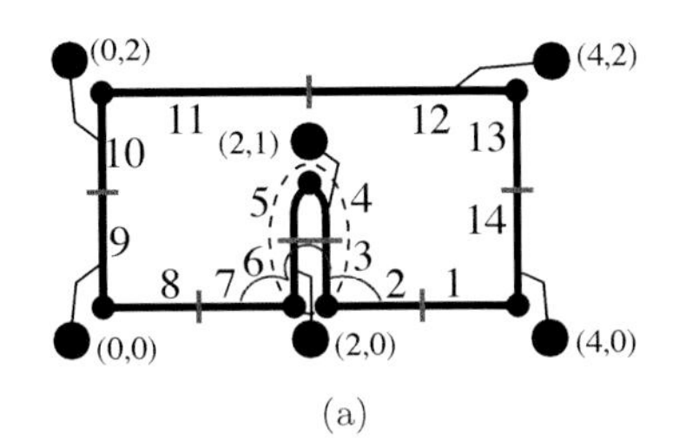

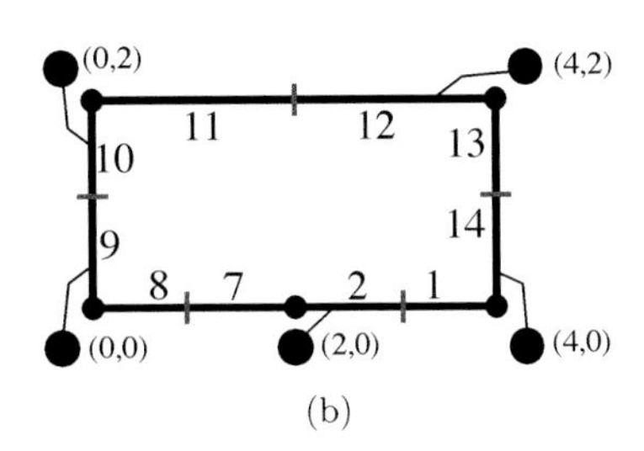

FIGURE 7.6
(a) An embedded 2-Gmap G.
(b) The 2-Gmap $G_{R_1}(\{3,4,5,6\})$ obtained from G by using the geometrical 1-removal operation of the dangling edge $\{3,4,5,6\}$.

Algorithm 74: `geomTriangulationNGMap(gm,d,i,p)`: geometrical triangulation of an i-cell for embedded n-Gmaps

---

Input: gm: an embedded n-Gmap;
  d $\in$ gm.Darts: a dart;
  i $\in \{1,\ldots,n\}$;
  p: a geometrical point.
Result: Make a triangulation of the i-cell c_i(d) around the point p.

```
1 d′ ← triangulationNGMap(gm,d,i);
2 setPoint(d′,p);
```

from these subdivisions are removed. The object shown in Fig. 7.8(f) is the final lantern.

Many other operations are available in MOKA. Among them, one is very useful for geometric modeling: the corefinement, which is a basis for Boolean operations. For example, this operation can create the windows of the lantern, as illustrated in Fig. 7.9. The two last steps of the previous construction method (subdivide the four main faces and remove the central faces) are replaced by the creation of a beam across the lantern (Fig. 7.9(a)), and by the subtraction operation between the lantern and this first beam to bore the lantern in a first direction (cf. Fig. 7.9(b)). A second beam is created in the second main direction (Fig. 7.9(c)), and the subtraction operation between the lantern and this second beam is applied to bore the lantern in the second direction (cf. Fig. 7.9(d)). Note that the resulting object is not exactly the same as the object constructed by the first method. Here, a 2-closed 3-Gmap with two handles is constructed, while a 3-Gmap with four 2-boundaries results from the first construction.

Another example is depicted in Fig. 7.10: the construction of a screw by applying the Boolean operations. In the first line, the intersection operation between the extrusion of a hexagon and an ellipse is applied: this creates the head of the screw. In the second line, firstly the union operation is applied to the head of the screw and a cylinder; secondly the subtraction operation is applied between the result of the union and a spiral. This produces the final screw shown bottom right.

The construction of a Klein bottle is illustrated in Fig. 7.11. First a square face and an extrusion path are created (Fig. 7.11(a)). A third path is also created, describing a scaling coefficient for the extrusion of the face along the extrusion path. The corresponding extruded object is shown in Fig. 7.11(b); it does not have a constant size along the path. This is not yet a Klein bottle since this is a surface with two 2-boundaries corresponding to the two extremities of the path. These boundaries are identified by applying the 2-sew on the four pairs of darts in these two boundaries, producing the Klein bottle given in

FIGURE 7.7
Example of basic objects available in MOKA.

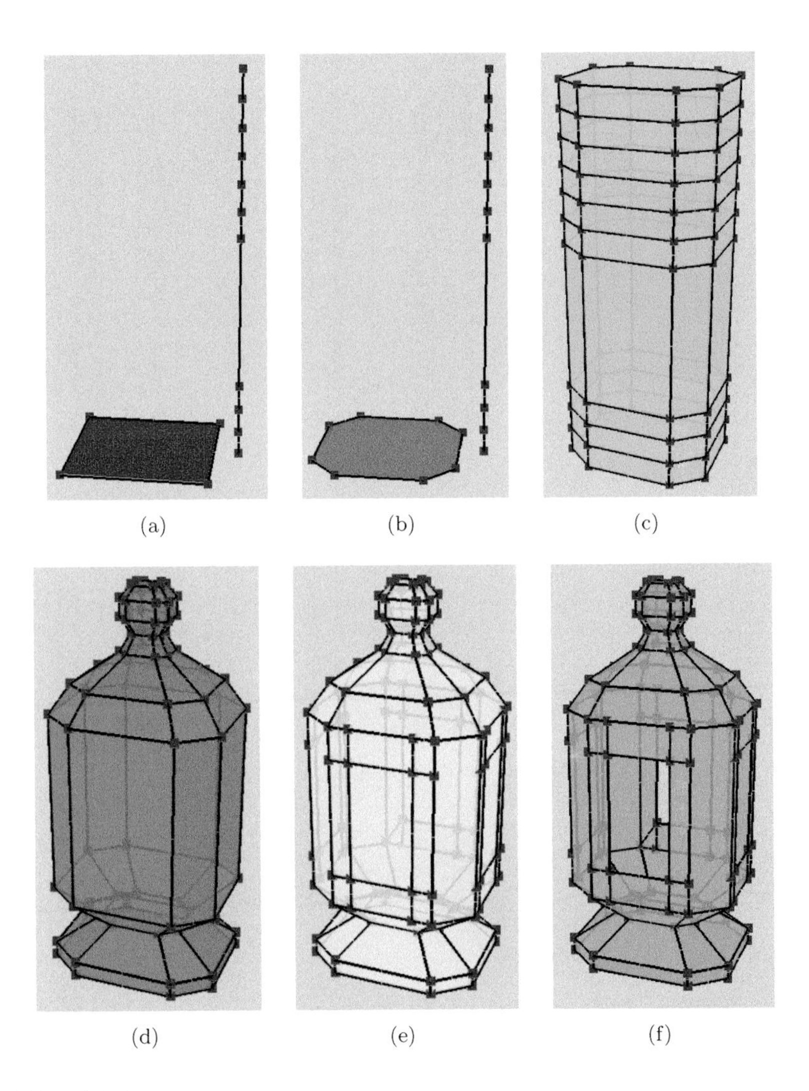

FIGURE 7.8
Construction of a lantern using MOKA.
(a) Creation of a square and a path of ten edges.
(b) Chamfering of the four vertices of the square.
(c) Extrusion of the face along the path.
(d) Translation and homothety of some points.
(e) Subdivision of the four big faces.
(f) Removal of the four central faces.

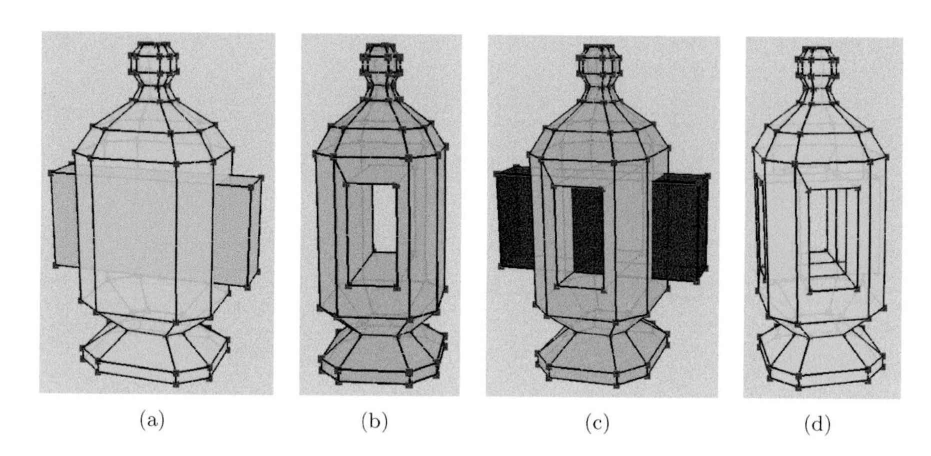

FIGURE 7.9
Construction of a lantern with corefining using MOKA.
(a) Creation of a beam across the lantern.
(b) Result of the subtraction of the beam from the lantern.
(c) Creation of a beam across the lantern in the second main direction.
(d) Result of the subtraction of the second beam from the lantern.

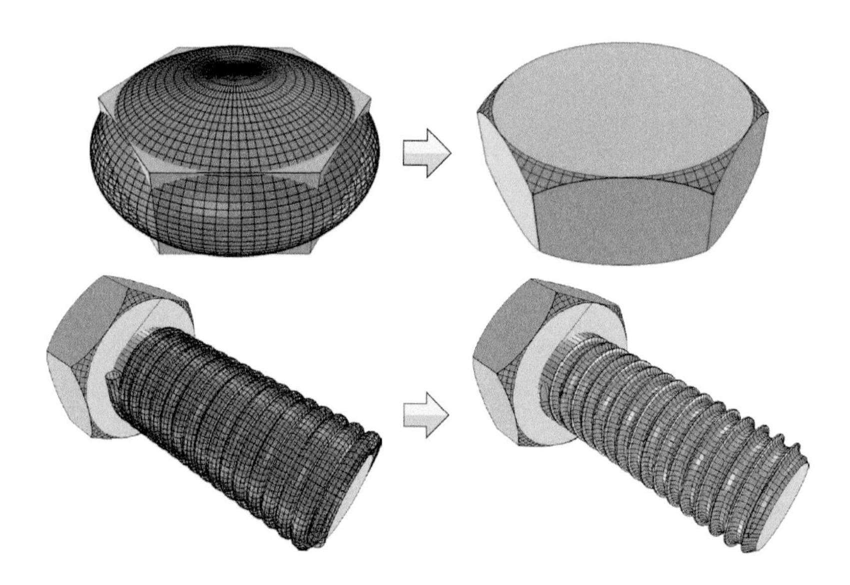

FIGURE 7.10
Construction of a screw with corefining using MOKA (images from [126]).

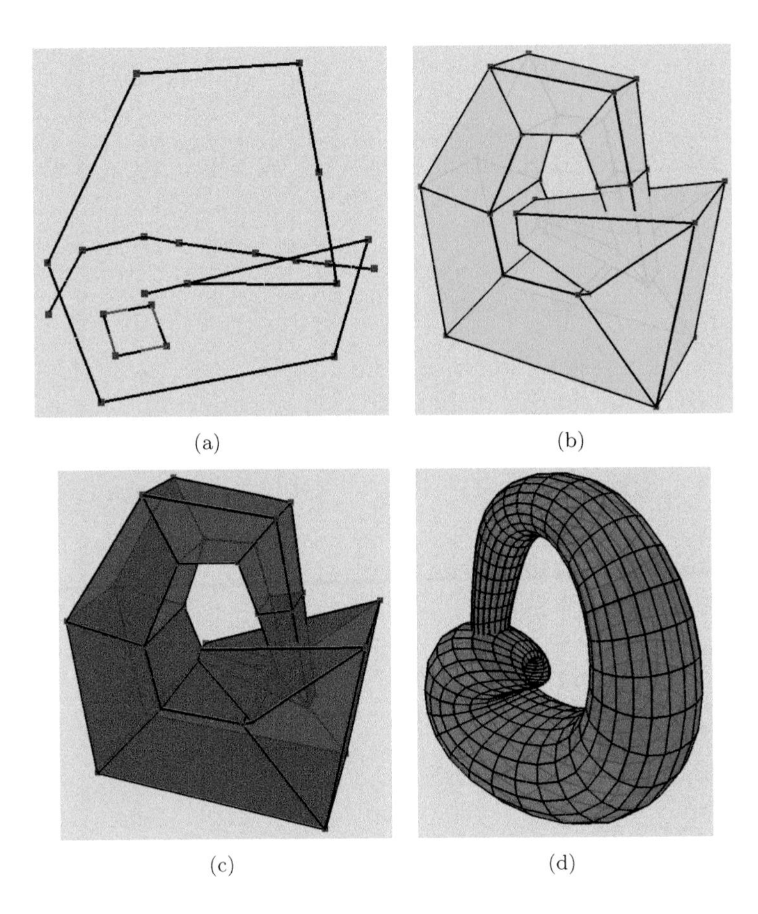

FIGURE 7.11
Construction of a Klein bottle using MOKA.
(a) Creation of a square, an extrusion path and a scaling path.
(b) Result of the extrusion.
(c) Result after the 2-sew applied to the four pairs of darts in the 2-boundaries.
(d) Final result obtained after a smoothing step.

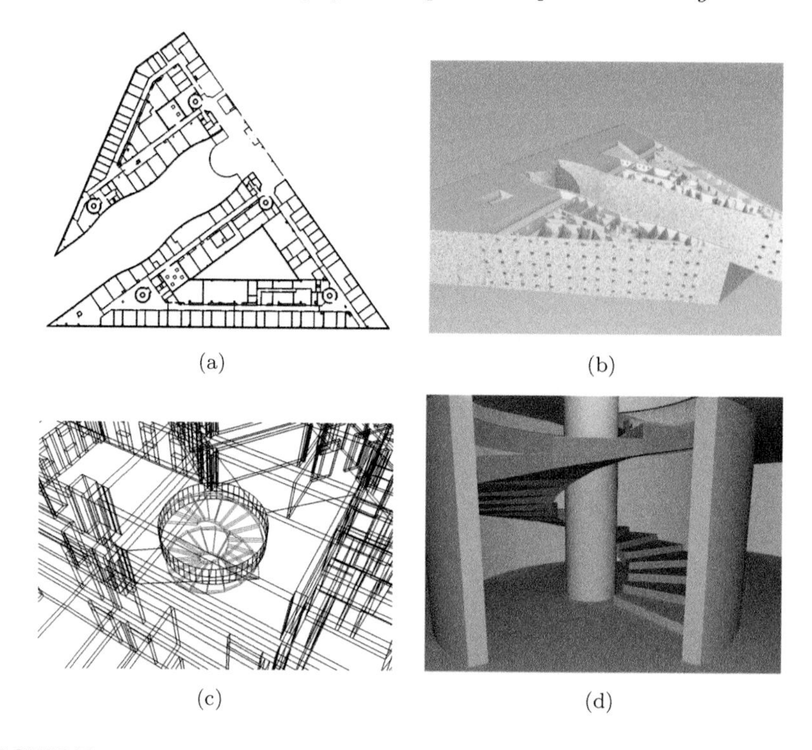

FIGURE 7.12
Use of Moka for building reconstruction (images from [134]).
(a) Initial data: a 2D Autocad plan of a building.
(b) Result of the reconstruction.
(c) Zoom on a spiral staircase.
(d) Rendering of a spiral staircase.

Fig. 7.11(c). The auto-intersection of the Klein bottle is only geometric: there is no vertex nor edge in this intersection. Indeed, it is not possible to embed a Klein bottle in $\mathbb{R}^3$ without auto-intersection. Lastly a smoothing operation using splines is applied to obtain the final result given in Fig. 7.11(d).

Moka was used for different projects, for which users need to describe real 3D objects, with possibly multi-incidence and mixing different types of cells. For example, Moka was used to describe physical objects in order to propose a topology-based physical simulation [177, 176], or to define a discrete pyramid representing objects in images with different discrete embeddings [72, 92].

The MokaArchi project [134, 136] is based on Moka for the reconstruction of 3D buildings from 2D autocad plans (see Fig. 7.12).

Moka was also used as the basic kernel for the construction of 3D geological models; and several specific operations were conceived for handling

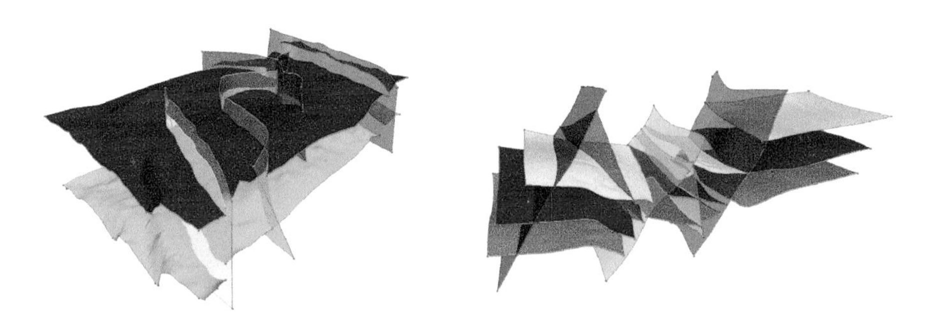

FIGURE 7.13
Use of MOKA for geological description (images from [126]).

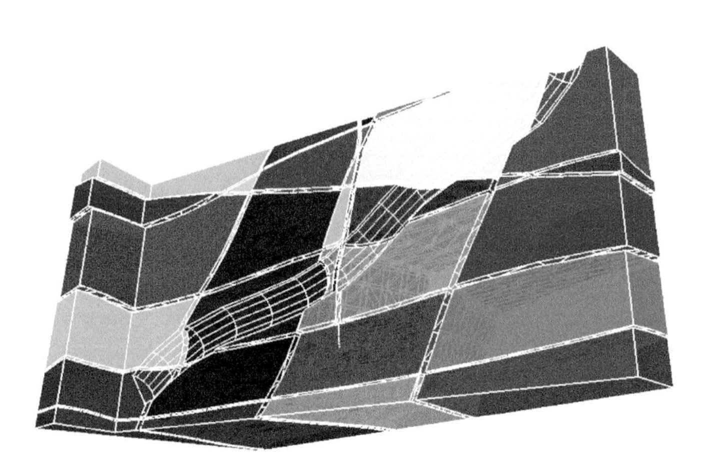

FIGURE 7.14
The corefining operation is applied in order to drill a well in the geological basement.

these models [29, 126] (see Fig. 7.13). For example, a well can be drilled in the geological basement by applying the corefining operation (see Fig. 7.14).

7.3 Image Processing

n-Gmaps and n-maps can be used to describe partitions of labeled images into regions and thus efficient image processing algorithms can be conceived. This fits into **structural image processing** methods, as n-Gmaps and n-maps can be considered as natural extensions of graphs for describing regions of labeled images. Indeed, in 2D, a 2-map or a 2-Gmap is a representation of a planar graph.

7.3.1 Preliminary Notions

Each pixel of a 2D image is associated with a *label*, a value from a finite set of labels. Two pixels (x, y) and (x', y') are 4-adjacent if $|x - x'| + |y - y'| = 1$, i.e. the two pixels share one edge of their boundaries. Two different pixels (x, y) and (x', y') are 8-adjacent if $max(|x - x'|, |y - y'|) = 1$, i.e. the two pixels share either an edge or a point of their boundaries. A list of pixels is a k-path (with $k = 4$ or 8) if each pair of successive pixels in the path are k-adjacent. A set of pixels is k-connected if any pair of pixels of the set are linked by a k-path of pixels of the set. A *region* in a labeled image is a maximal set of 4-connected pixels having same label. A lexicographic order exists on pixels: $(x, y) < (x', y')$ if $x < x'$ or ($x = x'$ and $y < y'$). This order is extended on regions: $R_i < R_j$ if $min(R_i) < min(R_j)$, $min(R)$ being the minimal pixel of region R.

These notions are illustrated in Fig. 7.15(a). Pixel a is 4-adjacent to pixels b, c, d and e, and 8-adjacent to pixels b, c, d, e, f, g, h and i. There are 6 regions in this image, labeled from R_1 to R_6. An additional region is considered, labeled R_0 and called the infinite region, which is the complementary of the image.

The *interpixel* framework [145, 143, 146, 141, 142] refers to pixels, linels which are unit segments separating each pair of 4-adjacent pixels and *pointels* which correspond to the extremities of linels (cd. Fig. 7.15(b)). Mainly, the interpixel framework make it possible to define a topology for 2D images, similar to the classical topology. For instance, curves can be defined as particular sets of linels and pointels, and properties can be defined, similar to that of curves in Euclidean space.

Given a 2D labeled image, it is often necessary to describe the regions and the different adjacency relations between these regions. Indeed, this information is required by many image processings, such as region merging used in bottom-up segmentation algorithms. The first data structure proposed to

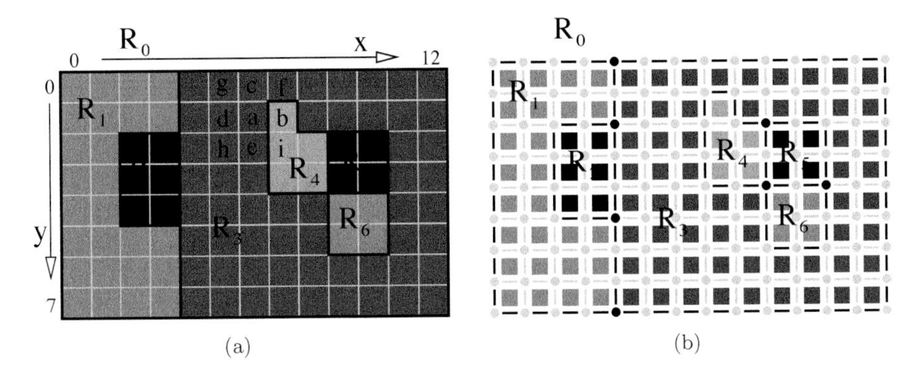

FIGURE 7.15
(a) A 2D labeled image. (b) The interpixel elements.

describe such information was the *region adjacency graph* (*RAG*): a vertex corresponds to each region of the image, and two vertices are linked by an edge if the two corresponding regions are adjacent [193]. The RAG has several advantages: this is a simple data structure (a graph) which can be defined for any dimension. Given a region, it is easy to retrieve all its adjacent regions. Merging two adjacent regions can be done in the RAG by contracting the edge linking the two corresponding vertices. However, there are also several drawbacks. First, a RAG does not take the multi-adjacency into account. When two regions are adjacent several times, there is only one edge in the corresponding RAG. Second, the regions adjacent to a given region are not ordered[3], and this order is often important for operations. Third, a RAG does not describe all the cells of the subdivision corresponding to a labeled image. In 2D, only 2-cells are described (which are regions), and some 1-cells (which are frontiers separating regions) corresponding to edges of the RAG.

Several solutions were proposed to solve these drawbacks, such as *dual graphs* [217, 152] for example. But this ad hoc solution is defined only in 2D and cannot be extended for higher dimension. In order to propose a generic definition, including the description of the multi-adjacency and the order relations, many works have concluded that a good solution consists in basing this definition on n-maps, and similar solutions have been proposed [95, 110, 32, 66]. Here, the solution defined in [66] for 2D labeled images is described, one of its main advantages is that it can be extended in 3D [59].

7.3.2 2D Topological Map

A 2-map is used in order to describe the subdivision corresponding to a given 2D labeled image. The edges of the 2-map describe the boundaries of the

[3]Although an order exists between 2D regions in a 2D Euclidean space.

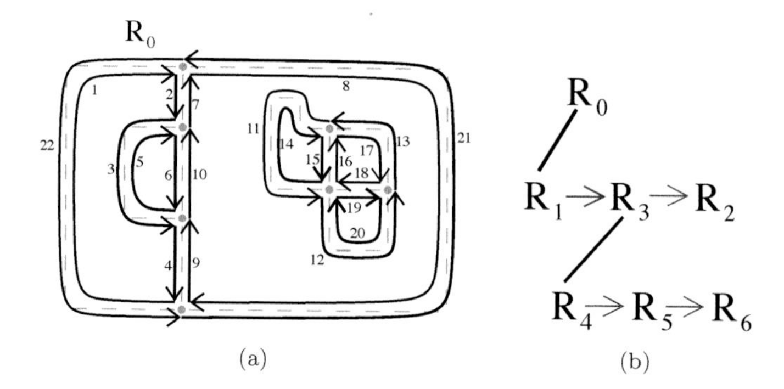

FIGURE 7.16
(a) 2-map describing the 2D labeled image of Fig. 7.15(a).
(b) Enclosure tree of regions associated with this 2-map. *Son* relations are drawn by black segments, and *samecc* relation by grey arrows.

regions (see Fig. 7.16(a)). Two adjacent regions share a part of their boundary: this part is described in the 2-map by an edge, which contains two darts. Indeed, thanks to the infinite region R_0, the 2-map is without boundary for all dimensions. In the example (cf. Fig. 7.16(a)), regions R_1 and R_2 are adjacent, and in the 2-map, edge $\{3, 5\}$ describes this adjacency relation. Moreover, when two regions are multi-adjacent (such as regions R_1 and R_3, which are adjacent twice), there are two corresponding edges in the 2-map (edges $\{2, 7\}$ and $\{4, 9\}$).

As 2-maps are ordered models, the order on the boundaries around regions is naturally taken into account. For example, starting from dart 1 which belongs to region R_1, applying β_1 relations produces the sequence of darts $(1, 2, 3, 4)$, which allows to iterate through the external boundary of R_1 in a clockwise direction. When a region has k holes (e.g. region R_3 has 1 hole), it has k internal boundaries, one for each hole. Starting from one dart belonging to an internal boundary, and applying β_1 relations, we iterate through the internal boundary in a counterclockwise direction. The sequence of darts $(11, 12, 13)$ is obtained for the internal boundary of region R_3. These two different directions are coherent (clockwise for external boundaries, counterclockwise for internal boundaries): when iterating through a boundary, the region is always to the right of the oriented boundary.

The 2-map describes the subdivision of the image in regions. Faces correspond to regions. Edges correspond to adjacency relations between regions, i.e. to parts of their common boundaries. Vertices correspond to adjacency relations between boundaries. Thanks to 2-map properties, all the incidence and adjacency relations between the cells are described in an ordered way. However there is a problem for regions with holes. In this case (as region R_3 in the example), the region has several boundaries (one external and k internals), and no link exists between these different boundaries in the 2-map: they correspond to distinct connected components.

To solve this problem, an enclosure tree of regions is added to describe all enclosed relations. Intuitively a region R_i is *enclosed* by a region R_j if R_i is surrounded by R_j. More precisely, R_i is enclosed by R_j if all the 8-paths starting from a pixel of R_i and going to a pixel of R_0 traverse region R_j. Region R_i is *directly enclosed* by region R_j if it is enclosed by R_j and if there is no region R_k such that R_i is enclosed by R_k and R_k is enclosed by R_j.

In our example, region R_4 is enclosed by region R_3 because all the paths starting from a pixel of R_4 traverse region R_3 to go to R_0. Region R_2 is not enclosed by region R_1, because some paths going from R_2 to R_0 exist which do not contain pixels of R_1. R_4 is enclosed by R_3 and by R_0, but it is only directly enclosed by R_3.

The "directly enclosed" relations are encoded by an *enclosure tree of regions*. This tree has one node for each region of the image, its root corresponds to the infinite region R_0. A region R_i is son of region R_j in the tree if R_i is directly enclosed by R_j. All the regions sons of R_i are grouped by 8-connected components thanks to the *samecc* relation: two regions belonging to the same connected component and having the same father belong to the same list of regions with the *samecc* relation. Only the smaller regions (smaller for the lexicographic order) sons of each 8-connected component are described as sons of R_i in the tree, other regions will be retrieved thanks to the *samecc* relation.

The enclosure tree of regions and the 2-map are linked. Each dart of the 2-map knows its region denoted by d.`Region`, and each region knows one of its darts denoted by R.`Representative`. This dart must belong to the external boundary of R and the corresponding edge must contain the smaller pointel of the external boundary of R (smaller in term of lexicographic order).

Figure 7.16(b) describes the enclosure tree of regions associated with the 2-map given in Fig. 7.16(a) (cf. Fig. 7.15(a) for the labeled image). The three regions R_1, R_2 and R_3 are directly enclosed in region R_0, and they belong to the same 8-connected component. The smaller region is R_1, which is son of R_0 in the tree, the two other regions R_2 and R_3 being linked with R_1 by the *samecc* relation. Similarly, the three regions R_4, R_5 and R_6 are directly enclosed in region R_3, and they belong to the same 8-connected component. The smaller region is R_4 which is son of R_3 in the tree, the two other regions R_5 and R_6 being linked with R_4 by the *samecc* relation.

All the boundaries of a given region R can be retrieved thanks to the enclosure tree of regions and to the links between the tree and the 2-map. The external boundary of R is given by $\langle\beta_1\rangle(R.\texttt{Representative})$. All internal boundaries are obtained by $\langle\beta_1\rangle(\beta_2(R'.\texttt{Representative}))$ for all regions R' son of R (since R' is the smaller region of an 8-connected component and thanks to the properties of `Representative`, we know that $\beta_2(R'.\texttt{Representative})$ belongs necessarily to one internal boundary of R, cf. Fig. 7.16).

To encode the enclosure tree of regions, the `Dart` data structure (given in Section 4.4 page 108) is modified by adding a pointer (called `Region`) to a structure describing the regions. This structure contains the different pointers to represent the tree (in a classical implementation: $father$, $brother$, son), plus the *samecc* pointer that gives the next region belonging to the same connected component, plus the pointer `Representative` that gives the representative dart of the region.

Generally additional information is associated with regions, which will be useful for different image processing operations: for example the mean color (the mean value of all the colors of the pixels belonging to the region), and the region size (its number of pixels). Note that depending on the need of the operations, some information can be associated with some specific cells: for example a gradient can be associated with each edge of the topological map (the sum of all the differences of colors of pixels separated by the edge).

The 2-map plus the enclosure tree of regions describes fully the topological information contained in the labeled image. Now the 2-map is embedded in order to describe also the geometrical information. 2D images correspond to discrete geometrical space: vertices of the 2-map correspond to pointels, edges to sets of linels and regions to sets of pixels. This embedding is described through a matrix of interpixel elements. This matrix contains a Boolean for each pointel and each linel of the given 2D image. The value of a linel is true if it separates two pixels belonging to two different regions. The value of a pointel is true if it has more than two true incident linels.

Each dart d of the 2-map is linked with a pair (p, l) denoted by d.`Emb`. p is the first pointel of the frontier associated with the edge containing dart d, and l is the direction of the first linel of this frontier (considering the frontier oriented clockwise or counterclockwise depending if d belongs to an external or internal boundary). This direction is a value in the set $\{0, 1, 2, 3\}$ corresponding respectively to the four possible directions *right*, *up*, *left* and *down*. The pointel p associated with a dart d has a true value in the interpixel matrix, except if the edge containing d is a loop. This case occurs for a directly enclosed region R having no other region in its 8-connected component (called also an isolated region, you can see an example in Fig. 7.24). In this case, the frontier separating R from its enclosed region is a cycle, and all the pointels of this cycle have two true incident linels, thus they are all false. Any pointel of the cycle can be linked with d and can be used as a starting point. This is the only case where the embedding of a dart is linked with a pointel off in the interpixel matrix (but a linel on). For all other configurations, the embedding of a dart is always a pointel and a linel which are on in the interpixel matrix.

To encode the embedding of the 2-map in the interpixel matrix, the `Dart` data structure (given in Section 4.4 page 108) is modified by adding a data called `Emb`, which is a structure describing the pair (p, l).

Figure 7.17(b) describes the interpixel embedding of the 2-map shown in Fig. 7.17(a). Some links between darts and pairs (p, l) are given in Fig. 7.17(c).

This embedding is implicit. The links between darts and interpixel elements do not describe the embedding of all the cells, but allow to compute them. Embedding of vertices is directly encoded as each dart is associated

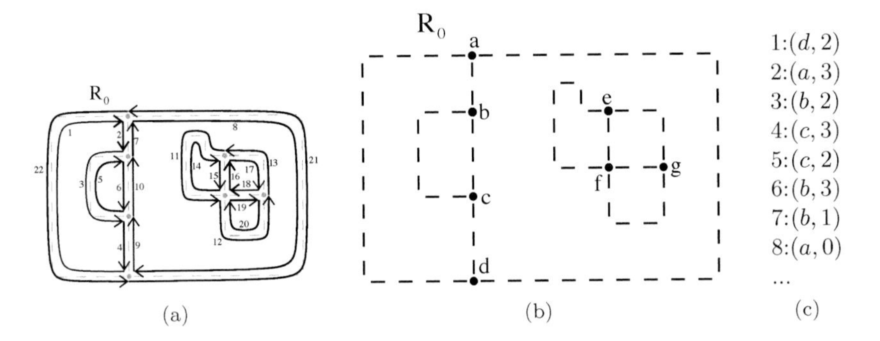

FIGURE 7.17
(a) 2-map.
(b) Its embedding in the interpixel matrix.
(c) Some links between darts and pairs (p, l).

with a pointel (darts belonging to the same vertex are necessarily associated with the same pointel). Edge embedding can be retrieved by iterating through true linels starting from the initial linel given by the pair (p, l), and following the true linels until obtaining either a true pointel or the initial linel (when the frontier is a cycle). At last, region embedding can be retrieved starting from a pixel inside the region, and applying a flood fill algorithm stopped by true linels. The initial pixel is retrieved thanks to the pair (p, l) associated with the external boundary of the region and taking into account that this boundary is oriented clockwise.

The model composed with the 2-map describing the 2D labeled image, together with the enclosure tree of regions, and its embedding in the interpixel matrix is called 2D *topological map*. Using these three components, it is possible to conceive efficient high level operations of 2D image processing.

7.3.3 Operations

7.3.3.1 Topological Map Construction

The first operation defined is the construction of a 2D topological map from a 2D labeled image. Algorithm 75 implements an incremental solution, which mainly consists in iterating through all the pixels of the image line by line, in creating a square associated with the current pixel and, depending on the local configuration, in removing the two edges to the left and to the up of the current pixel and the vertex to the left and up corner of this pixel. During the scan line of the image, *last* denotes the dart to the left of the current pixel and *up* the dart to the up. An important invariant is the fact that the 2-map describing the pixels smaller than the current pixel (according to lexicographic order) is already build. This invariant is required to 2-sew the square corresponding to the current pixel to the former 2-map.

Algorithm 75: Incremental construction of a 2D topological map

Input: I: a 2D labeled image of $n_1 \times n_2$ pixels.
Output: The 2D topological map corresponding to I.

```
1  cm ← createNMap(2);
2  last ← Build the upper border of the image in the 2-map cm;
3  emb ← an interpixel matrix of size (n1 + 1, n2 + 1) initialized to false;
4  for j ← 0 to n2 do
5      for i ← 0 to n1 do
6          up ← computeUpFromLast(cm,last);
7          tmp ← createSquare(cm,last,up,(i, j),n1);
8          V ← c0(last);
9          if pixel (i − 1, j) belongs to the same region than pixel (i, j) then
10             V ← V \ c1(last);
11             removeNMap(cm,last,1);
12         else switch on linel ((i, j), 3) in emb;
13         if pixel (i, j − 1) belongs to the same region than pixel (i, j) then
14             V ← V \ c1(up);
15             removeNMap(cm,up,1);
16         else switch on linel ((i, j), 0) in emb;
17         if V = {v0, . . .} ≠ ∅ then
18             if isRemovableNMap(cm,v0,0) then
19                 if v0.Betas[1] ≠ v0 then removeNMap(cm,v0,0);
20             else switch on pointel (i, j) in emb;
21         last ← tmp;
22 Remove all the darts connected with last;
23 tree ← Compute the enclosure tree of regions;
24 return (cm, emb, tree)
```

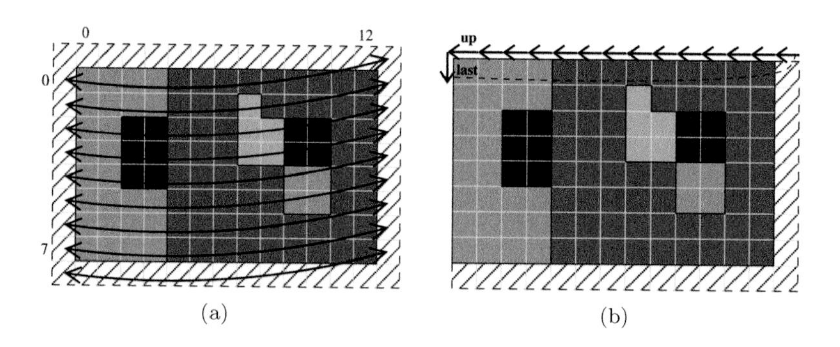

FIGURE 7.18
(a) 2D labeled image folded on a cylinder.
(b) Initial border created in the 2-map before to start the construction.

In order to take into account the boundaries with the infinite region, the image is extended by one pixel all around it, these additional pixels belonging to the infinite region. In order to avoid special cases for these pixels and thus to simplify the algorithm, each last pixel of a line is identified with the first pixel of the next line (see Fig. 7.18(a)). Making these identifications is equivalent to fold the image on a cylinder with a shift on the lines, so that we can now iterate through the entire image by only advancing the current pixel in x direction.

In order to satisfy our invariant for the first pixel of the image (pixel $(0,0)$), an upper border is first built in the 2-map (cf. Fig. 7.18(b)). It is composed by one dart to the left of the first pixel, and all the darts to the up of the pixels in the first line of the image. This border is extended for one pixel to the right of the image to consider the infinite region. To simulate the folding on a cylinder, β_1 links the dart to the left of the first pixel and the dart to the up of the last pixel of the first line (represented by the dashed curve in the figure). It is enough to add one pixel in the column to the right of the image because, thanks to the cylinder, the last pixel of a line is also the first pixel of the next line. The dart to the left of the first pixel is the first *last* dart, ready to be 2-sewn with the square describing pixel $(0,0)$. Thanks to this border, each pixel of the image can be processed without special case for first and last columns and lines.

In the main loop of the construction algorithm (line 4 and 5) we consider pixel (i,j). First a square is created and 2-sewn with darts *last* and *up*. This is done in Algorithm 76, where first four darts are created and linked by β_1 (lines 1 to 5). Then dart $res[0]$ is 2-sewn with dart *last*, and dart $res[1]$ with dart *up* (lines 6 and 7). Lastly the embedding of the four new darts is initialized given (x,y) the position of the up-left pointel. To take into account the bending of the image on a cylinder, the position of pointels are computed modulo the width of the image plus one. The function returns dart $res[2]$ which is the dart to the right of the new square, i.e. the next last when processing the next pixel of the image.

Note that during the main loop of the construction algorithm, only *last* dart is kept since *up* dart can be computed from *last* thanks to Algorithm 77 (where mainly we turn around the vertex containing dart *last* until obtaining a 2-free dart).

In the next step of the construction algorithm, the two edges containing the darts *last* and *up* and the vertex containing dart *last* are considered in order to test if they must be removed or kept. The edge containing *last* must be removed if pixel $(i-1,j)$ belongs to the same region than the current pixel. Indeed in this case, there is no frontier between these two pixels. Note that if $i=0$, pixel $(-1,j)$ belongs to the infinite region R_0. Thus, if the current pixel belongs also to the infinite region (when $j=n_2$), both pixels belong to the same region; otherwise, they belong to two different regions. If $i=n_1$, pixel (i,j) belongs to the infinite region and both pixels belong to the same region only if pixel $(i-1,j)$ belongs also to the infinite region (when $j=n_2$).

Algorithm 76: `createSquare(cm,last,up,(x,y),w)`: create a new square 2-sewn with darts `last` and `up`

Input: `cm`: a 2D topological map;
`last`: the dart to the left of the current pixel;
`up`: the dart to the up of the current pixel;
$(\mathtt{x},\mathtt{y})$: the coordinates of the up-left pointel of the square;
`w`: the width of the image.

Result: Create a new square 2-sewn with darts `last` and `up`, with $(\mathtt{x},\mathtt{y})$ as coordinates of the up-left pointel of the square (modulo $\mathtt{w}+1$ in x-axis).

Output: The dart to the right of the new square.

1 **for** i $\leftarrow$ 0 **to** 3 **do**
2   *res*[i] $\leftarrow$ `createDartNMap(cm)`;
3   **if** i $>$ 0 **then**
4   *res*[i].Betas[0] $\leftarrow$ *res*[i $-$ 1]; *res*[i $-$ 1].Betas[1] $\leftarrow$ *res*[i];
5 *res*[0].Betas[0] $\leftarrow$ *res*[3]; *res*[3].Betas[1] $\leftarrow$ *res*[0];
6 *res*[0].Betas[2] $\leftarrow$ last; last.Betas[2] $\leftarrow$ *res*[0];
7 *res*[1].Betas[2] $\leftarrow$ up; up.Betas[2] $\leftarrow$ *res*[1];
8 $x' \leftarrow (\mathtt{x}+1)\%(\mathtt{w}+1)$; $y' \leftarrow \mathtt{y}+((\mathtt{x}+1)/(\mathtt{w}+1))$;
9 *res*[0].Emb $\leftarrow ((\mathtt{x},\mathtt{y}+1),1)$; *res*[1].Emb $\leftarrow ((\mathtt{x},\mathtt{y}),0)$;
10 *res*[2].Emb $\leftarrow ((x',y'),3)$; *res*[3].Emb $\leftarrow ((x',y'+1),2)$;
11 **return** *res*[2];

Algorithm 77: `computeUpFromLast(cm,last)`: get the dart up to the current pixel given the dart to its left

Input: `cm`: a 2D topological map;
`last`: the dart to the left of the current pixel.

Output: up the dart up to the current pixel.

1 up $\leftarrow$ last.Betas[0];
2 **while** *not* isFreeNMap(cm,up,2) **do**
3   up $\leftarrow$ up.Betas[2].Betas[0]
4 **return** up;

Similarly, the edge containing *up* must be removed if pixel $(i, j-1)$ belongs to the same region than the current pixel. If $j = 0$, pixel $(i, j-1)$ belongs to the infinite region, and both pixels belong to the same region if pixel (i, j) belongs also to the infinite region (and the same for $j = n_2$). An edge is removed by applying the 1-removal operation given in Section 6.2.2 page 205. When an edge is not removed (lines 12 and 16), the corresponding linel is switched on in the interpixel matrix because it separates two pixels belonging to two different regions.

Lastly, we test if the vertex containing dart $last$ needs to be removed. For that, before to process edges, the set of darts in $c_0(last)$ are stored in V (line 8). When an edge is removed, the darts of the edge are removed from V (lines 10 and 14). Indeed, since these darts are removed, they no longer belong to the vertex incident to $last$. After processing the edges, if V is empty (line 17), there is no more vertex to process. Otherwise, there is at least one dart $v_0 \in V$. If $c_0(v_0)$ is removable and if dart v_0 is not a loop, the vertex has to be removed. If the vertex is removable but the edge is a loop, the vertex must be kept; otherwise, this involves to remove also the edge and thus to lose one boundary. Lastly if the vertex is not removable, pointel (i, j), which is the pointel corresponding to vertex $c_0(v_0)$, is switched on. Indeed, in this case more than two linels are incident to this pointel.

This is the end of the processing of pixel (i, j). The $last$ dart is moved on the dart to the right of the last created square, which is the next $last$ dart for the pixel $(i+1, j)$. Note that, thanks to the folding on the cylinder, there is no special processing for the last pixel of a line: the next $last$ of a line is equal to the first $last$ of the next line.

During the construction of the 2-map, information associated with regions and with the different cells can be computed. For example, to compute the mean color of each region, we sum up the color of each pixel added to a given region, and we increment its number of pixels. After the construction of the topological map, the mean color is directly computed by dividing the sum of colors by the number of pixels. As another example, to compute the gradient of each edge, it is possible to initialize the gradient of the edge containing dart $last$ by the absolute value of the difference of colors of the two pixels separated by this edge (and similarly for the edge containing dart *up*), and to initialize the length of these two edges to 1. Then, when a vertex is removed, the two gradients (and the two lengths) of the two incident edges are added. At the end of the construction, each edge is associated with the sum of the gradients of all of its linels and with its length. Thanks to these two values, the gradient of each edge can be directly computed.

Figure 7.19 illustrates the first steps of our construction algorithm for the 2D image used in our previous example. The initial border is given in Fig. 7.18(b). The 2-map after the creation of the square corresponding to pixel $(0, 0)$ is depicted in Fig. 7.19(a). This pixel does not belong to the same region than pixel $(-1, 0)$ nor than pixel $(0, -1)$ which belong both to R_0, thus the two edges are not removed. Then the vertex $c_0(last)$ is tested: since it is

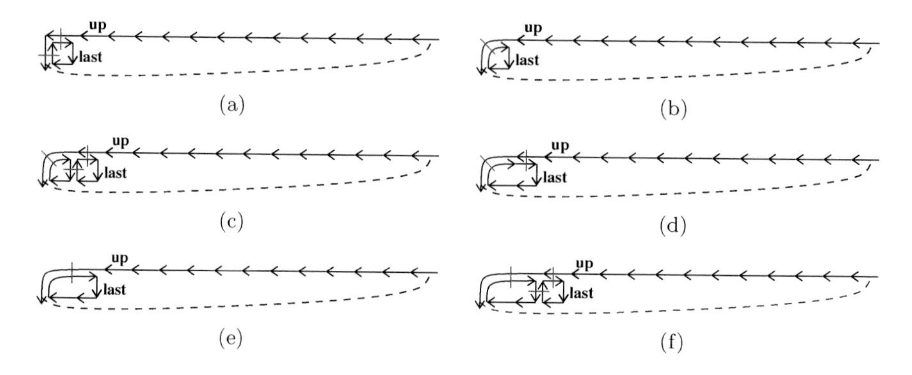

FIGURE 7.19
(a) After the creation of the square for pixel $(0, 0)$.
(b) After processing pixel $(0, 0)$.
(c) After the creation of the square for pixel $(1, 0)$.
(d) After processing the two edges.
(e) After processing pixel $(1, 0)$.
(f) After the creation of the square for pixel $(2, 0)$.

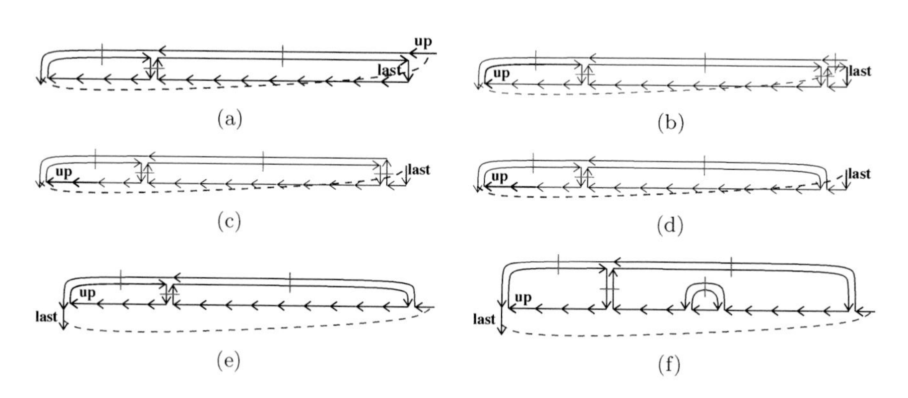

FIGURE 7.20
(a) After processing pixel $(12, 0)$.
(b) After the creation of the square for pixel $(13, 0)$.
(c) After processing the two edges.
(d) After processing pixel $(13, 0)$.
(e) Same configuration with *last* dart drawn before the first pixel of the next line thanks to the modulo for x coordinates.
(f) After processing pixel $(13, 1)$.

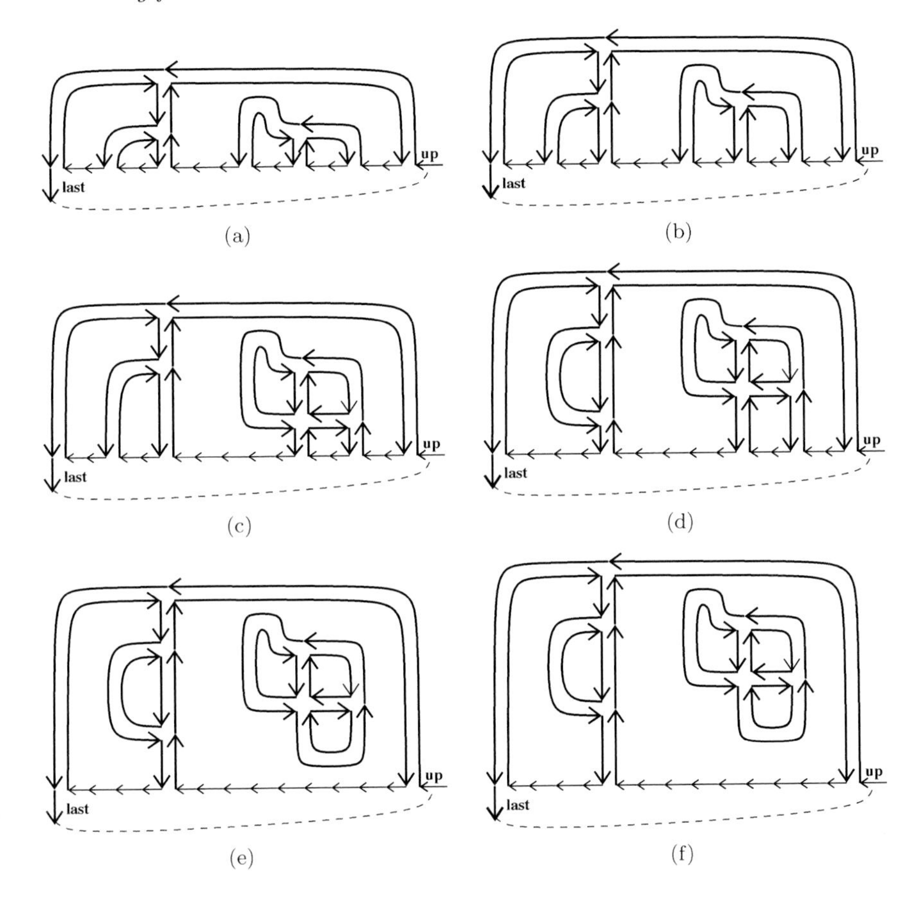

FIGURE 7.21
(a) After processing pixel $(13, 2)$.
(b) After processing pixel $(13, 3)$.
(c) After processing pixel $(13, 4)$.
(d) After processing pixel $(13, 5)$.
(e) After processing pixel $(13, 6)$.
(f) After processing pixel $(13, 7)$.

removable and dart *last* is not a loop, the vertex is removed, producing the 2-map given in (b). Then the square corresponding to pixel $(1, 0)$ is created (see (c)). The edge between pixel $(0, 0)$ and $(1, 0)$ is removed because the two pixels belong to the same region (see (d)). Then the vertex satisfies the conditions and is thus removed (see (e)). Lastly (f) shows the 2-map after the creation of the square corresponding to pixel $(2, 0)$.

Figure 7.20(a) shows the 2-map after processing all the pixels up to pixel $(12, 0)$. The next pixel is $(13, 0)$, the last pixel of the first line. As for other pixels, a square describing this pixel is created (see (b)). Since the last pixel of

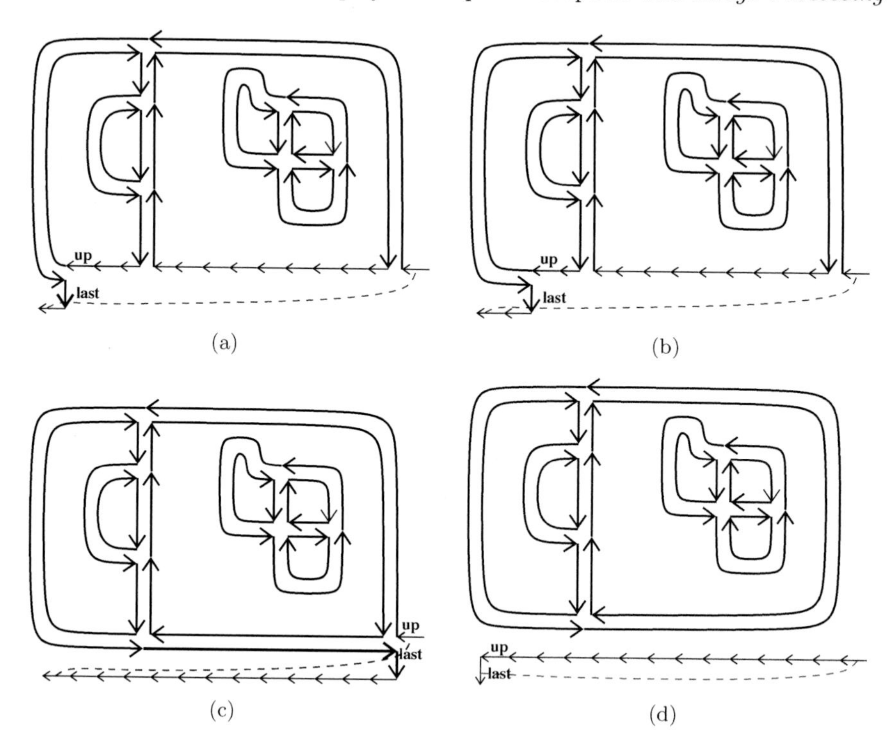

FIGURE 7.22
(a) After processing pixel (0, 8).
(b) After processing pixel (1, 8).
(c) After processing of pixel (12, 8).
(d) After processing pixel (13, 8).

a line belongs to the infinite region, it does not belong to the same pixel than its pixel to its left (except for the last line), and it belongs to the same region than the pixel to its up. Thus the edge containing *last* is kept and the edge containing *up* is removed, producing the 2-map shown in (c). Then the vertex containing *last* is removed (cf. (d)). This 2-map is equivalent to the one drawn in (e) due to the folding of the image in a cylinder. Indeed the two pointels associated with darts *last* and $\beta_1(last)$ have $n_1 + 1$ as x coordinate. Thanks to the modulo computation (cf. Algorithm 76 line 8), they are moved to $x = 0$ and to the next line. The obtained configuration, regarding pixel $(0, 1)$, is similar to the initial configuration of the initial border for pixel $(0, 0)$. There is a dart to the up of each pixel of the second line, and a dart to the left of the first pixel. The pixels of the second line can thus be processed. The 2-map obtained after processing of all the pixels of the second line is shown in (f).

The different 2-maps obtained after processing each last pixel of the different lines of the image are shown in Fig. 7.21. The 2-map is progressively built, while the initial border goes progressively down. After having considered all the pixels of the image, the 2-map given in Fig. 7.21(f) is obtained.

The last step consists in processing the pixels of the last line which belong to the infinite region. As illustrated in Fig. 7.22, the initial border is progressively disconnected from the 2-map describing the image. After processing the last pixel of the image (pixel $(13, 8)$ in our example), the 2-map describing the image is extracted. Since this last pixel belongs to the same regions than its left and up regions, both edges containing darts *last* and *up* are removed and thus the border is disconnected from the rest of the 2-map. This border is no longer useful, and it can be removed (line 22). At last, the enclosure tree of regions is computed (line 23 of the extraction algorithm). The algorithm is not described here (cf. [66]). The topological map is then computed, i.e. its three parts: the 2-map, its embedding and the enclosure tree of regions.

The complexity of Algorithm 75 is linear in the number of pixels of the image. Indeed, all these pixels are processed once, and all the operations applied in the loop are either atomic, or linear in the number of darts of the considered cells (vertices and edges), and no cell is considered twice.

The construction algorithm can be modified for different uses, by replacing the test if pixels belong to same region (lines 9 and 13). For instance, given a 2D image, a pre-segmentation can be computed while simultaneously constructing the corresponding topological map. Pixels are compared according to the distance between their colors: they belong to the same region if this distance is smaller than a threshold given by the user. This modified version can be useful to decrease the memory space occupation of the topological map when processing big images.

7.3.3.2 Region Merging

Algorithm 78 describes the merging operation of two regions in a topological map. This operation takes as input a topological map plus a dart `d` which belongs to an edge which separates the two regions to merge. It uses as a basic tool the edge removal operation (given in Algorithm 53 page 212). However, as regions can be multi-adjacent, it is necessary to iterate through all the darts belonging to $c_2(\mathtt{d})$ (i.e. the same face containing `d`) and to remove all the edges separating the two regions to merge. Moreover, when an edge is removed, all linels associated with this edge must be switched off (line 5 of the algorithm).

After the foreach loop, all the edges separating the two regions are removed and their corresponding embeddings are switched off. Now, each removable vertex incident to a non loop edge has to be removed. Indeed after the removal of an edge, a vertex which was not removable can become removable. In this case, the vertex must be removed, so that each edge corresponds exactly to an adjacency relation between regions. The last operation is the updating of the enclosure tree of regions. Indeed, the merging of the two regions can modify the enclosed relations. Note that these last two steps concern only the two modified regions and thus modifications are local ones. So, only the vertices incident to the removed edges are checked, as the regions containing and enclosed by the two merged regions.

Algorithm 78: `mergeRegions(cm,d)`: merge two regions in a topological map

Input: `cm`: a 2D topological map;
`d`: a dart.

Result: `cm` is modified by merging the two regions `d.Region` and `d.Betas[2].Region`.

```
1 R2 ← d.Betas[2].Region;
2 merge information of d.Region and R2;
3 foreach dart d' in c2(d) do
4     if d'.Betas[2].Region = R2 then
5         switch off all the linels between d'.Emb and d'.Betas[2].Emb;
6         removeNMap(cm,d',1);
7 remove each removable vertex incident to a non loop edge;
8 update the enclosure tree of regions of cm;
```

Let us consider the topological map of our example (given in Fig. 7.16) where the merge algorithm is applied to dart 2, which separates regions R_1 and R_3. In the main loop, edges $\{2, 7\}$ and $\{4, 9\}$ are removed, which both belong to common boundary of these two regions. For the embedding, the five linels corresponding to these two edges are switched off. This is achieved by starting from the pair (p, l) given by `d'.Emb`, and following the path of linels until obtaining the second extremity of the path given by `d'.Betas[2].Emb`. All the linels in the path have exactly one successor thus there is no ambiguity when following the linels. In our example, when edge $\{2, 7\}$ is removed, we start from the pair 2.`Emb` $= (a, 3)$. This linel is switched off, as the next one (the second linel of the edge). Then pointel b is obtained which is the pointel of the dart $7 = \beta_2(2)$: the end of the path is reached. Similarly, when edge $\{4, 9\}$ is removed, its three linels are switched off.

This produces the 2-map shown in Fig. 7.23(a), which is not a topological map because some adjacency relations are represented by more than one edge. Region $R'_1 = R_1 \cup R_3$ is now adjacent once to region R_0 and once to region R_2 but each of these two adjacency relations is described by two edges in the 2-map.

To solve this problem, the four vertices incident to the removed edges will be checked during the next step of the algorithm. Vertices $\{3, 6\}$ and $\{8, 22\}$ are removed, because they are removable and incident to non loop edges. After these removals the two other vertices $\{1, 21\}$ and $\{5, 10\}$ are incident to loops: thus they are not removed (otherwise a boundary between two different regions would be completely removed). This produces the 2-map shown in Fig. 7.24. Note that when a vertex is removed, the embedding is updated by switching to off its corresponding pointel. In our example, the two pointels a

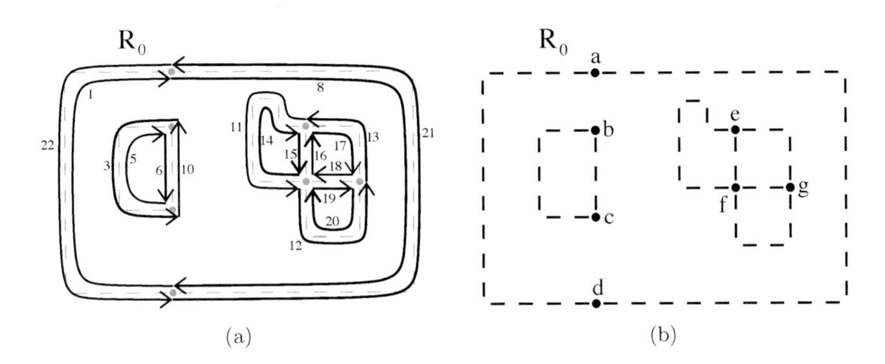

FIGURE 7.23
(a) 2-map obtained from the 2-map of Fig. 7.16(a) after the removal of the two edges between regions R_1 and R_3.
(b) Its embedding in the interpixel matrix.

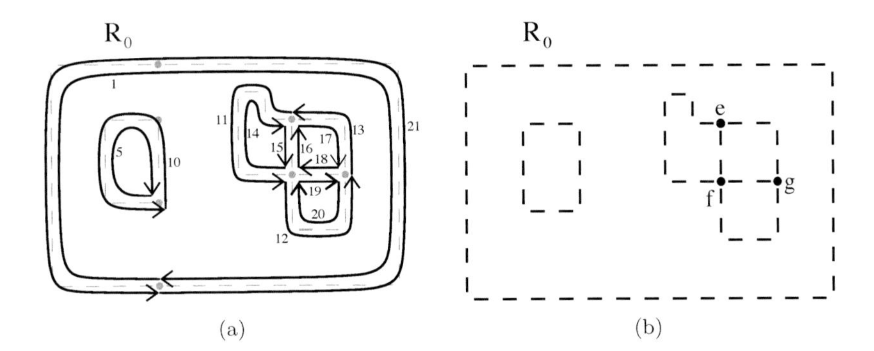

FIGURE 7.24
(a) 2-map obtained from the 2-map of Fig. 7.23(a) after the removal of removable vertices incident to non loop.
(b) Its embedding in the interpixel matrix.

and b are switched off, because the corresponding vertices are removed, and the two pointels c and d are switched off, because the corresponding vertices are incident to loops. For instance, $1.$`Emb` $= (d, 2)$ and $21.$`Emb` $= (d, 0)$ and d is now off.

The last step of the merging algorithm is the updating of the enclosure tree of regions (the algorithm is not provided here). The first modification consists in removing the node in the tree corresponding to the second region, and in moving all the sons of this node as son of the node corresponding to the first region. Indeed, the two regions are now merged into one region, which contains all the regions enclosed initially in the two merged regions.

Then the enclosed relations are updated when the merging of the two regions creates new enclosed regions, or removes enclosed regions. In our example, the merging of regions R_1 and R_3 has created a new enclosed region. Indeed, R_2 is now enclosed in R'_1 while R_2 was not enclosed neither in R_1 nor in R_3 before the merging. All sons of R'_1 can be reconstructed in the tree by iterating through all of its darts, and by grouping these darts by connected components relative to $\langle\beta_1\rangle$. Each connected component corresponds to a boundary of the new region: the first is the external boundary, the others are internal boundaries. For each internal boundary, we iterate through the corresponding connected component, reaching thus all the regions enclosed in the new region and belonging to the same connected component.

7.3.3.3 Image Segmentation

The region merging operation is the basic tool for the generic image segmentation method, described in Algorithm 79. This algorithm follows a bottom-up approach: it consists in merging adjacent regions according to a given oracle. This is a direct translation of similar algorithms defined on region adjacency graphs. First (line 1), a list of darts is constructed, having one dart per each edge of the topological map. Then all the darts in this list are processed so each edge is processed exactly once. For each dart `d`, `Oracle` says if the edge containing `d` must be removed, and when this is the case the `mergeRegions` algorithm is applied in order to merge the two regions around the given edge.

Algorithm 79: 2D image segmentation with topological map

Input: `cm`: a 2D topological map describing a labeled image;
`Oracle`: an oracle.

Result: `cm` is modified to describe the segmentation of the labeled image according to `Oracle`.

```
1 List ← one dart per edge of cm;
2 foreach dart d in List do
3     if Oracle(d) then
4         mergeRegions(cm,d);
```

This segmentation algorithm is generic, because it is guided by `Oracle` which can be any user defined function. In the following, different oracles are defined for gray level images, but similar functions can be defined for color images. A first simple oracle is defined in Algorithm 80, by using the mean gray level of each region, and allowing to merge two regions if the distance between their two mean gray levels is smaller than a given threshold τ defined by the user.

Algorithm 80: `criterionMeanGray(d)`: merging criterion based on mean gray level

Input: `d`: a dart.
Result: `true` iff `d.Region` and `d.Betas[2].Region` must be merged.
1 $R_1 \leftarrow$ `d.Region`; $R_2 \leftarrow$ `d.Betas[2].Region`;
2 $dist \leftarrow |R_1.\text{mean\_gray} - R_2.\text{mean\_gray}|$;
3 **return** $dist < \tau$;

This criterion is computed by adding the sum of all the pixel gray levels and the size of each region to the information associated with each region. Thanks to these two values, the mean gray level of each region is computed in constant time. These two values can be computed directly during the construction algorithm (Algorithm 75). Moreover they can be updated incrementally in constant time when two regions are merged (line 2 of Algorithm 78).

The main advantage of this criterion is its simplicity. However it does not produce very good segmentation results. A better solution is proposed in [105] using external and internal contrasts. $e.ext$ is the external contrast associated with each edge e of the topological map. It is the minimal gray level difference between two 4-adjacent pixels separated by a linel belonging to edge e. $R.int$ is the internal contrast associated with each region R of the topological map. It is the maximal gray level difference between two pixels belonging to the minimal spanning tree of pixels in R. Intuitively, $R.int$ is the maximal value separating two pixels in R according to minimum paths linking these pixels (minimum in term of gray value differences).

The external contrast can be initialized during the construction algorithm. When a new edge is created, its external contrast is initialized as the gray value difference of the two pixels separated by the edge. When a vertex is removed, the two incident edges are merged and the external contrast of the resulting edge is updated by keeping the minimum value of the two initial edges. The internal contrast can also be initialized and updated during the segmentation algorithm. Indeed, $R.int$ is equal to 0 if all the pixels in R have the same gray level value, and $R_1'.int = e.ext$ when two regions R_1 and R_2 are merged along edge e, e being the edge with the smallest external contrast separating the two regions.

Thanks to these contrasts, we can decide if two regions must be merged or not using Algorithm 81. Given a dart `d` belonging to the edge with the smallest external contrast separating two regions R_1 and R_2, the algorithm tests if the external contrast of the edge is smaller than the minimum internal contrast of the two merged regions. Instead of directly consider the internal contrast, a value given by the function $f(R)$ is added, which allows to weight the internal contrasts when they are not significant. The authors of [105] propose to use $f(R) = k/|R|$, where $|R|$ is the number of pixels in R and k is a parameter defined by the user allowing to tune the method.

Algorithm 81: `criterionContrast(d)`: merging criterion based on contrasts

Input: `d`: a dart belonging to the edge with the smallest external contrast separating regions `d.Region` and `d.Betas[2].Region`.
Result: `true` iff `d.Region` and `d.Betas[2].Region` must be merged.
1 $R_1 \leftarrow$ `d.Region`; $R_2 \leftarrow$ `d.Betas[2].Region`;
2 **return** $c_1(d).ext \leq min(R_1.int + f(R_1),\ R_2.int + f(R_2))$;

In order to use this criterion, a preliminary step is added in the segmentation algorithm, allowing to sort all the edges regarding their external contrast. So, the list of darts is sorted just after its construction (after line 1 of Algorithm 79). Then, in the foreach loop, these edges will be processed by increasing external contrast order.

A third merging criterion is defined in Algorithm 82, but contrary to the two previous ones, a geometrical criterion is used instead of a colorimetric one. Two regions are merged if the size of the smaller one is smaller than a threshold ς defined by the user. This operation can be used after a segmentation process, to clean up the result of the segmentation by removing all the small regions, which can often be considered as noise. Additional information is the size of each region, stored in all regions, initialized during the construction algorithm and updated during the region merging method.

Algorithm 82: `criterionSize(d)`: merging criterion based on regions size

Input: `d`: a dart.
Result: `true` iff `d.Region` and `d.Betas[2].Region` must be merged.
1 $R_1 \leftarrow$ `d.Region`; $R_2 \leftarrow$ `d.Betas[2].Region`;
2 $minsize \leftarrow min(R_1.size, R_2.size)$;
3 **return** $minsize < \varsigma$;

Some results of the segmentation algorithm using 2D topological map are presented in Fig. 7.25 for three 2D images: *fruits*, *bike* and *hut*, with sizes 800×600 pixels. These images are segmented according to the mean gray level criterion, producing the images given in the left column. Then these images are segmented again, but now with the size of regions criterion, producing the images given in the right column. The size of regions criterion allows us to

FIGURE 7.25
Image segmentation using the mean gray level criterion (left column), then the size of regions criterion (right column).
(a) $\tau = 20$. (b) $\varsigma = 20$. (c) $\tau = 35$. (d) $\varsigma = 20$. (e) $\tau = 30$. (f) $\varsigma = 20$.

TABLE 7.1
Information on the different topological maps obtained by segmentation using mean gray level and size criteria. *Original* is the topological map constructed from the nonsegmented image; *Segmented* is the topological map obtained from the segmentation using the mean gray level criterion; *Small regions* is the final topological map obtained from the segmentation using the size of regions criterion.

	Original		Segmented		Small regions	
Image	#Darts	#Regions	#Darts	#Regions	#Darts	#Regions
Fruits	1,661,114	384,798	21,092	5,005	2,292	475
Bike	1,782,074	423,081	32,254	7,944	1,406	304
Hut	1,807,908	435,105	43,668	10,940	4,054	878

clean up the results of the first segmentation method. These results illustrate one interest of our image segmentation algorithm which can start from any image partition (i.e. from an image already segmented or not). This allows to chain up several steps using different criteria.

The results obtained with the same images by using the criterion based on contrasts are shown in Fig. 7.26. This second criterion preserves better the boundaries of the object (this is for example visible for the fruits).

The two tables 7.1 and 7.2 describe the numbers of darts and of regions of the corresponding topological maps obtained in our experiments. The numbers of darts and regions of the topological map extracted from a nonsegmented image are important (414 000 regions, and 1,750,000 darts in average). After the segmentation step, there are 7,600 regions and 25,500 darts in average. Then the segmentation based on the size of regions produces topological maps having small numbers of darts and regions (470 regions and 1,940 darts in average).

Other operations have been defined on 2D topological map: we can cite for example a robust region filling algorithm [65], a deformable partition method preserving the topology of the whole partition [74], a topological criterion allowing to control the evolution of the Betti numbers during the segmentation [100] or a method to approximate the digital polygonal curves of an image [69]. Indeed, the main advantage of topological map is to fully describe the topology of the image partition, while allowing easily to add geometric and colorimetric information. Moreover efficient operations can be defined for modifying the partition. For these reasons, they are a very good basic tool for efficient image processing methods.

Moreover, topopological maps and corresponding operations introduced in this chapter have been extended in 3D [59]. Lastly, a hierarchical representation based on topological maps has been proposed [44, 45, 200] in order to

(a) (b) (c) (d) (e) (f)

FIGURE 7.26
Image segmentation using the contrasts criterion (left column), then the size of regions criterion (right column).
(a) $k = 17000$. (b) $\varsigma = 20$. (c) $k = 17000$. (d) $\varsigma = 20$. (e) $k = 17000$. (f) $\varsigma = 20$.

TABLE 7.2
Information on the different topological maps obtained by segmentation using contrasts and size criteria. *Original* if the topological map constructed from the nonsegmented image; *Segmented* is the topological map obtained after the segmentation using the contrasts criterion; *Small regions* is the final topological map obtained after the segmentation using the size of regions criterion.

	Original		Segmented		Small regions	
Image	#Darts	#Regions	#Darts	#Regions	#Darts	#Regions
Fruits	1,661,114	384,798	11,228	4,485	1,050	320
Bike	1,782,074	423,081	25,360	10,013	1,320	401
Hut	1,807,908	435,105	19,448	7,480	1,532	446

represent different segmentations of a same image at different levels. All these works illustrate the interest of n-maps for image processing.

8

Cellular Structures as Structured Simplicial Structures

So far we have been concerned with the definitions of n-Gmaps and n-maps, construction operations and embedding definitions for applications in geometric modeling, computational geometry, discrete geometry, computer graphics, image processing and analysis. But note that only *examples* of geometric objects which can be associated with n-Gmaps and n-maps have been provided. An important question remains: *what are the geometric objects* which correspond to n-Gmaps and n-maps, and more generally to cellular structures, including incidence graphs?

For instance:

- assume an incidence graph contains an edge incident to three vertices: is it possible to associate a geometric object with it[1]?
- assume a 2-dimensional incidence graph or a 2-Gmap describes a subdivided torus. Add a 3-dimensional cell which has the torus as boundary (by adding a 3-dimensional cell to the incidence graph, which is incident to all the faces of the torus or by increasing the dimension of the 2-Gmap). What is the corresponding object? a full torus[2]?

In fact, it is always possible to associate a simplicial analog *with any cellular structure. This simplicial analog is a triangulation: it is made of simplices, i.e. vertices, edges, triangles, tetrahedra, etc., and it formally describes the topology of the object which is associated with the cellular structure.*

But it looks strange to define the interpretation of a cellular structure as a simplicial object. Why not establish a direct relation between cellular structures and "subdivided objects"[3]?

- classical objects in topology are simplicial complexes (cf. Section 8.1.1) and CW-complexes [174, 1, 115] (cf. Section 8.1.2). The cells of these objects are *homeomorphic to balls*, and we have seen above that cellular structures exist, which cells do not correspond to balls. Moreover, as far as we know, no combinatorial characterization of balls exists, i.e. it is

[1] Yes

[2] No

[3] Note that "subdivided objects" are so far not formally defined!

not possible to check a cellular structure in order to know if (a subpart of) it corresponds to a ball: so, given a cellular structure, it is not even possible to decide whether a CW-complex can be directly associated with it.

- CW-complexes exist, which cannot be triangulated. But all cellular structures (as n-Gmaps or incidence graphs) have a simplicial analog. That means that only triangulable objects are taken into account by these cellular structures. This corresponds to the fact that these cellular structures are combinatorial structures (defined by a set of discrete objects on which discrete applications act), and that only simplicial objects (and objects derived from simplicial ones) have a combinatorial characterization (cf. Section 8.1.1 and Section 8.1.2).

The intuitive idea of simplicial analogs is related to the classical notion of *barycentric triangulation*. Let O be a k-dimensional "subdivided object" in $\mathbb{R}^n$, such that all its cells are convex ones (as a consequence, any cell is homeomorphic to a ball). The barycentric triangulation of O is constructed incrementally in the following way (cf. Fig. 8.1):

- "triangulate" all vertices: it simply consists in replacing all vertices of O by vertices numbered 0 (note that 0 is the dimension of the vertices);
- triangulate all edges. Given an edge, it consists in adding a new vertex numbered 1 at the barycenter of the edge (1 being the dimension of the edge), and in replacing the edge by its triangulation related to this new vertex, i.e. by two edges both incident to this new vertex and to an extremity of the initial edge;
- triangulate all faces. Given a face, it consists in adding a new vertex numbered 2 at the barycenter of the face (2 being the dimension of the face), and in replacing the face by its triangulation related to this new vertex, i.e. by a set of triangles, each incident to a vertex numbered 0, to a vertex numbered 1 and to a vertex numbered 2;
- iterate this process for all i-cells, for i increasing from 3 to k; the last triangulation produces the barycentric triangulation of O.

Note that any *main* i-dimensional simplex of the barycentric triangulation (i.e. a simplex which is not incident to a higher-dimensional simplex) is incident to $i+1$ vertices, numbered from 0 to i. More generally:

- there is a bijection between initial i-dimensional cells and vertices numbered i;
- the interior of any initial i-dimensional cell corresponds to the associated vertex numbered i, and all j-dimensional simplices ($1 \leq j \leq i$) which are incident to this vertex and such that their incident vertices are numbered by integers lower than or equal to i.

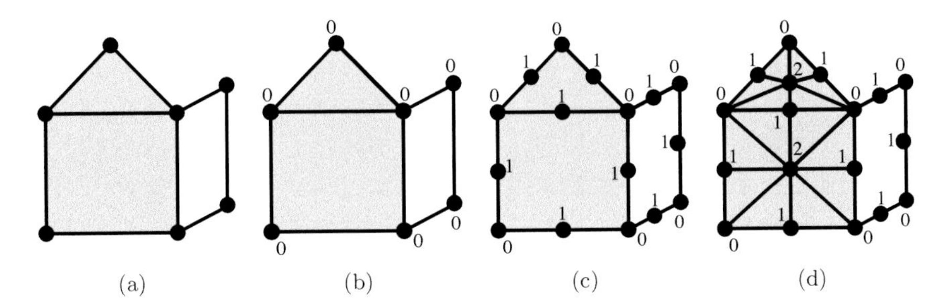

FIGURE 8.1
(a) Initial 2-dimensional geometric object.
(b) Barycentric triangulation of 0-cells.
(c) Barycentric triangulation of 1-cells.
(d) Barycentric triangulation of 2-cells, producing the barycentric triangulation of the initial object.

So, given a geometric object, satisfying some geometric properties, it is possible to associate a geometric simplicial analog. Now, geometry will be omitted; more generally, we will see how to associate a simplicial (combinatorial) analog with *any* cellular structure studied in this book (n-Gmap or incidence graph). The vertices of the simplicial analog will also be numbered, providing a representation of the initial cells; when the cellular structure can be associated with a geometric object with convex cells in $\mathbb{R}^n$, the simplicial analog of the cellular structure corresponds to the barycentric triangulation of the geometric object.

First, two main classes of simplicial (combinatorial) structures will be studied, together with the relations between theses classes. Then, the simplicial analogs of cellular (combinatorial) structures will be defined. At last, several consequences will be deduced from the correspondences between simplicial and cellular structures.

8.1 Simplicial Structures

8.1.1 Abstract Simplicial Complexes

This section is mainly based upon [1].

Abstract simplicial complexes

Definition 76 (abstract simplicial complex) *An abstract simplicial complex K is a set of nonempty subsets of a given finite set V such that:*

1. $\{v\} \in K$ *for every* $v \in V$;
2. $\tau \subseteq \sigma \in K \implies \tau \in K$.

The elements of K are called *simplices.* Let $\sigma \in K$ such that $|\sigma| = i+1$; σ is called an (abstract) i-dimensional simplex, or i-simplex. The elements of V are called *vertices*, and any vertex $v \in V$ is assimilated with the 0-simplex $\{v\} \in K$.

Let $\sigma \in K$. $\tau \subseteq \sigma$ is a *face* of σ: if $\tau = \sigma$, then it is the *principal* face of σ, or *main* face, else it is a *proper* face of σ. The *boundary* of σ is the set of its proper faces; it is an abstract simplicial complex. The *star* (resp. *proper star*) of σ is the set of simplices of which σ is a face (resp. a proper face): so, σ is contained in its star, but not in its proper star. A *principal*, or *main*, simplex of K is such that its proper star is empty.

Example. Let $V_{ex} = \{v_1, v_2, v_3, v_4, v_5, v_6\}$, and $K_{ex} = \{\{v_1\}, \{v_2\}, \{v_3\}, \{v_4\}, \{v_5\}, \{v_6\}, \{v_1, v_2\}, \{v_1, v_3\}, \{v_2, v_3\}, \{v_2, v_6\}, \{v_3, v_4\}, \{v_3, v_5\}, \{v_3, v_6\}, \{v_4, v_5\}, \{v_5, v_6\}, \{v_1, v_2, v_3\}, \{v_2, v_3, v_6\}, \{v_3, v_4, v_5\}\}$. The boundary of $\{v_1, v_2, v_3\}$ is $\{\{v_1\}, \{v_2\}, \{v_3\}, \{v_1, v_2\}, \{v_1, v_3\}, \{v_2, v_3\}\}$. The star of $\{v_3\}$ (resp. $\{v_2, v_3\}$) is $\{\{v_3\}, \{v_1, v_3\}, \{v_2, v_3\}, \{v_3, v_4\}, \{v_3, v_5\}, \{v_3, v_6\}, \{v_1, v_2, v_3\}, \{v_2, v_3, v_6\}, \{v_3, v_4, v_5\}\}$ (resp. $\{\{v_2, v_3\}, \{v_1, v_2, v_3\}, \{v_2, v_3, v_6\}\}$). Note that K_{ex}, as any abstract simplicial complex, is completely defined by its main simplices $\{v_5, v_6\}, \{v_1, v_2, v_3\}$, $\{v_2, v_3, v_6\}, \{v_3, v_4, v_5\}$, due to property 2 of Def. 76.

Functions can be defined on abstract simplicial complexes.

Definition 77 (abstract simplicial map) *Let K and L be abstract simplicial complexes. An abstract simplicial map $f : K \to L$ is a map from the vertices of K to the vertices of L which satisfies: if $v_0, v_1, \ldots, v_i$ are vertices of a simplex in K, then $f(v_0), f(v_1), \ldots, f(v_i)$ are vertices of a simplex in L. If f is bijective, f is called an isomorphism, and K and L are said to be isomorphic.*

Example. An abstract simplicial map exists between the abstract simplicial complex K_{ex} defined in the previous example and the abstract simplicial complex K'_{ex} defined as follows: let $V'_{ex} = \{v'_1, v'_2, v'_3, v'_4, v'_6\}$, and $K'_{ex} = \{\{v'_1\}, \{v'_2\}, \{v'_3\}, \{v'_4\}, \{v'_6\}, \{v'_1, v'_2\}, \{v'_1, v'_3\}, \{v'_2, v'_3\}, \{v'_2, v'_6\}, \{v'_3, v'_4\}, \{v'_3, v'_6\}, \{v'_1, v'_2, v'_3\}, \{v'_2, v'_3, v'_6\}\}$. This abstract simplicial map associates v_1 with v'_1, v_2 with v'_2, v_3 with v'_3, v_4 with v'_4, v_5 with v'_3, v_6 with v'_6,

Simplicial Complexes

Let K be an abstract simplicial complex: K is a combinatorial object. A geometric object can be associated with K, it is the *geometric realization* of K:

cf. Fig. 8.2(a) (resp. Fig. 8.2(b)), representing the geometric realization of K_{ex} (resp. K'_{ex}), defined in the previous examples. More precisely, the geometric realization of an abstract simplicial complex is a *simplicial complex*, as defined below.

The points $P_0, P_1, \ldots, P_i$ of $\mathbb{R}^n$ are said to be *linearly independent* if the vectors $P_1 - P_0, P_2 - P_0, \ldots, P_i - P_0$ are linearly independent. The *line segment* between two points x and y in $\mathbb{R}^n$ is the set $\{z = tx + (1-t)y | t \in [0,1]\}$. Let $A \subseteq \mathbb{R}^n$: A is *convex* if $x, y \in A$ implies that the line segment between x and y is contained in A. The *convex hull* of $B \subseteq \mathbb{R}^n$ is the intersection of all convex sets containing B.

Definition 78 (simplex) *Let $i \geq 0$. An i-dimensional simplex σ, or i-simplex, is the convex hull of $i+1$ linearly independent points $P_0, P_1, \ldots, P_i$.*

σ is denoted $P_0 P_1 \ldots P_i$. The points P_j are the *vertices* of σ. Let $h \leq i$: τ is an h-dimensional *face* of σ if it is an h-simplex which vertices are vertices of σ. Every point w of σ can be written uniquely in the form:

$$w = \sum_{j=0}^{i} \lambda_j P_j, \lambda_j \in [0,1] \forall j, \sum_{j=0}^{i} \lambda_j = 1$$

The λ_j are the *barycentric coordinates* of w. w is an *interior* point of σ if $\lambda_j > 0$ for all j; else it is a *boundary* point of σ. The *interior* (resp. *boundary*) of σ is the set of all its interior (resp. boundary) points.

Definition 79 (simplicial complex) *A simplicial complex is a finite set of simplices in some $\mathbb{R}^n$ satisfying:*

1. *$\sigma \in K \implies$ all faces of σ belong to K;*
2. *if $\sigma, \tau \in K$, then either $\sigma \cap \tau = \emptyset$, either $\sigma \cap \tau$ is a common face of σ and τ.*

The subset $|K| = (\bigcup_{\sigma \in K} \sigma)$ of $\mathbb{R}^n$ is the *underlying space* of K. The dimension of K is defined to be -1 if $K = \emptyset$, and the maximum of the dimensions of the simplices of K otherwise.

Functions can be defined on simplicial complexes: cf. Fig. 8.2.

Definition 80 (simplicial map) *Let K and L be simplicial complexes. A simplicial map $f : K \to L$ is a map from the vertices of K to the vertices of L which satisfy: if $P_0, P_1, \ldots, P_i$ are vertices of a simplex in K, then $f(P_0), f(P_1), \ldots, f(P_i)$ are vertices of a simplex in L. If f is bijective, f is called an isomorphism, and K and L are said to be isomorphic.*

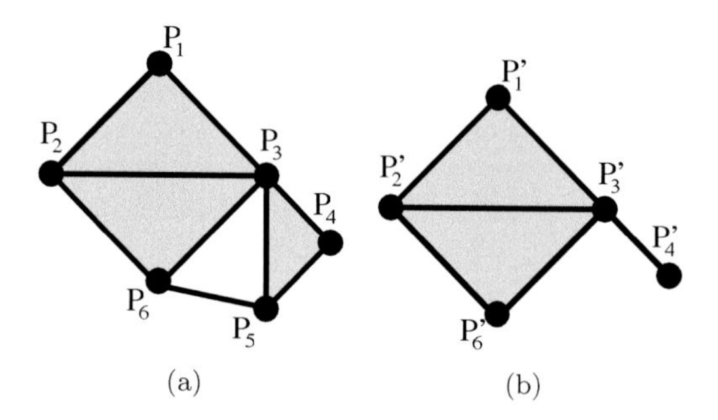

FIGURE 8.2
(a) A simplicial complex.
(b) A simplicial complex, image of the simplicial complex of (a) by the simplicial map which associates P_1 and P_1', P_2 and P_2', P_3 and P_3', P_4 and P_4', P_5 and P_3', P_6 and P_6'.

Simplicial maps induce continuous maps on the underlying spaces. Let $f : K \to L$ be a simplicial map. Define $|f| : |K| \to |L|$ as follows: if $x \in |K|$, then x belongs to the interior of a unique simplex $P_0P_1 \dots P_i$ of K; let $x = \sum_{j=0}^{i} \lambda_j P_j$, where λ_j are the barycentric coordinates of x, and define $|f|(x) = \sum_{j=0}^{i} \lambda_j f(P_j)$. $|f|$ is continuous; if it is a homeomorphism, it is called a linear homeomorphism. Note that $|f|$ is a linear homeomorphism if and only if f is an isomorphism.

Geometric realization

Every simplicial complex K determines an abstract simplicial complex $A(K) = \{\{P_0, P_1, \dots, P_i\} | P_0P_1 \dots P_i \text{ is an } i\text{-simplex of } K\}$.

Conversely:

Definition 81 (geometric realization of an abstract simplicial complex) *Let K be an abstract simplicial complex. A geometric realization of K is a pair $(\varphi, GR(K))$, where $GR(K)$ is a simplicial complex and φ is a bijection between the vertices of K and the vertices of $GR(K)$, such that $\{v_0, v_1, \dots, v_i\}$ is an i-simplex of K if and only if $\varphi(v_0)\varphi(v_1)\dots\varphi(v_i)$ is an i-simplex of $GR(K)$.*

Every abstract simplicial complex has a unique geometric realization, up to isomorphism.

Note also that a simplicial map between simplicial complexes induces an abstract simplicial map between the corresponding abstract simplicial complexes; conversely, an abstract simplicial map between two abstract simplicial complexes induces a simplicial map between their geometric realizations.

Operations

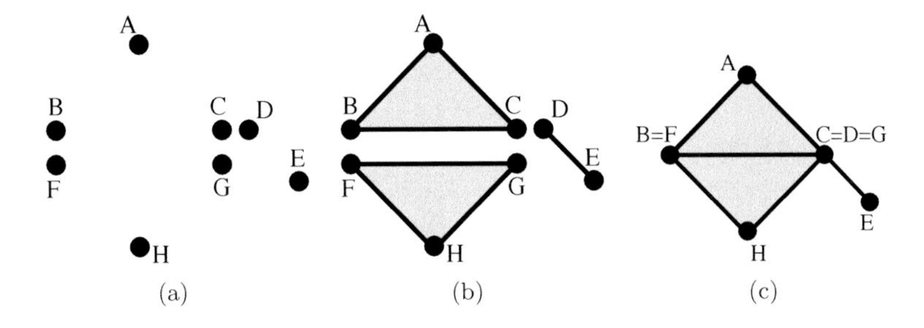

FIGURE 8.3
(a) Adding vertices.
(b) Several cones: for instance, 1-simplex $\{D, E\}$ is the cone of 0-simplex $\{D\}$ according to vertex E; 2-simplex $\{A, B, C\}$ is the cone of 1-simplex $\{A, B\}$ according to vertex C.
(c) Several identifications of vertices, namely B with F, C with D, then $C = D$ with G.

Every abstract simplicial complex can be constructed, starting from an empty abstract simplicial complex corresponding to an empty set of vertices, by the three following operations (cf. Fig. 8.3). Let K be an abstract simplicial complex and V be its set of vertices:

1. adding a vertex to V and the corresponding 0-simplex to K;

2. creating a cone: given a vertex v, and a subcomplex K' of K, this operation consists, for any i-simplex $\sigma = \{v_0, v_1, \dots, v_i\}$ of K', to add to K the $(i + 1)$-simplex $\{v_0, v_1, \dots, v_i, v\}$ and all its faces, excluding simplices which still belong to K. A particular case is the following. An *isolated* i-simplex σ is a main i-simplex together with its boundary, such that all its faces are only incident to σ and to its faces (i.e. the simplex is a connected component). Any isolated i-simplex can be constructed, starting from an isolated $(i-1)$-simplex τ, by adding a new vertex v and by making a cone between τ and v. So, any isolated i-simplex can be constructed by applying i times this operation, starting from an isolated 0-simplex.

3. identifying two vertices v and v': this operation consists in replacing v and v' by a new vertex v", sometimes denoted $v = v'$, in V and in all simplices of K (note that an abstract simplicial map exists between the initial and the resulting abstract simplicial complexes). It is clear that something is changed only if $v \neq v'$. Moreover, since abstract simplices are sets of vertices, and abstract simplicial complexes are sets of simplices, identifying two vertices can lead to:

   - degenerate a simplex: it is the case when v and v' both belong to a simplex. For instance, identifying v_3 with v_5 in the abstract simplicial complex K_{ex} corresponding to Fig. 8.2(a) leads to the degeneracy of 1-simplex $\{v_3, v_5\}$ into 0-simplex $\{v_3 = v_5\}$, and of 2-simplex $\{v_3, v_4, v_5\}$ into 1-simplex $\{v_3 = v_5, v_4\}$;

- identify other simplices. Obvioulsy, 0-simplices $\{v\}$ and $\{v'\}$ are identified into $\{v = v'\}$. Other simplices can be identified: for instance, identifying v_3 with v_5 in the abstract simplicial complex K_{ex} corresponding to Fig. 8.2(a) leads to the identification of $\{v_3, v_4\}$ with $\{v_5, v_4\}$, of $\{v_3, v_6\}$ with $\{v_5, v_6\}$.

8.1.2 Semi-Simplicial Sets

This section is mainly based upon [174].

Semi-Simplicial Sets

Definition 82 (semi-simplicial set) *An* n-dimensional semi-simplicial set $S = (K, (d_j)_{j=0,\dots,n})$ *is defined by (cf. Fig. 8.4):*

- $K = (\overset{n}{\underset{i=0}{\cup}} K_i)$, *where* K_i *is a finite set of elements called* i*-simplices;*
- $\forall j \in \{0, \dots, n\}$, face operator $d_j : K \longrightarrow K$ *is s.t.*[4]*:*
  - $\forall i \in \{1, \dots, n\}, \forall j \in \{0, \dots, i\}$, $d_j : K_i \longrightarrow K_{i-1}$*;* $\forall j > i$, d_j *is undefined on* K_i*, and no face operator is defined on* K_0*;*
  - commutation property of face operators*:* $\forall i \in \{2, \dots, n\}$, $\forall \sigma \in K_i$, $\forall j, k \in \{0, \dots, i\}$, $d_k(d_j(\sigma)) = d_{j-1}(d_k(\sigma))$ *for* $k < j$*.*

Let σ be an i-simplex of S. Simplex τ is a k-face of σ if a sequence of face operators $d_{j_{i-1}}, \cdots, d_{j_k}$ exists such that $\tau = d_{j_k}(\cdots d_{j_{i-1}}(\sigma))$ (note that, due the commutation property of face operators, any sequence of face operators is equivalent to a unique sequence of face operators such that the indices of the operators decrease, i.e. $j_{i-1} > \dots > j_k$). If the sequence is empty, then $\tau = \sigma$ and τ is the *principal*, or *main*, face of σ; else τ is a *proper* face of σ. The boundary of σ is the set of its proper faces (sometimes together with the related face operators: in this case, the boundary is a semi-simplicial set). The star (resp. proper star) of σ is the set of simplices of which σ is a face (resp. proper face). A *principal*, or *main*, simplex is such that its proper star is empty. An oriented graph can be associated with any semi-simplicial set: it is thus easy to extend some notions related to graphs, as that of connected component for instance.

Note that, contrary to abstract simplicial sets, semi-simplicial sets allow multi-incidence between simplices. For instance, an edge can be incident twice to a vertex (it is a loop); more generally, an i-simplex can be incident to less

[4] $d_0, \cdots, d_i$ associate its $(i + 1)$ $(i - 1)-$faces with each $i-$simplex.

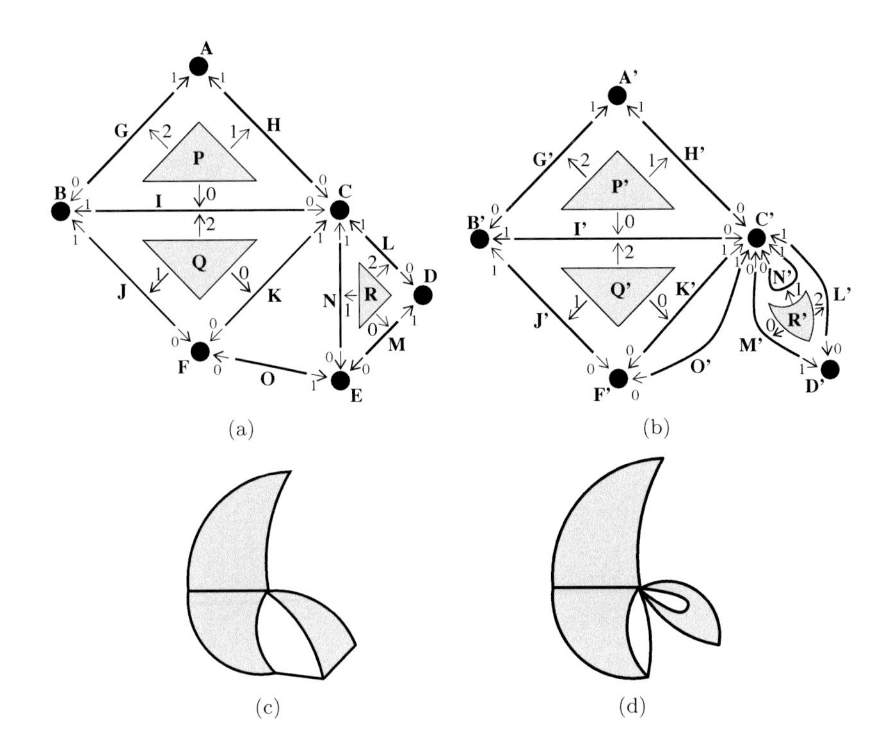

FIGURE 8.4
(a) A semi-simplicial set. An i-simplex is represented as a geometric i-simplex, and face operators are represented by arrows (d_j corresponds to arrows numbered j). B is a 0-face of P, since $d_0(d_2(P)) = d_1(d_0(P)) = B$. The boundary of simplex P contains simplices A, B, C, G, H, I. The star of simplex B contains simplices B, G, P, I, Q, J.
(b) Another semi-simplicial set, such that a semi-simplicial set morphism exists between the semi-simplicial sets depicted on (a) and (b). Note that N' is incident twice to C', and that 2-simplex R' is incident to 0-simplices C' and D'. Note also that L' and M' have the same boundary, as K' and O'.
(c) (resp. (d)) Geometric realization of the semi-simplicial set depicted on (a) (resp. (b)).

than $i+1$ vertices. Moreover, distinct i-simplices can be incident to the same vertices; they can even share the same boundary.

Functions can be defined on semi-simplicial sets.

Definition 83 (semi-simplicial set morphism) *Let $S = (K, (d_j)_{j=0,\dots,n})$ and $S' = (K', (d'_j)_{j=0,\dots,n})$ be semi-simplicial sets. A morphism $f : S \to S'$ is a map from the simplices of K to the simplices of K' which satisfy:*

- $\forall i, 0 \leq i \leq n, f : K_i \to K'_i$;
- $\forall i, 0 < i \leq n, \forall \sigma \in K_i, \forall j, 0 \leq j \leq i, f(d_j(\sigma)) = d'_j(f(\sigma))$.

If f is bijective, f is called an isomorphism, and S and S' are said to be isomorphic.

A semi-simplicial set morphism exists between the two semi-simplicial sets depicted on Fig. 8.4(a) and (b), which associates any simplex Z with simplex Z', except E which is associated with C'.

CW-complexes
This paragraph is based on [180].

Let $\bar{B}^n$ (resp. B^n) denotes an n-dimensional closed (resp. open) ball. In this paragraph, an n-dimensional *open cell* denotes a space which is homeomorphic to B^n.

Definition 84 (*CW*-complex) *A CW-complex is a space X and a collection of disjoint open cells $\{e_\alpha\}$ whose union is X, such that:*

1. *X is Hausdorff[5];*
2. *for each open n-cell e_α of the collection, there exists a continuous map $f_\alpha : \bar{B}^n \to X$ that maps B^n homeomorphically onto e_α and carries $\partial\bar{B}^n$, i.e. the boundary of $\bar{B}^n$, onto a finite union of open cells, each dimension less than n;*
3. *a set A is closed in X if $A \cap \bar{e_\alpha}$ is closed in $\bar{e_\alpha}$ for each α, where $\bar{e_\alpha}$ denotes the closure of e_α.*

Schematically, a *CW*-complex is a set of balls "glued" together by continuous *attaching functions.*

Geometric realization
CW-complexes exist, which cannot be directly associated with semi-simplicial sets; more generally, and as said above, *CW*-complexes exist, which are even not triangulable.

[5] i.e. distinct points of X have disjoint neighborhoods.

Conversely, *the geometric realization of a semi-simplicial set is a CW-complex*. Let $S = (K, (d_j)_{j=0,\dots,n})$ be a semi-simplicial set. For $i = 0$ to n, for any i-simplex σ, associate a closed i-dimensional ball with σ, and define the attaching function f_σ in such a way that it "satisfies" the face operators, i.e. it carries the boundary of the ball onto the union of cells corresponding to $d_j(\sigma)$, for $0 \leq j \leq i$.

Any semi-simplicial set has a unique geometric realization, up to isomorphism (cf. Fig. 8.4(c) and (d)). Note also that a semi-simplicial set morphism between semi-simplicial sets induces a continuous function between the corresponding geometric realizations.

So, a simplex of a semi-simplicial set can have a nonconvex shape. In practice, semi-simplicial sets can be embedded using Bézier triangular spaces [155], for instance.

Operations

Any semi-simplicial set can be constructed, starting from an empty semi-simplicial set, by the three following operations (cf. Fig. 8.5). Let $S = (K = (\bigcup_{i=0}^{n} K_i), (d_j)_{j=0,\dots,n})$ be an n-dimensional semi-simplicial set:

1. adding a new 0-simplex to K_0;
2. creating a cone: let v be a 0-simplex of K_0, and K' be a subset of K such that $(K', (d_j/K'))$ is a semi-simplicial set. Let K" be a set of simplices, such that a bijection φ exists between K' and K". More precisely, for $0 \leq j \leq n$, φ associates j-simplices of K' and $(j+1)$-simplices of K".

   Then, for any j-simplex τ of K":

   - if $j = 1$, $d_1(\tau) = \varphi^{-1}(\tau)$ and $d_0(\tau) = v$;
   - else $d_j(\tau) = \varphi^{-1}(\tau)$ and for $0 \leq k < j$, $d_k(\tau) = \varphi(d_k(\varphi^{-1}(\tau)))$.

   As for abstract simplicial complexes, it is possible to derive an operation for constructing an isolated i-simplex, starting from an isolated $(i-1)$-simplex (remember that an isolated i-simplex is an i-simplex together with its boundary; it is a connected component of the semi-simplicial set, and it is incident to $i+1$ distinct 0-simplices);
3. identifying two simplices. Let σ_1 and σ_2 be two i-simplices, which have the same boundary, i.e. $\forall j, 0 \leq j \leq i$, $d_j(\sigma_1) = d_j(\sigma_2)$ (the condition is trivially satisfied if $i = 0$). Then σ_1 and σ_2 are replaced by a new i-simplex μ, such that:

   - $\forall j, 0 \leq j \leq i, d_j(\mu) = d_j(\sigma_1)$;
   - $\forall \tau \in K_{i+1}, \forall j, 0 \leq j \leq i+1$, if $d_j(\tau) = \sigma_1$ or $d_j(\tau) = \sigma_2$, then $d_j(\tau)$ is modified in order to be equal to μ.

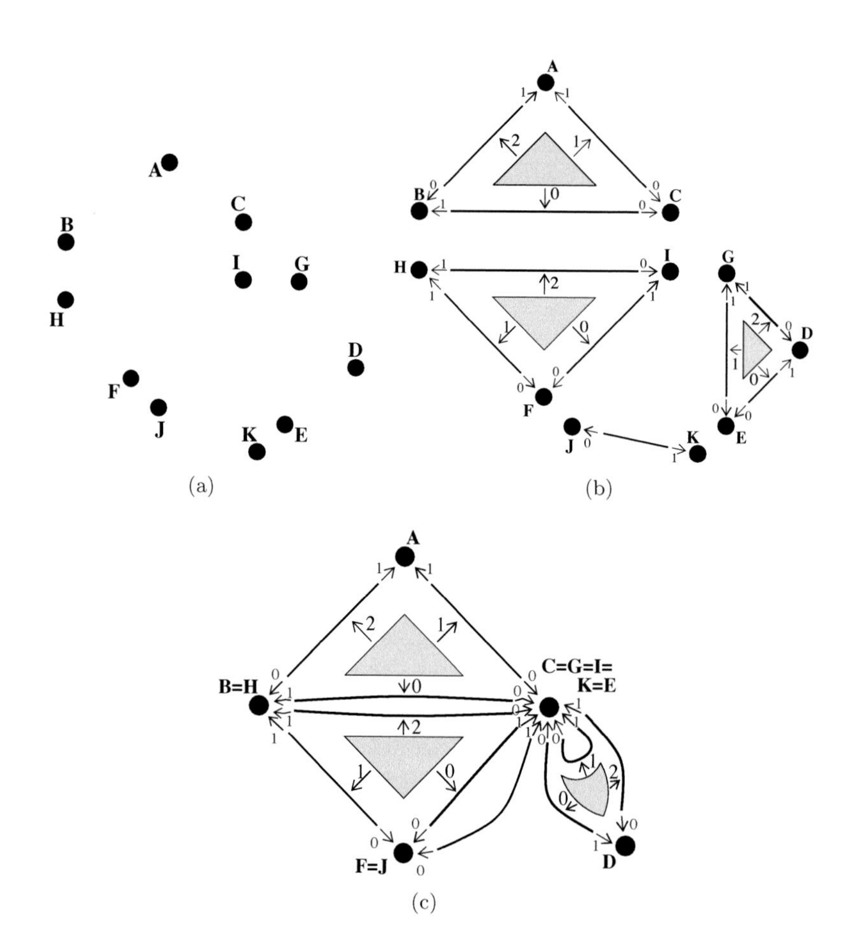

FIGURE 8.5
(a) Adding 11 vertices to an empty semi-simplicial set.
(b) Creating 1- and 2-simplices as cones.
(c) Identifying vertices. Next, the semi-simplicial set depicted on Fig. 8.4(b) is obtained by identifying two 1-simplices.

Nothing is really changed when $\sigma_1 = \sigma_2$. When $\sigma_1 \neq \sigma_2$, contrary to abstract simplicial complexes, there are not "side" effects: all simplices remain in K, except σ_1 and σ_2 which are replaced by μ: no other simplices are identified, and no simplex is degenerated. Note that a semi-simplicial set morphism exists between the initial and the resulting semi-simplicial sets. At last, note also that this operation can be generalized. Given any two i-simplices, the generalized operation first identifies the boundaries of the simplices by applying several times the basic identification operation, and then identifies the i-simplices themselves.

8.1.3 Conversions of Simplicial Structures

Semi-simplicial sets are a "more general" structure than abstract simplicial complexes: as said above, semi-simplicial sets allow multi-incidence between simplices, but abstract simplicial complexes do not. That means that a semi-simplicial set can be associated with any abstract simplicial complex, but the converse is not true. More precisely:

1. Let K be an abstract semi-simplicial set defined on the set of vertices V. A semi-simplicial set can be associated with K in the following way. Define an order on the vertices of V. Associate with any simplex (which is a set of vertices) the *sequence* of its vertices according to the order defined on V, and associate a simplex in the semi-simplicial set with this sequence. Define now the face operators in the following way: let σ be an i-simplex associated with a sequence $(v_0, \ldots, v_j, \ldots, v_i)$; for any $j, 0 \leq j \leq i$, $d_j(\sigma)$ is the $(i-1)$-simplex associated with the sequence $(v_0, \ldots, \hat{v_j}, \ldots, v_i)$, where $\hat{v_j}$ denotes that v_j is removed. For instance, the semi-simplicial set corresponding to the abstract simplicial complex depicted on Fig. 8.2(a), where the order of the vertices corresponds to the order of the points, i.e. from P_1 to P_6, is depicted on Fig. 8.4(a). Note that the commutation property of face operators is satisfied by this construction. Note also that several nonisomorphic semi-simplicial sets can be associated with a single abstract simplicial complex, depending on the order chosen on V.

2. Conversely, let S be a semi-simplicial set, such that:

   - any i-simplex σ is incident to $(i+1)$ distinct 0-simplices; in other words, the set $V(\sigma) = \{d_0(d_1(\ldots \hat{d_j}(\ldots d_i(\sigma)))), 0 \leq j \leq i\}$ contains $(i+1)$ 0-simplices;
   - for any $i, 1 \leq i \leq n$, for any distinct i-simplices σ and τ, the sets of vertices incident to σ and τ, i.e. $V(\sigma)$ and $V(\tau)$, are distinct.

   In this case, an abstract simplicial complex K can be associated with S: the set of vertices associated with K is the set of 0-simplices of S, and the simplices of K are the sets of vertices associated with the simplices of S.

At last, note that a simplex of an abstract simplicial complex is not oriented: it is a set of vertices, and no order is defined on this set.

On the contrary, a simplex of a semi-simplicial is implicitly oriented. For instance, let σ be an i-simplex incident to $i+1$ distinct vertices, and let $v_j = d_0(d_1(\ldots \hat{d_j}(\ldots d_i(\sigma))))$ for $0 \leq j \leq i$. The sequence $(v_0, \ldots, v_j, \ldots, v_i)$ corresponds to an orientation[6] of σ. This orientation notion can be generalized for i-simplices incident to less than $i+1$ vertices. But if the simplices are oriented, the orientability notion has no meaning in general for semi-simplicial sets, since the orientability notion is defined for a *subclass* of semi-simplicial sets: cf. Section 8.2.

8.2 Numbered Simplicial Structures and Cellular Structures

See for instance [39, 165, 85].

8.2.1 Numbered Simplicial Structures

Numbered semi-simplicial sets

Definition 85 (numbered semi-simplicial set) *A numbered semi-simplicial set (S, ν) is an n-dimensional semi-simplicial set S together with a function $\nu : K_0 \to [0, n]$, where K_0 is the set of 0-simplices of S, such that:*

- *for any main 0-simplex σ of S, $\nu(\sigma) = 0$;*
- *for any $i, 1 \leq i \leq n$, for any main i-simplex σ of S, the set of integers associated by ν with the set of 0-simplices incident to σ is $[0, i]$.*

A consequence of this definition is that any i-simplex is incident to $(i+1)$ distinct vertices: cf. Fig. 8.6(a). Moreover, it is possible to define the face operators $(d_j)_{j=0,\ldots,n}$ in such a way that, for any $i, 0 \leq i \leq n$ and for any i-simplex σ, $\nu(v_0) < \ldots < \nu(v_j) < \ldots < \nu(v_i)$, where $v_j = d_0(d_1(\ldots \hat{d_j}(\ldots d_i(\sigma))))$ for $0 \leq j \leq i$; note that if σ is a main simplex, $\nu(v_j) = j$, for $0 \leq j \leq i$. If $i = 0$, σ is said to be numbered $(\nu(\sigma))$, else σ is numbered $(\nu(v_0), \ldots, \nu(v_i))$. Note that semi-simplicial sets exist, which cannot be numbered: cf. Fig. 8.6(b).

An *i-dimensional cell*, or *i-cell*, is a 0-simplex numbered (i), together with all incident j-simplices numbered $(\nu_0, \ldots, \nu_{j-1}, i)$, $1 \leq j \leq i$: cf. Fig. 8.6(c).

---

[6] For instance, take an n-sided polygon P, incident to vertices $P_1, \ldots, P_n$. The sequence of vertices $S = (P_1, \ldots, P_n)$ defines an orientation of P; any sequence S' obtained from S by applying an even number of transpositions corresponds to the same orientation; any sequence S" obtained from S by applying an odd number of transpositions corresponds to the inverse orientation.

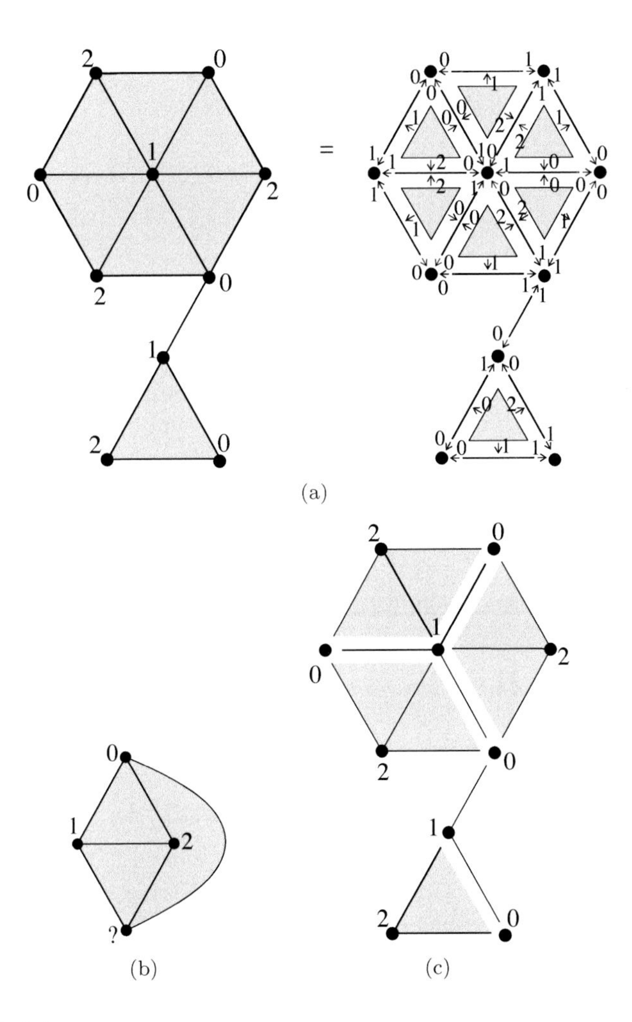

FIGURE 8.6
(a) A numbered semi-simplicial set. Note that the numbering of face operators corresponds to the numbering of 0-simplices.
(b) A semi-simplicial set which cannot be numbered.
(c) Cells of the numbered semi-simplicial set of (a).

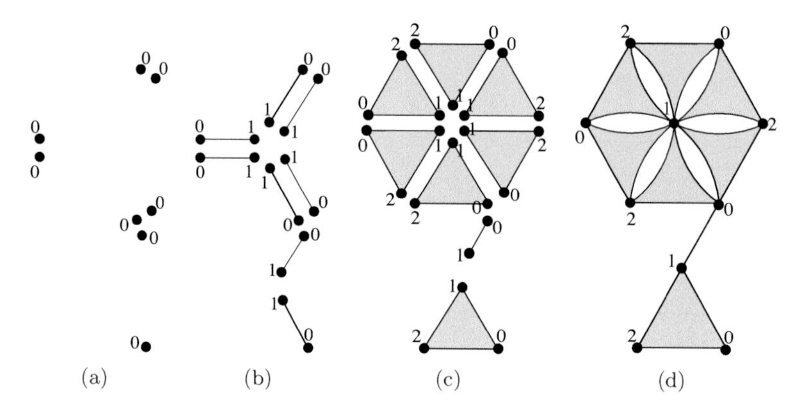

FIGURE 8.7
(a) Adding 0-simplices.
(b) Creating isolated 1-simplices by cones on 0-simplices.
(c) Creating isolated 2-simplices by cones on 1-simplices.
(d) Identification of 0-simplices. The numbered semi-simplicial set of Fig. 8.6(a) can now be constructed by identifying some 1-simplices.

Any numbered semi-simplicial set can be constructed by the three following operations, starting from an empty semi-simplicial set (cf. Fig. 8.7):

- adding a new 0-simplex numbered (0);
- creating an isolated i-simplex numbered $(0, \ldots, i)$, by adding a new 0-simplex numbered (i) and making a cone between this 0-simplex and an isolated $(i-1)$-simplex;
- identifying two i-simplices σ_1 and σ_2, if they satisfy the following condition: either $i = 0$ and $\nu(\sigma_1) = \nu(\sigma_2)$, either $i \neq 0$ and σ_1 and σ_2 have the same boundary. This operation can be generalized: given two i-simplices σ_1 and σ_2 which have the same numbering, the generalized operation first identifies the boundaries of σ_1 and σ_2, by applying several times the basic identification operation, and then identifies σ_1 and σ_2 themselves.

Numbered simplicial quasi-manifolds

Definition 86 (numbered simplicial quasi-manifold) *An n-dimensional numbered simplicial quasi-manifold is a numbered semi-simplicial set which can be constructed by the following operations, starting from an empty semi-simplicial set (cf. Fig. 8.8(a)):*

- *adding isolated n-simplices;*

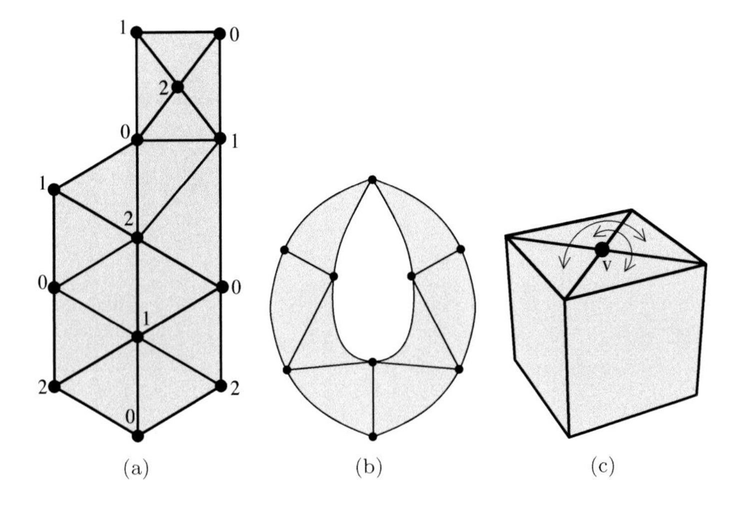

FIGURE 8.8
(a) A 2-dimensional numbered simplicial quasi-manifold: it can be constructed by adding isolated 2-simplices numbered $(0, 1, 2)$, and by identifying 1-simplices, in such a way that at most two 2-simplices are incident to any 1-simplex. Note that cells (together with their boundaries) can exist, which are not numbered simplicial quasi-manifolds: for instance, edges exist, made of three 1-simplices numbered $(0, 1)$ incident to a 0-simplex numbered 1.
(b) A pseudo-manifold.
(c) A cube. The upper face is subdivided into four triangles which share vertex v, corresponding to the center of the initial face of the cube. Identify two by two the four triangles, as depicted by the two arrows (this is not possible in the usual 3-dimensional space). The resulting quasi-manifold does not correspond to a manifold, since the neighborhood of v is a cone on a torus.

- *identifying $(n-1)$-simplices and their boundaries, in such a way that at most two n-simplices are incident to any $(n-1)$-simplex.*

This notion corresponds to a particular case of the well-known notion of *pseudo-manifold* (cf. Fig. 8.8(b)). A pseudo-manifold X is the geometric realization of an n-dimensional semi-simplicial set S such that:

- S is *homogeneous*, i.e. any k-simplex of S is a k-face of an n-simplex, $0 \leq k \leq n$;
- every $(n-1)$-simplex is a face of at most two n-simplices[7];

[7]taking multi-incidence into account: that means that, if τ is an $(n-1)$-simplex, at most two distinct pairs (σ_1, i_1) and (σ_2, i_2) exist, such that $0 \leq i_1, i_2 \leq n$ and $d_{i_1}(\sigma_1) = d_{i_2}(\sigma_2) = \tau$.

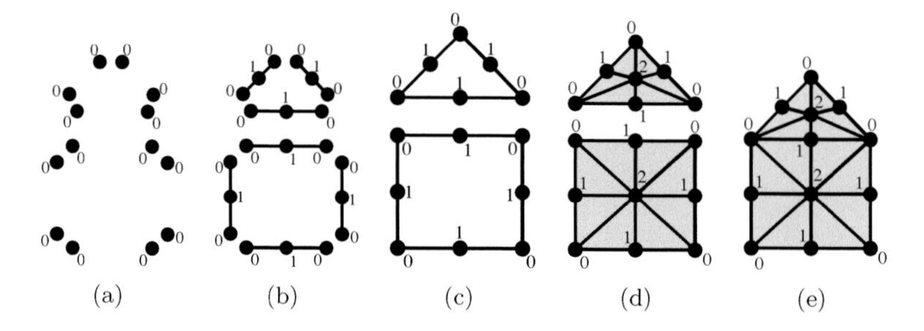

FIGURE 8.9
(a) A 0-dimensional quasi-manifold.
(b) Cellular cones.
(c) Identification of vertices.
(d) Cellular cones.
(e) Identification of edges.

- for any distinct n-simplices σ and σ', a sequence of n-simplices $\sigma_1, \ldots, \sigma_p$ exists in S, such that $\sigma_1 = \sigma$, $\sigma_p = \sigma'$, and for $1 \leq i < p$, an $(n-1)$-simplex exists, which is a common face of σ_i and σ_{i+1}.

It is easy to prove that numbered simplicial quasi-manifolds satisfy the three properties above.

Cellular quasi-manifolds

Definition 87 (cellular quasi-manifold) *An n-dimensional cellular quasi-manifold S is an n-dimensional numbered simplicial quasi-manifold such that (cf. Fig. 8.9):*

- *if $n = 0$, then S is a set of 0-simplices numbered (0), which are partitioned into connected components consisting of one or two 0-simplices;*
- *if $n > 0$, then S can be constructed by the two following operations:*
  1. creating cellular cones *on an $(n-1)$-dimensional cellular quasi-manifold S'. This operation consists in making a cone on each connected component of S', the new vertex associated with the connected component being numbered n;*
  2. identifying two $(n-1)$-dimensional cells. *Let c_1 and c_2 be two $(n-1)$-dimensional cells of S, such that each $(n-1)$-simplex of c_1 (resp. c_2) is incident to only one n-simplex, and such that an isomorphism ϕ exists between c_1 and c_2:*
     * *if $c_1 \neq c_2$, the operation consists in identifying the $(n-1)$-simplices of c_1 and c_2 (and their boundaries) according to ϕ;*

$*$ *if* $c_1 = c_2$, ϕ *is an automorphism, and the operation can be applied if* $\phi = \phi^{-1}$ *and* ϕ *is not the identity on* c_1. *In this case also, the operation consists in identifying the* $(n-1)$*-simplices of* c_1 *(and their boundaries), according to* ϕ. c_1 *is said to be* bent on itself.

A connected component of a 0-dimensional cellular quasi-manifold made by two vertices corresponds to 0-sphere. A cellular cone on such a connected component creates an edge incident to two vertices. Curves are constructed by identifying extremity vertices of such edges. Cellular cones on closed curves produce faces, which can be glued together by identifying edges. We get here the construction of paper surfaces as presented in chapter 2. More formally, the geometric realization of any 2-dimensional cellular quasi-manifold is a surface, i.e. a 2-dimensional manifold. This property is not true for higher dimensions. 3-dimensional cellular quasi-manifolds exist, such that their geometric realizations are not manifolds: cf. Fig. 8.8(c) and Section 2.4.

Note that a cell is defined as a cone on a cellular quasi-manifold (this is the answer to a question in chapter 2: *what is a cell?*). Note also that the interior of a cell is always the interior of a cellular quasi-manifold, whatever the way cells are glued together. This is due to the fact that *cells are identified* together, and not simply parts of cells (as for numbered simplicial quasi-manifolds). At last, the geometric realization of a cellular quasi-manifold is always a pseudo-manifold, since a cellular quasi-manifold is a numbered simplicial quasi-manifold.

8.2.2 Simplicial Interpretation of n-Gmaps

Numbered simplicial quasi-manifolds and premaps

Let $Q = (S, \nu)$ be a numbered simplicial quasi-manifold, where $S = (K = (\bigcup_{i=0}^{n} K_i), (d_j)_{j=0,\dots,n})$. Remember that Q can be constructed by adding isolated n-simplices, and by identifying $(n-1)$-simplices, such that any $(n-1)$-simplex is incident to at most two n-simplices. Since the numbering of any n-simplex is $[0, n]$, the numbering of any $(n-1)$-simplex is $[0, n] - \{i\}$ for some $i, 0 \leq i \leq n$. The construction of Q is then completely represented by the set of n-simplices and $(n+1)$ *adjacency functions* denoted for instance $a_i : K_n \to K_n, 0 \leq i \leq n$, such that, for any n-simplex σ:

- either $a_i(\sigma) = \sigma$, and σ is the single n-simplex incident to its $(n-1)$-face numbered $[0, n] - \{i\}$;

- either $a_i(\sigma) = \sigma' \neq \sigma$, and σ and σ' are the two n-simplices incident to the same $(n-1)$-face numbered $[0, n] - \{i\}$.

Note that any a_i is an involution.

Let us now define *n-premaps*.

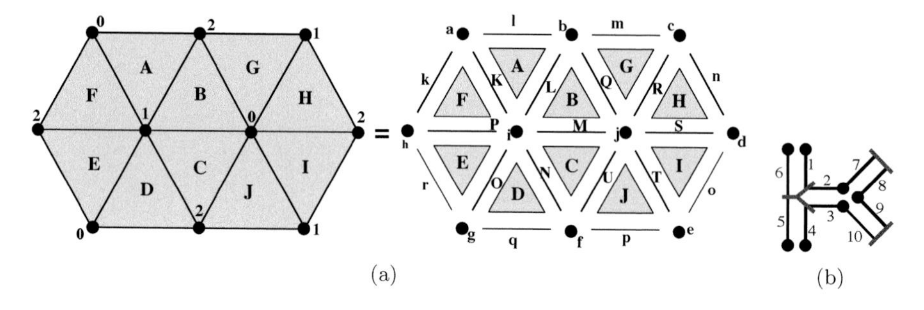

FIGURE 8.10
(a) A 2-dimensional numbered simplicial quasi-manifold.
(b) The corresponding premap.

Definition 88 (n-premap) *An n-dimensional premap, with $0 \leq n$, is an $(n+2)$-tuple $P = (D, \alpha_0, \ldots, \alpha_n)$ where:*

1. *D is a finite set of darts;*
2. *$\forall i \in \{0, \ldots, n\}$, α_i is an involution on D.*

It is clear that $(K_n, a_0, \ldots, a_n)$ as defined above is an n-premap: cf. Fig. 8.10. So, any numbered simplicial quasi-manifold can be associated with a unique n-premap. For instance in Fig. 8.10, dart 1 (resp. 2, 3, 4, 5, 6, 7, 8, 9, 10) is associated with 2-simplex A (resp. B, C, D, E, F, G, H, I, J). E and F share a 1-simplex numbered $(1, 2)$, thus $\alpha_0(5) = 6$ and $\alpha_0(6) = 5$; E is the single 2-simplex incident to its 1-face numbered $(0, 2)$, so $\alpha_1(5) = 5$; E and D share a 1-simplex numbered $(0, 1)$, thus $\alpha_2(5) = 4$ and $\alpha_2(4) = 5$.

Conversely, let $P = (D, \alpha_0, \ldots, \alpha_n)$ be an n-premap. Let I be any nonempty sequence $(j_0, \ldots, j_i) \subset [0, n]$: associate with the set of orbits $\langle \alpha_0, \ldots, \hat{\alpha_{j_0}}, \ldots, \hat{\alpha_{j_i}}, \ldots, \alpha_n \rangle$, denoted $\langle\rangle_{N \setminus I}$, a set of simplices numbered $(j_0, \ldots, j_i)$.

For instance in Fig. 8.10:

- A (resp. $B, C, D, E, F, G, H, I, J$) is associated with $\langle\rangle(1) = \langle \hat{\alpha_0}, \hat{\alpha_1}, \hat{\alpha_2} \rangle(1) = \langle\rangle_{[0,2] \setminus \{0,1,2\}}(1)$ (resp. $\langle\rangle(2)$, $\langle\rangle(3)$, $\langle\rangle(4)$, $\langle\rangle(5)$, $\langle\rangle(6)$, $\langle\rangle(7)$, $\langle\rangle(8)$, $\langle\rangle(9)$, $\langle\rangle(10)$);
- 1-simplices numbered $(1, 2)$, namely L (resp. N, P, m, n, o, p), correspond to $\langle \alpha_0 \rangle(1) = \langle\rangle_{[0,2] \setminus \{1,2\}}(1)$ (resp. $\langle \alpha_0 \rangle(3)$, $\langle \alpha_0 \rangle(5)$, $\langle \alpha_0 \rangle(7)$, $\langle \alpha_0 \rangle(8)$, $\langle \alpha_0 \rangle(9)$, $\langle \alpha_0 \rangle(10)$); 1-simplices numbered $(0, 2)$, namely k (resp. l, q, r, Q, S, U), correspond to $\langle \alpha_1 \rangle(6) = \langle\rangle_{[0,2] \setminus \{0,2\}}(6)$ (resp. $\langle \alpha_1 \rangle(1), \langle \alpha_1 \rangle(4), \langle \alpha_1 \rangle(5), \langle \alpha_1 \rangle(2), \langle \alpha_1 \rangle(8), \langle \alpha_1 \rangle(3)$); 1-simplices numbered $(0, 1)$, namely K (resp. M, O, R, T), correspond to $\langle \alpha_2 \rangle(1) = \langle\rangle_{[0,2] \setminus \{0,1\}}(1)$ (resp. $\langle \alpha_2 \rangle(2), \langle \alpha_2 \rangle(4), \langle \alpha_2 \rangle(7), \langle \alpha_2 \rangle(9)$);

- 0-simplices numbered (2), namely b (resp. d, f, h), correspond to $\langle\alpha_0, \alpha_1\rangle(1) = \langle\rangle_{[0,2]\setminus\{2\}}(1)$ (resp. $\langle\alpha_0, \alpha_1\rangle(8), \langle\alpha_0, \alpha_1\rangle(3), \langle\alpha_0, \alpha_1\rangle(5)$); 0-simplices numbered (1), namely c (resp. e, i), correspond to $\langle\alpha_0, \alpha_2\rangle(7) = \langle\rangle_{[0,2]\setminus\{1\}}(7)$ (resp. $\langle\alpha_0, \alpha_2\rangle(9), \langle\alpha_0, \alpha_2\rangle(1)$); 0-simplices numbered (0), namely a (resp. g, j), correspond to $\langle\alpha_1, \alpha_2\rangle(1) = \langle\rangle_{[0,2]\setminus\{0\}}(1)$ (resp. $\langle\alpha_1, \alpha_2\rangle(4), \langle\alpha_1, \alpha_2\rangle(2)$).

The face operators are defined in the following way: let $I = (j_0, \ldots, j_l, \ldots, j_i)$ be such that $i \geq 1$, d be a dart, and σ be the simplex associated to $\langle\rangle_{[0,n]\setminus I}(d)$; then $d_l(\sigma)$ is the $(i-1)$-simplex associated with $\langle\rangle_{[0,n]\setminus(I-\{j_l\})}(d)$. For instance in Fig. 8.10:

- $d_0(A) = L$, since A is associated with $\langle\rangle_{[0,2]\setminus\{0,1,2\}}(1) = \langle\rangle(1)$ and L is associated with $\langle\rangle_{[0,2]\setminus(\{0,1,2\}-\{0\})}(1) = \langle\alpha_0\rangle(1)$; $d_1(A) = l$, since l is associated with $\langle\rangle_{[0,2]\setminus(\{0,1,2\}-\{1\})}(1) = \langle\alpha_1\rangle(1)$; $d_2(A) = K$, since K is associated with $\langle\rangle_{[0,2]\setminus(\{0,1,2\}-\{2\})}(1) = \langle\alpha_2\rangle(1)$; note that this is coherent with the fact that, for any $i, 0 \leq i \leq n$, α_i corresponds to an "adjacency relation" between n-simplices sharing a common face obtained by d_i;

- $d_0(K) = i$ (resp. $d_1(K) = a$), since K is associated with $\langle\rangle_{[0,2]\setminus\{0,1\}}(1) = \langle\alpha_2\rangle(1)$ and i (resp. a) is associated with $\langle\rangle_{[0,2]\setminus\{1\}}(1) = \langle\alpha_0, \alpha_2\rangle(1)$ (resp. $\langle\rangle_{[0,2]\setminus\{0\}}(1) = \langle\alpha_1, \alpha_2\rangle(1)$); $d_0(l) = b$ (resp. $d_1(l) = a$), since l is associated with $\langle\rangle_{[0,2]\setminus\{0,2\}}(1) = \langle\alpha_1\rangle(1)$ and b (resp. a) is associated with $\langle\rangle_{[0,2]\setminus\{2\}}(1) = \langle\alpha_0, \alpha_1\rangle(1)$ (resp. $\langle\rangle_{[0,2]\setminus\{0\}}(1) = \langle\alpha_1, \alpha_2\rangle(1)$); $d_0(L) = b$ (resp. $d_1(L) = i$), since L is associated with $\langle\rangle_{[0,2]\setminus\{1,2\}}(1) = \langle\alpha_0\rangle(1)$ and b (resp. i) is associated with $\langle\rangle_{[0,2]\setminus\{2\}}(1) = \langle\alpha_0, \alpha_1\rangle(1)$ (resp. $\langle\rangle_{[0,2]\setminus\{1\}}(1) = \langle\alpha_0, \alpha_2\rangle(1)$);

At last, any 0-simplex corresponding to an orbit $\langle\rangle_{[0,n]\setminus\{i\}} = \langle\alpha_0, \ldots, \hat{\alpha_i}, \ldots, \alpha_n\rangle$ is numbered i (and this is coherent with the numbering of face operators).

So, any n-premap can be associated with a unique n-dimensional numbered simplicial quasi-manifold.

In conclusion, *n-dimensional numbered simplicial quasi-manifolds are equivalent to n-premaps*, and the conversion processes described above are inverse to each other.

Cellular quasi-manifolds and n-Gmaps

The conversion processes described above still apply for n-dimensional cellular quasi-manifolds and n-Gmaps. So, *n-dimensional cellular quasi-manifolds are equivalent to n-Gmaps.*

More precisely, a 0-dimensional cellular quasi-manifold is a set of components, each component contains one or two vertices numbered 0. Each connected component of a 0-Gmap $G = (D, \alpha_0)$ contains one or two darts: so, the orbits $\langle\rangle_{[0]\setminus\{0\}} = \langle\rangle$ correspond in the first case to one vertex numbered 0, and in the second case to two vertices numbered 0.

Assume an $(n-1)$-dimensional cellular quasi-manifold Q is equivalent to an $(n-1)$-Gmap $G = (D, \alpha_0, \ldots, \alpha_{n-1})$. Let Q' be the n-dimensional cellular quasi-manifold defined by constructing cellular cones for all connected components of Q; let $G' = (D, \alpha_0, \ldots, \alpha_{n-1}, \alpha_n = Id)$, where Id is the identity on D, be the n-Gmap deduced from G by increasing its dimension. Then Q' is equivalent to G'. In fact (cf. Fig. 8.11):

- any i-simplex σ numbered $\{j_0, \ldots, j_i\}$ in Q is a simplex of Q'. Let d be a dart, such that σ corresponds to orbit $O = \langle\alpha_0, \ldots, \hat{\alpha_{j_0}}, \ldots, \hat{\alpha_{j_i}}, \ldots, \alpha_{n-1}\rangle(d)$ in G; σ corresponds also to orbit $O' = \langle\alpha_0, \ldots, \hat{\alpha_{j_0}}, \ldots, \hat{\alpha_{j_i}}, \ldots, \alpha_{n-1}, \alpha_n\rangle(d)$ in G', and O and O' are equal since $\alpha_n = Id$;

- a unique $(i+1)$-simplex σ', numbered $\{j_0, \ldots, j_i, n\}$, is associated with σ in Q': σ' corresponds to the orbit $\langle\alpha_0, \ldots, \hat{\alpha_{j_0}}, \ldots, \hat{\alpha_{j_i}}, \ldots, \alpha_{n-1}, \hat{\alpha_n}\rangle(d)$ in G', which clearly is equal to O;

- at last, for each connected component of Q, a unique 0-simplex numbered n is created in Q', corresponding to the unique orbit $\langle\alpha_0, \ldots, \alpha_{n-1}\rangle$ in G' associated with the corresponding connected component in G.

Assume an n-dimensional cellular quasi-manifold Q is equivalent to an n-Gmap $G = (D, \alpha_0, \ldots, \alpha_{n-1}, \alpha_n)$. Let c_1 and c_2 be two $(n-1)$-cells, incident respectively to $(n-1)$-simplices σ_1 and σ_2, such that $\sigma_1 \neq \sigma_2$, σ_1 (resp. σ_2) is incident to a unique n-simplex, and an isomorphism ϕ associates c_1 and c_2, with $\phi(\sigma_1) = \sigma_2$ (and $\phi = \phi^{-1}$ if $c_1 = c_2$). Let d_1 (resp. d_2) be the dart of D corresponding to the n-simplex incident to σ_1 (resp. σ_2): note that $\langle\alpha_n\rangle(d_1) = \{d_1\}$ corresponds to σ_1 (resp. $\langle\alpha_n\rangle(d_2) = \{d_2\}$ corresponds to σ_2). Let Q' be the n-dimensional cellular quasi-manifold defined by identifying c_1 and c_2 according to ϕ; let $G' = (D, \alpha_0, \ldots, \alpha_{n-1}, \alpha'_n)$ be the n-Gmap deduced from G by n-sewing d_1 and d_2. Then Q' is equivalent to G'. In fact:

- isomorphism ϕ induces an isomorphism ϕ' between orbits $\langle\alpha_0, \ldots, \alpha_{n-2}\rangle(d_1)$ and $\langle\alpha_0, \ldots, \alpha_{n-2}\rangle(d_2)$ (or an automorphism such that $\phi' = \phi'^{-1}$ on orbit $\langle\alpha_0, \ldots, \alpha_{n-2}\rangle(d_1)$ if ϕ is an automorphism on c_1);

- identifying σ_1 and σ_2 (and all simplices corresponding by ϕ) corresponds to link d_1 and d_2 by α'_n (and all darts corresponding by ϕ'), since $\langle\alpha'_n\rangle(d_1) = \{d_1, d_2\}$.

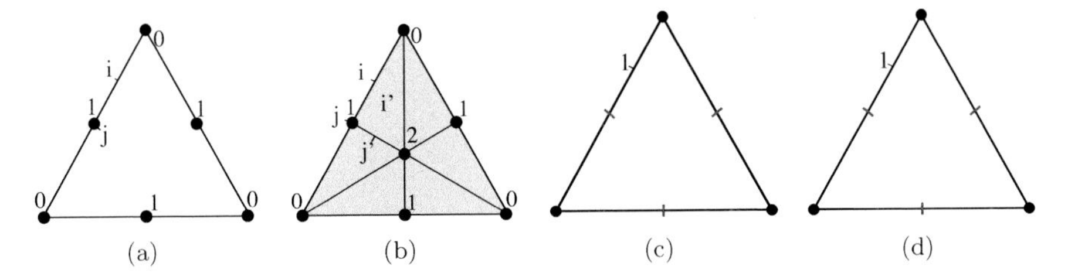

FIGURE 8.11
(a) A connected 1-dimensional cellular quasi-manifold.
(b) The connected 2-dimensional cellular quasi-manifold obtained by a cellular cone.
(c) The 1-Gmap corresponding to (a).
(d) The 2-Gmap corresponding to (b). i (resp. j) in (a) corresponds to orbit $\langle\rangle(1)$ (resp. $\langle\alpha_0\rangle(1)$) in (c). i (resp. j) in (b) corresponds to orbit $\langle\alpha_2\rangle(1)$ (resp. $\langle\alpha_0, \alpha_2\rangle(1)$) in (d). i' (resp. j'), associated with i (resp. j), corresponds to orbit $\langle\rangle(1)$ (resp. $\langle\alpha_0\rangle(1)$) in (d). The connected component of (a) (resp. of (b)) corresponds to orbit $\langle\alpha_0, \alpha_1\rangle(1)$ in (c) (resp. to orbit $\langle\alpha_0, \alpha_1, \alpha_2\rangle(1)$ in (d)); 0-simplex v numbered (2) in (b) corresponds to orbit $\langle\alpha_0, \alpha_1\rangle(1)$ in (d).

So, n-Gmaps correspond to cellular quasi-manifolds and premaps correspond to numbered simplicial quasi-manifolds. The interior of any cell of any cellular quasi-manifold corresponds to the interior of a cellular quasi-manifold; this is not the case for any cell of a numbered simplicial quasi-manifold. This corresponds to the fact that, given any n-Gmap, the involutions which indices differ from at least 2 commute; this is not the case for any premap.

Orientable quasi-manifolds and (hyper)maps

Let Q be an n-dimensional numbered simplicial quasi-manifold, and let $P = (D, \alpha_0, \dots, \alpha_n)$ be the associated premap. Remember that simplices are implicitly oriented within semi-simplicial sets. For instance, look at the 2-dimensional simplicial quasi-manifold in Fig. 8.12: assume 2-simplices are oriented according to their numbering, i.e. their orientations corresponds to the sequence $(0, 1, 2)$. So, the orientation of a 2-simplex is the inverse of the orientations of the 2-simplices which share a 1-simplex with it. This property can be extended for any dimension. That means that, given an n-simplex σ corresponding to a dart d in P, its orientation is inverse from the n-simplices corresponding to darts $\alpha_i(d)$, for $0 \leq i \leq n$.

Let Q be a connected quasi-manifold without boundary, i.e. any $(n-1)$-simplex is incident to two distinct n-simplices; this means that P is connected too, and that all involutions in P are without fixed points. Q is *orientable* if it is possible to orient all n-simplices in a coherent way. This can be checked in the following way. Choose an n-simplex with its implicit orientation; for

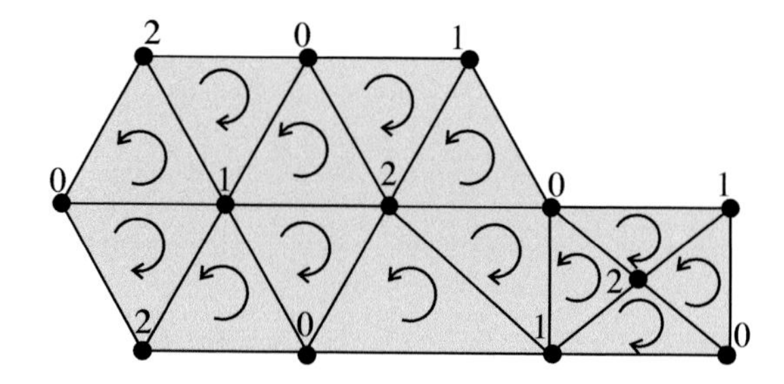

FIGURE 8.12
In a numbered simplicial quasi-manifold, the implicit orientation of an n-simplex is inverse from the implicit orientations of the n-simplices which share an $(n-1)$-simplex with it.

the simplices which share an $(n-1)$-simplex with it, choose the inverse of their implicit orientation; iterate this process, i.e. traverse the whole quasi-manifold by adjacency. If all simplices are coherently oriented, Q is orientable (and oriented), else Q is not orientable. Another way consists in choosing also an n-simplex with its implicit orientation, and to traverse all simplices which must have this orientation, i.e. the simplices adjacent to the simplices adjacent to the initial simplex, and then iterate this process. At the end, either the half of the simplices of Q have been traversed, and Q is orientable (the set of traversed simplices corresponds to one orientation, the remaining simplices correspond to the other orientation), either all simplices of Q have been traversed and Q is not orientable.

This process can be applied for the associated connected premap P. Let $H = (D, \alpha_1 \circ \alpha_0, \ldots, \alpha_n \circ \alpha_0)$ be the *hypermap of the orientations of* P. H contains at most two connected components: either H contains two connected components, and P is orientable (as Q), either H contains one connected component, and P is not orientable (as Q).

Conversely, let $H = (D, \beta_1, \ldots, \beta_n)$ be a connected n-hypermap. $H^{-1} = (D, \beta_1^{-1}, \ldots, \beta_n^{-1})$ is the hypermap *inverse* of H. Let $H' = (D', \beta'_1, \ldots, \beta'_n)$, such that $D \cap D' = \emptyset$, and an isomorphism ϕ exists, which maps H onto the inverse of H' (in other words, H' is a copy of H^{-1}). Define $P = (D \cup D', \pi_0, \ldots, \pi_n)$, with:

- $\pi_{0|D} = \phi$; $\pi_{0|D'} = \phi^{-1}$;
- for $1 \leq i \leq n$, $\pi_{i|D} = \beta'_i \circ \phi$; $\pi_{i|D'} = \beta_i \circ \phi^{-1}$.

P is a connected orientable premap, such that its hypermap of the orientations contains two connected components, which are H and H'.

A particular case is the relation between n-Gmaps and n-maps, which satisfy similar properties. Let $G = (D, \alpha_0, \ldots, \alpha_n)$ be a connected n-Gmap, such that all involutions are without fixed points, and let C be its associated cellular quasi-manifold. Let $M = (D, \alpha_1 \circ \alpha_0, \ldots, \alpha_n \circ \alpha_0)$: it is the n-map of the orientations of G. M contains at most two connected components. It M contains two connected components, G is said orientable, and C is orientable, else G is said not orientable, and C is not orientable.

Conversely, let M be a connected n-map. It is possible to construct, using M and an n-map inverse of M, a connected orientable n-Gmap, such that its map of the orientations contains two connected components, which are the two initial maps.

All these notions can be extended for n-dimensional numbered simplicial quasi-manifolds with boundaries (resp. cellular quasi-manifolds with boundaries), n-premaps (resp. n-Gmaps) and n-hypermaps with partial permutations (resp. n-maps with partial permutations).

Some remarks about cellular quasi-manifolds

So, n-Gmaps corresponds to cellular quasi-manifolds, and their cells correspond to the interior of quasi-manifolds. A key point is the fact that quasi-manifolds can be combinatorially defined: they are more general than manifolds, but as far as we know, no other class has been defined such that:

- it contains manifolds;
- it can be combinatorially defined;
- and it is strictly contained in the set of quasi-manifolds.

So, as far as we know, the set of quasi-manifolds is the "smallest" set which contains manifolds and which can be combinatorially defined, i.e. such that the combinatorial definition of the structure takes directly into account the expected properties. Of course, it is possible to define subclasses by adding properties which can combinatorially defined and computed: for instance, n-Gmaps without multi-incidence can be defined, which are equivalent to n-surfaces, i.e. a subset of incidence graphs (cf. Section 8.2.4)[8]; n-Gmaps can be restricted in order that the homology of all cells is that of balls (cf. [4]); but the representation itself, as a set of darts on which involutions act, is not modified.

Such remarks can also be stated for cells of quasi-manifolds. Such a cell is more "general" than a simplex, which can also be combinatorially defined; but it is necessary for some applications in geometric modeling or image analysis to

[8]Similarly, the definition of incidence graphs makes it possible to represent cells which are not quasi-manifolds; n-surfaces are defined as incidence graphs satisfying some properties; but the representation itself, as a set of cells on which incidence relations act, is not modified.

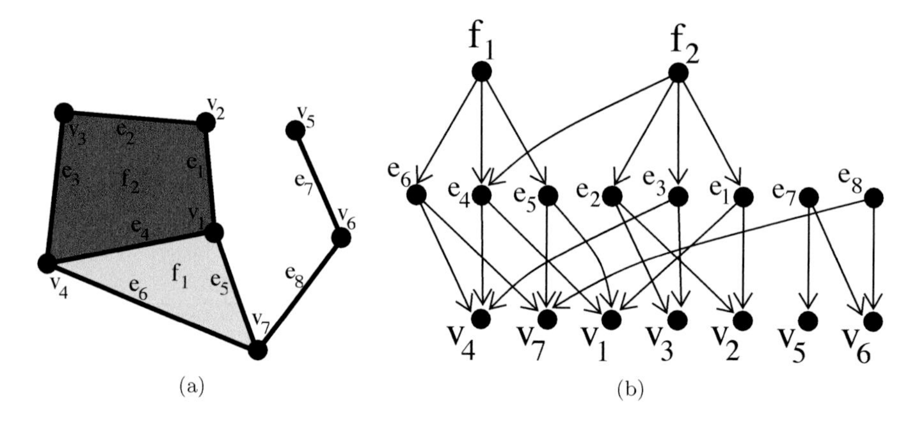

FIGURE 8.13
(a) A 2-dimensional complex.
(b) The corresponding incidence graph.

handle more "general" cells than simplices[9]. For instance, boolean operations (union, intersection, difference), even applied to simplicial "objects", produce nonsimplicial "objects".

Several applications handle "objects" with *convex* cells (i.e. excluding multi-incidence between cells), but note that:

- convexity is not a combinatorial property;
- for several applications, cells can be nonconvex ones, for instance when cells are embedded as (parts of) free-form spaces (e.g. splines).

So, as far as we know, cells of quasi-manifolds make the "smallest" subset such that the structure of cells is not regular, as simplices or products of simplices, and which can be combinatorially characterized.

8.2.3 Simplicial Interpretation of Incidence Graphs

Remember the definition of incidence graphs given in chapter 2: an incidence graph is a set of vertices, integers and oriented edges, such that (cf. Fig. 8.13):

- each vertex of the graph is associated with an integer, called its dimension; a vertex associated with i is called an i-cell;
- for any $i \geq 1$, each i-cell is linked by oriented edges with at least one $(i-1)$-cell (and there is at most one edge between an i-cell and an $(i-1)$-cell).

---

[9] Or other cells which can be combinatorially characterized, as cubes or more generally simploids, which are products of simplices.

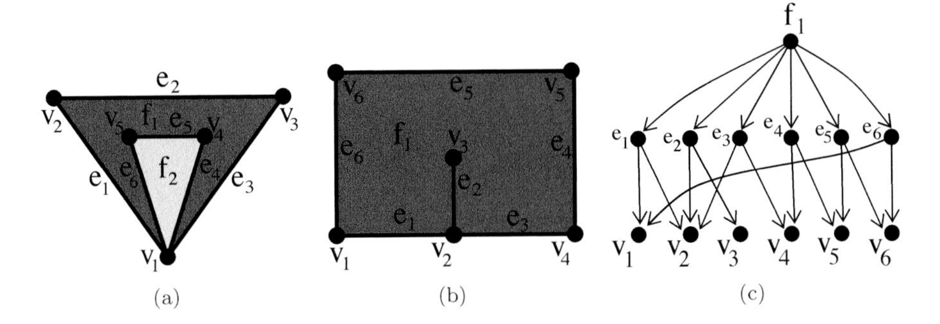

FIGURE 8.14
(a) Two corners of face f_1, namely (e_1, v_1, e_6) and (e_3, v_1, e_4) share vertex v_1: face f_1 is adjacent to itself through this vertex.
(b) A surface subdivision in which face f_1 is incident twice to edge e_2.
(c) The corresponding incidence graph.

We claimed that such incidence graphs cannot take multi-incidence into account, but it seems possible to deduce incidence graphs from subdivisions in which multi-incidence occurs: cf. Fig. 8.14. It is thus necessary to precisely define the geometric objects which can be unambiguously associated with such incidence graphs.

The key notion here is the notion of cell-tuple [39]. Let c_i be a main i-cell, i.e. a cell which is not the extremity of an edge in the incidence graph: for instance, f_1, f_2, e_7, e_8 in Fig. 8.13 are main cells. Let $(c_0, \ldots, c_i)$ be a sequence of cells, such that:

- for any $j, 0 \leq j \leq i$, c_j is a j-cell;
- for any $j, 0 \leq j < i$, there is an edge between c_{j+1} and c_j.

Associate now an i-simplex $\{c_0, \ldots, c_i\}$ with $(c_0, \ldots, c_i)$; if you generalize this process for all main cells, you get the main simplices of an abstract simplicial complex. Moreover, this abstract simplicial complex can be numbered, by associating its dimension with each vertex of the complex, since any vertex of the complex corresponds to a cell of the incidence graph: cf. Fig. 8.15.

Look now at Fig. 8.15(b), in which is depicted the cellular quasi-manifold corresponding to Fig. 8.14(b). This cellular quasi-manifold corresponds to a 2-Gmap, but it is not an abstract simplicial complex: several edges share the same boundaries; for instance there are two edges incident to vertices corresponding to v_2 and f_1: so, it is not possible to associate an incidence graph with the subdivision depicted on Fig. 8.14(b).

Look now at the incidence graph depicted on Fig. 8.14(c): this incidence graph does not describe the topology of the surface depicted on Fig. 8.14(b). If you construct the associated abstract simplicial complex, you will see that

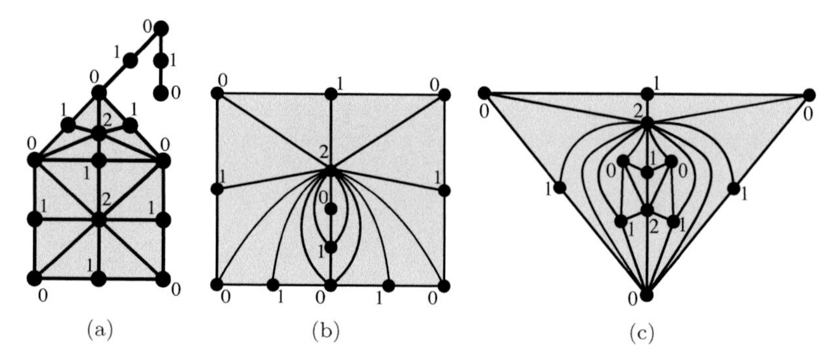

FIGURE 8.15
(a) The numbered abstract simplicial complex corresponding to the incidence graph of Fig. 8.13.
(b) The cellular quasi-manifold corresponding to the subdivision depicted in Fig. 8.14(b): it is not an abstract simplicial complex, since several edges share the same boundaries.
(c) The cellular quasi-manifold corresponding to the subdivision depicted in Fig. 8.14(a).

only one 2-simplex is incident to vertex v_3, i.e. $\{v_3, e_2, f_1\}$, and this does not correspond to the cellular quasi-manifold depicted on Fig. 8.15(b).

This is also the case for the subdivision depicted on Fig. 8.14(a). If you construct the incidence graph which seems naturally corresponds to this subdivision, and then the associated abstract simplicial complex, you will see that only one 1-simplex links the 0-simplices associated with v_1 and f_1; but there are two such 1-simplices in the cellular quasi-manifold associated with the subdivision: cf. Fig. 8.15(c), and once again, these two 1-simplices share the same boundary, so it is not an abstract simplicial complex: as before, this denotes multi-incidence.

In conclusion, any incidence graph has a simplicial analog, which is a (numbered) abstract simplicial complex. But if you try to intuitively deduce an incidence graph from a given subdivision, be careful with multi-incidence! That is why incidence graphs are often used for representing subdivisions with convex cells, avoiding thus multi-incidence: such subdivisions are *regular CW-complexes* [39, 139]: cells are homeomorphic to balls (due to the convexity property), and no multi-incidence occurs (for the same reason).

8.2.4 n-Gmaps and Incidence Graphs

Definitions and properties

n-surfaces make a subclass of incidence graphs, which are close to n-

Gmaps. Some preliminary definitions are needed in order to define n-surfaces: cf. Fig. 8.16.

Let G be an incidence graph, and let c^i be a vertex of G. The closure of c^i is the subgraph of G such that c^i is its root; in other words, it contains c^i, all vertices c^k such that a path from c^i to c^k exists in G (so $k < i$), and all edges of G linking these vertices. The boundary of c^i is the closure of c^i minus c^i and its incident edges. Let G^- be the graph deduced from G by reversing all edges; the star (resp. proper star) of c^i is the subgraph of G which corresponds to the closure of c^i (resp. the boundary of c^i) in G^-. The neighborhood of c^i in G is the subgraph of G which contains all vertices of the star and the boundary of x, and all edges linking these vertices. The strict neighborhood of c^i is the neighborhood of c^i minus c^i and all its incident edges, plus all edges (c^{i+1}, c^{i-1}) such that edges (c^{i+1}, c^i) and (c^i, c^{i-1}) exist in G.

Definition 89 (n-surfaces) *[84, 85]*

1. *An incidence graph is a* 0*-surface if it is composed by exactly two vertices* c^0 *and* c'^0*, and no edges;*
2. *an incidence graph* G *is an* n*-surface,* $n > 0$*, if it is connected and for any vertex* c^i*, the strict neighborhood of* c^i *is an* $(n-1)$*-surface.*

Remember that a 0-sphere is a set of two points: the structure of a 0-sphere corresponds thus to a 0-surface. The property characterizing an n-surface is a local property which has to be satisfied for any vertex of the graph, i.e. for any cell of the associated subdivision. The correspondence with n-Gmaps will be explained below.

Note that the dimension of all main cells is n. Moreover, n-surfaces satisfy the following *switch* property: cf. Figs. 8.16(f) and (g). Let G be an n-surface, to which two vertices c^{n+1} and c^{-1} are added, and edges which link c^{n+1} with any vertex numbered n, and any vertex numbered 0 with c^{-1}. Then for any cells x^i and y^j, $b(x^i) \cap s(y^j)$ *is either empty or made of exactly two elements*, where $b(x^i)$ (resp. $s(y^j)$) is the set of vertices numbered $i-1$ in the boundary of x^i (resp. the set of vertices numbered $j+1$ in the star of y^j).

Conversion between n*-surfaces and connected* n*-Gmaps without boundary and without multi-incidence*

Let S be an n-surface. Let D be the set of all $(n+1)$-cell tuples $(c_0, \ldots, c_n)$, i.e. such that $|(c_0, \ldots, c_n)| = n+1$: in other words, an $(n+1)$-cell tuple corresponds to a main simplex in the associated numbered semi-simplicial set. Let $t = (c_0, \ldots, c_i, \ldots, c_n) \in D$, with $0 \leq i \leq n$: due to the switch property, a unique cell tuple $t' = (c_0, \ldots, c'_i, \ldots, c_n)$ exists in D, which shares all cells with t except c_i; then define $switch_i(t) = t'$ and $switch_i(t') = t$. It has been shown that $G = (D, switch_0, \ldots, switch_n)$ is a connected n-Gmap which satisfies [5]:

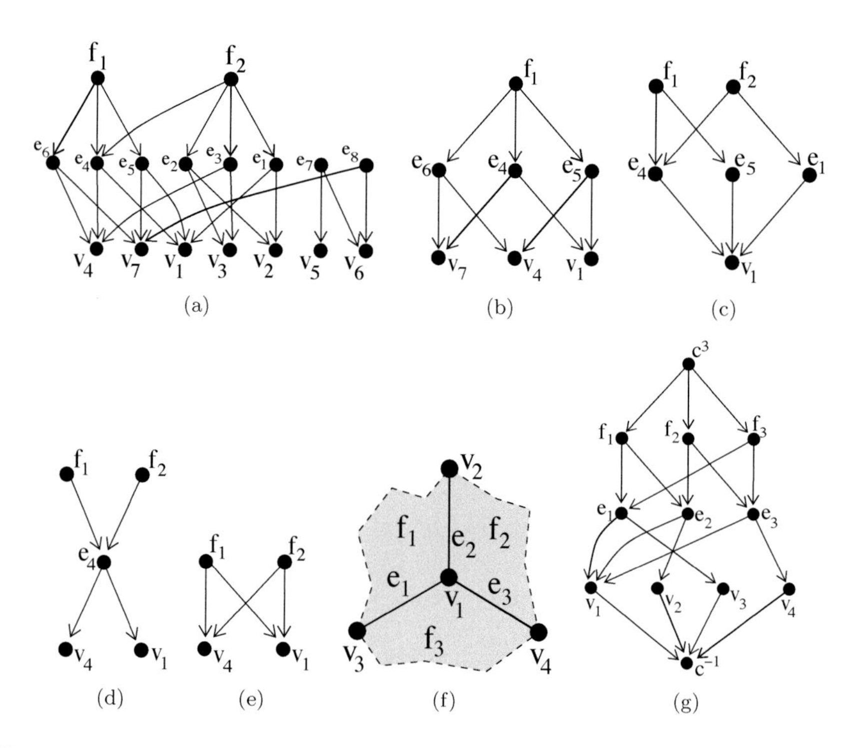

FIGURE 8.16
(a) The incidence graph describing the 2D object given in Fig. 8.13(a).
(b) The closure of f_1.
(c) The star of v_1.
(d) The neighborhood of e_4.
(e) The strict neighborhood of e_4.
(f) A part of a surface subdivision.
(g) The part of the incidence graph which corresponds to (f), illustrating the switch property. For instance, (e_1, f_1, c^3) (resp. $(v_1, e_1, f_1), (c^{-1}, v_1, e_1)$) corresponds to (e_1, f_3, c^3) (resp. $(v_1, e_2, f_1), (c^{-1}, v_3, e_1)$).

1. G is without boundaries, i.e. all involutions $switch_i$ are without fixed points;

2. G is *without multi-incidence*, i.e.[10]

$$\forall d \in D, \forall I \subseteq N = \{0, \ldots, n\}, \langle\rangle_{N \setminus I}(d) = \bigcap_{i \in I} \langle\rangle_{N \setminus \{i\}}(d).$$

The equivalence between n-surfaces and connected n-Gmaps without boundary and without multi-incidence leads to the fact that the simplicial analog of any n-surface is an n-dimensional cellular quasi-manifold. The local property characterizing n-surfaces, namely the strict neighborhood of any cell is an $(n-1)$-surface, corresponds to the fact that any i-cell of an n-Gmap is an $(n-1)$-Gmap.

8.3 Some Consequences

8.3.1 Chain of Maps

Definition and basic operations

For several applications, it can be useful to handle other classes of "objects", for instance subdivided "objects" which cells are quasi-manifolds but which are not quasi-manifolds themselves. This leads to the definition of *chains of maps*:

Definition 90 (chain of maps) *[104] An n-dimensional chain of maps is a tuple $C = ((G^i)_{i=0,\ldots,n}, (\sigma^i)_{i=1,\ldots,n})$ such that (cf. Fig. 8.17(a) and (b)):*

1. *$\forall i,\ 0 \leq i \leq n,\quad G^i = (D^i, \alpha_0^i, \ldots, \alpha_{i-1}^i, \alpha_i^i = \omega)$ is an i-dimensional generalized map such that ω is undefined on D^i;*

2. *$\forall i,\ 1 \leq i \leq n,\quad \sigma^i : D^i \longrightarrow D^{i-1}$;*
   for $i \geq 2$, σ^i satisfies, for any dart d of D^i:

   (a) *for any $j, 0 \leq j \leq i-2$, $\sigma^i(\alpha_j^i(d)) \in \{\sigma^i(d), \alpha_j^{i-1}(\sigma^i(d))\}$ (in other words, σ^i is a function which associates an orbit $\langle \alpha_0^i, \cdots, \alpha_{i-2}^i \rangle$ of G^i with (a subset of) an orbit $\langle \alpha_0^{i-1}, \cdots, \alpha_{i-2}^{i-1} \rangle$ of G^{i-1});*

   (b) *$\sigma^{i-1} \circ \sigma^i \circ \alpha_{i-1}^i(d) = \sigma^{i-1} \circ \sigma^i(d)$.*

[10]This property simply expresses the fact that two distinct simplices are incident to distinct sets of vertices in the associated cellular quasi-manifold, i.e. that this cellular quasi-manifold is a numbered abstract simplicial complex; so, it corresponds to an incidence graph.

Any connected component of an i-Gmap is an i-dimensional cell, or $i-$cell: that is why $\alpha_i^i = \omega$ is undefined, in order to describe only the interior of the cell. The cells are linked by *face operators* σ^i.

Chains of maps can be constructed in the following way: take isolated cells and their boundaries, and then identify cells (cf. Fig. 8.17(e) and (f)). More precisely, note first that a chain of maps can be associated with any n-Gmap (cf. Fig. 8.17(c) and (d) and the conversion process below). So, an isolated i-cell c_i is the chain of maps which corresponds to an i-Gmap $G^{c_i} = (D^{c_i}, \alpha_0^{c_i}, \ldots, \alpha_{i-1}^{c_i}, \alpha_i^{c_i} = Id)$ containing a unique connected component. The *canonical boundary* of G^{c_i} is $(D^{c_i}, \alpha_0^{c_i}, \ldots, \alpha_{i-1}^{c_i})$. The chain of maps associated with G^{c_i} contains thus an i-cell and its canonical boundary (cf. Fig. 8.17(e)). It is then easy to define an operation for identifying two i-cells $G^{c_i} = (D^{c_i}, \alpha_0^{c_i}, \ldots, \alpha_{i-1}^{c_i}, \alpha_i^{c_i} = \omega)$ and $G^{c_i'} = (D^{c_i'}, \alpha_0^{c_i'}, \ldots, \alpha_{i-1}^{c_i'}, \alpha_i^{c_i'} = \omega)$ such that:

1. a function $\varphi : D^{c_i} \rightarrow D^{c_i'}$ exists, such that for any $d \in D^{c_i}$, for any $j, 0 \leq j \leq i-1$, $\varphi(\alpha_j^{c_i}(d)) \in \{\varphi(d), \alpha_j^{c_i'}(\varphi(d))\}$; in other words, a correspondence exists between the cells, which have thus common elements in their structures;

2. for any $d \in D^{c_i}$, $\sigma^i(d) = \sigma^i(\varphi(d))$. In other words, the two cells share the same boundary.

The operation consists in modifying σ^{i+1} for any dart $d' \in G^{i+1}$ such that $\sigma^{i+1}(d') = d \in G^{c_i}$ in order that $\sigma^{i+1}(d') = \varphi(d)$, and in removing G^{c_i}.

Subclasses of chains of maps have been defined: for instance, when for any $i \geq 1$, σ^i is a isomorphism between an orbit $\langle \alpha_0^i, \ldots, \alpha_{i-2}^i \rangle$ of the boundary of an i-cell and an $(i-1)$-cell, there is a strict correspondence between the structure of i-cells and their boundaries. Such properties can be used in order to define optimized data structures.

Conversion between n-Gmap and n-dimensional chains of maps

Let $G = (D, \alpha_0, \ldots, \alpha_n)$ be an n-Gmap. The n-dimensional chain of maps $C = ((G^i = (D^i, \alpha_0^i, \ldots, \alpha_{i-1}^i, \omega))_{i=0,\ldots,n}, (\sigma^i)_{i=1,\ldots,n})$ can be associated with G, where (cf. Figs. 8.17(c) and (d)):

1. for any $i, 0 \leq i \leq n$, there is a bijection φ^i between the set of orbits $\langle \alpha_{i+1}, \ldots, \alpha_n \rangle$ of G and D^i. For instance in Figs. 8.17(c) and (d):
    - all darts of D^2 in (d) correspond to orbits $\langle\rangle$ in (c);
    - all darts of D^1 in (d) correspond to orbits $\langle \alpha_2 \rangle$ in (c); for instance a (resp. k) corresponds to $\langle \alpha_2 \rangle(1)$ (resp. $\langle \alpha_2 \rangle(11)$);
    - all darts of D^0 in (d) correspond to orbits $\langle \alpha_1, \alpha_2 \rangle$ in (c); for instance, A (resp. B) corresponds to $\langle \alpha_1, \alpha_2 \rangle(1)$ (resp. $\langle \alpha_1, \alpha_2 \rangle(2)$);

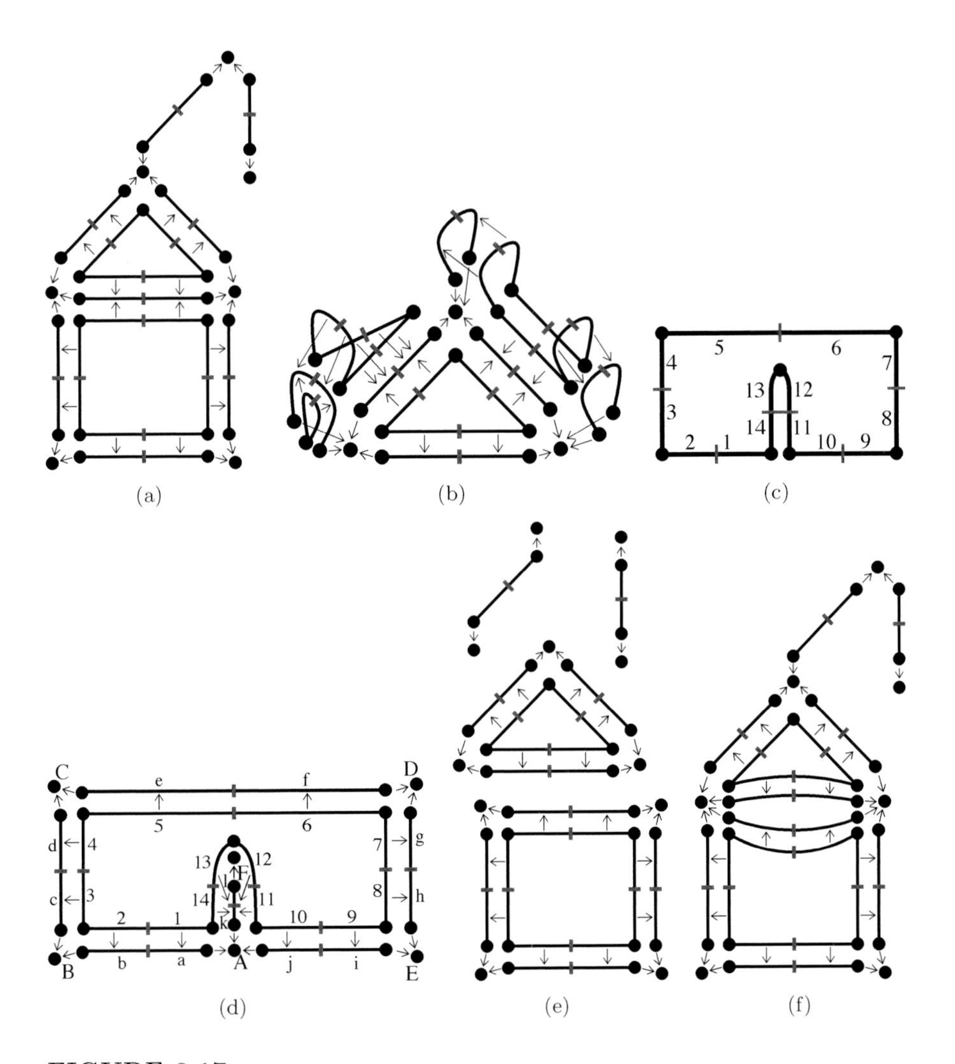

FIGURE 8.17
(a) The chain of maps corresponding to the incidence graph of Fig. 8.13.
(b) A chain of maps with multi-incidence.
(c) The 2-Gmap corresponding to the cellular quasi-manifold depicted in Fig. 8.15(b).
(d) The chain of maps corresponding to the 2-Gmap of (c).
(e) Isolated cells.
(f) Identifying 0-cells. The identification of two 1-cells will produce the chain of maps depicted in (a).

2. for any $i, 0 \leq i \leq n$, for any $j, 0 \leq j \leq i-1$, for any dart $d \in D^i$, $\alpha_j^i(d) = \varphi^i(\langle\alpha_{i+1}, \ldots, \alpha_n\rangle(\alpha_j(d')))$, where $d' \in D$ and $\varphi^i(\langle\alpha_{i+1}, \ldots, \alpha_n\rangle(d')) = d$. For instance in Figs. 8.17(c) and (d):
   - involutions α_0^2 and α_1^2 in (d) correspond to α_0 and α_1 in (c), since $\langle\rangle(d)$ is equivalent to d;
   - $\alpha_0^1(a) = b$ since a corresponds to $\langle\alpha_2\rangle(1)$ and b corresponds to $\langle\alpha_2\rangle(\alpha_0(1))$; similarly, $\alpha_0^1(k) = l$ since k corresponds to $\langle\alpha_2\rangle(11)$ and l corresponds to $\langle\alpha_2\rangle(12)$, where $12 = \alpha_0(11)$;
3. for any $i, 1 \leq i \leq n$, for any dart $d \in D^i$, $\sigma^i(d) = \varphi^i(\langle\alpha_i, \ldots, \alpha_n\rangle(d'))$, where $d' \in D$ and $\varphi^i(\langle\alpha_{i+1}, \ldots, \alpha_n\rangle(d')) = d$. For instance in Figs. 8.17(c) and (d):
   - $\sigma^2(1) = a$ in (d), since a corresponds to $\langle\alpha_2\rangle(1)$ in (c);
   - $\sigma^1(a) = A$ (resp. $\sigma^1(l) = F$) in (d) since a corresponds to $\langle\alpha_2\rangle(1)$ and A corresponds to $\langle\alpha_1, \alpha_2\rangle(1)$ (resp. l corresponds to $\langle\alpha_2\rangle(12)$ and F corresponds to $\langle\alpha_1, \alpha_2\rangle(12)$) in (c).

Simplicial interpretation of chains of maps

A simplicial analog can be associated with any chain of maps $C = ((G^i = (D^i, \alpha_0^i, \ldots, \alpha_{i-1}^i, \omega))_{i=0,\ldots,n}, (\sigma^i)_{i=1,\ldots,n})$. The conversion is easily deduced from that of n-Gmaps. For any $i, 0 \leq i \leq n$, the simplicial interpretation of G^i is that of any i-Gmap, but it describes the interior of i-cells, since $\alpha_i^i = \omega$ is undefined; so, all orbits involving α_i^i do not exist, meaning that all simplices in the simplicial interpretation of G^i are incident to a vertex numbered i. All face operators linking simplices which are internal to an i-cell are defined as for the simplicial interpretation of any i-Gmap. Face operators linking simplices issued from different cells correspond to face operators σ^i. More precisely, let i be such that $0 \leq i \leq n$:

1. let J be any nonempty sequence $\{j_0, \ldots, j_k\} \subset [0, i]$, such that $j_k = i$; associate a set of simplices numbered $(j_0, \ldots, j_k = i)$ with the set of orbits $\langle\rangle_{[0,i]\setminus J}$ of G^i;
2. assume $k > 0$. Let l be such that $0 \leq l \leq k-1$, d be a dart of D^i and σ be the simplex associated with $\langle\rangle_{[0,i]\setminus\{j_0,\ldots,j_l,\ldots,j_k=i\}}(d)$. Then $d_l(\sigma)$ is the simplex associated with $\langle\rangle_{[0,i]\setminus\{j_0,\ldots,\hat{j_l},\ldots,j_k=i\}}(d)$. $d_k(\sigma)$ is the simplex associated with $\langle\rangle_{[0,j_{k-1}]\setminus\{j_0,\ldots,j_{k-1}\}}(\sigma^{j_{k-1}+1}(\ldots(\sigma^i(d))))$;
3. each vertex corresponding to an orbit $\langle\rangle_{[0,i]\setminus\{i\}}$ of G^i is numbered i.

It is also possible to define conversion processes between equivalent subclasses of chains of maps and incidence graphs, as for connected n-Gmaps without boundary and without multi-incidence and n-surfaces.

Note also that other subclasses of combinatorial maps can be defined, according to the topological properties which have to be satisfied (cf. [104] for instance).

8.3.2 Euler-Poincaré Characteristic for Combinatorial Maps

The Euler-Poincaré characteristic is a well-known *topological invariant*[11]: if two subdivisions are homeomorphic, then their Euler-Poincaré characteristics are equal. The converse assertion is used in practice, i.e. if the Euler-Poincaré characteristics of two subdivisions are different, it is sure that the subdivisions are not homeomorphic.

The Euler-Poincaré characteristic of an n-dimensional simplicial "object" is the alternate sum of the number of i-simplices, for $0 \leq i \leq n$, i.e.:

$$\chi = \sum_{i=0}^{n} (-1)^i |K_i|$$

where K_i is the set of i-simplices.

8.3.2.1 For n-Gmaps

The definition of the Euler-Poincaré characteristic of an n-Gmap can be easily deduced from the simplicial interpretation of n-Gmaps, since an orbit for i involutions corresponds to an $(n-i)$-simplex.

Definition 91 (Euler-Poincaré characteristic of an n-Gmap) *Let $G = (D, \alpha_0, \ldots, \alpha_n)$ be an n-Gmap, with $n \geq 0$. Its Euler-Poincaré characteristic $\chi(G)$ is:*

$$\chi(G) = \sum_{i=0}^{n} (-1)^{(n-i)} \sum_{0 \leq k_0 < \ldots < k_{i-1} \leq n} \#\langle \alpha_{k_0}, \ldots, \alpha_{k_{i-1}} \rangle$$

where $\#\langle \alpha_{k_0}, \ldots, \alpha_{k_{i-1}} \rangle$ is the number of orbits $\langle \alpha_{k_0}, \ldots, \alpha_{k_{i-1}} \rangle$.

For instance:

- when $n = 0$, $\chi(G) = \#\langle\rangle = |D|$, i.e. it is equal to the number of vertices in the associated cellular quasi-manifold;
- when $n = 1$, $\chi(G) = -\#\langle\rangle + \#\langle \alpha_1 \rangle + \#\langle \alpha_0 \rangle$, i.e. it is equal to the numbers of vertices numbered (0) and (1) minus the number of edges of the associated cellular quasi-manifold. When G is connected and without 0 and 1-boundary, i.e. it corresponds to a closed polygonal curve, $\#\langle \alpha_0 \rangle = \#\langle \alpha_1 \rangle = \#\langle\rangle/2$; thus $\chi(G) = 0$, it is the Euler-Poincaré characteristic of a circle. When G is connected, without 0-boundary but with 1-boundaries, i.e. it corresponds to an open polygonal curve, $\#\langle \alpha_0 \rangle = \#\langle\rangle/2$ and $\#\langle \alpha_1 \rangle = 1 + \#\langle\rangle/2$; thus $\chi(G) = 1$, it is the Euler-Poincaré characteristic of a segment;

[11] Several topological invariants have been defined and studied, orientability for instance.

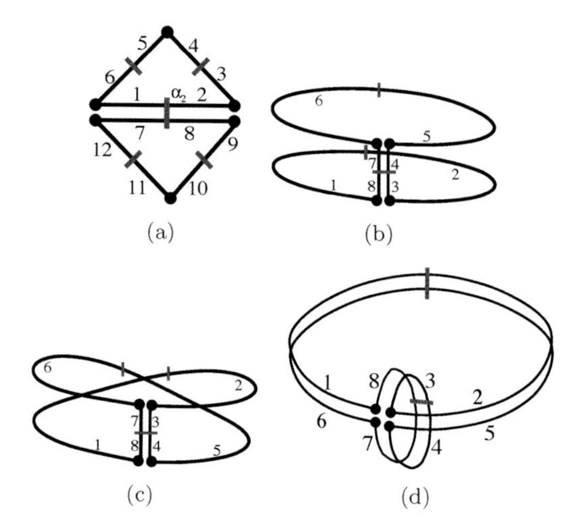

FIGURE 8.18
Examples of 2-Gmaps and their Euler-Poincaré characteristics.
(a) G_1 describes a disk: $\chi(G_1) = 1$.
(b) G_2 describes an annulus: $\chi(G_2) = 0$.
(c) G_3 describes a Möbius strip: $\chi(G_3) = 0$.
(d) G_4 describes a torus: $\chi(G_4) = 0$.

- when $n = 2$, $\chi(G) = \#\langle\rangle - \#\langle\alpha_0\rangle - \#\langle\alpha_1\rangle - \#\langle\alpha_2\rangle + \#\langle\alpha_0, \alpha_1\rangle + \#\langle\alpha_0, \alpha_2\rangle + \#\langle\alpha_1, \alpha_2\rangle$, i.e. it is equal to the number of triangles minus the number of edges (numbered $(1, 2), (0, 2), (0, 1)$) plus the number of vertices (numbered $(2), (1), (0)$). Four examples of 2-Gmaps are depicted in Fig. 8.18, together their Euler-Poincaré characteristics. For instance, for G_1 depicted in Fig. 8.18(a): $\#\langle\rangle = 12$, $\#\langle\alpha_0\rangle = \#\langle\alpha_1\rangle = 6$, $\#\langle\alpha_2\rangle = 10$, $\#\langle\alpha_0, \alpha_1\rangle = 2$, $\#\langle\alpha_0, \alpha_2\rangle = 5$ and $\#\langle\alpha_1, \alpha_2\rangle = 4$. Thus $\chi(G_1) = 12 - 6 - 6 - 10 + 2 + 5 + 4 = 1$: this is the Euler-Poincaré characteristic of a disk. Note that the Euler-Poincaré characteristics of an annulus, of a Möbius strip and of a torus are equal;
- when $n = 3$, $\chi(G) = -\#\langle\rangle + \#\langle\alpha_0\rangle + \#\langle\alpha_1\rangle + \#\langle\alpha_2\rangle + \#\langle\alpha_3\rangle - \#\langle\alpha_0, \alpha_1\rangle - \#\langle\alpha_0, \alpha_2\rangle - \#\langle\alpha_0, \alpha_3\rangle - \#\langle\alpha_1, \alpha_2\rangle - \#\langle\alpha_1, \alpha_3\rangle - \#\langle\alpha_2, \alpha_3\rangle + \#\langle\alpha_0, \alpha_1, \alpha_2\rangle + \#\langle\alpha_0, \alpha_1, \alpha_3\rangle + \#\langle\alpha_1, \alpha_2, \alpha_3\rangle$.

For a 2-Gmap having no 0 nor 1-boundary, $\#\langle\alpha_0\rangle = \#\langle\alpha_1\rangle = \#\langle\rangle/2$. Thus the previous formula $\chi(G) = \#\langle\rangle - \#\langle\alpha_0\rangle - \#\langle\alpha_1\rangle - \#\langle\alpha_2\rangle + \#\langle\alpha_0, \alpha_1\rangle + \#\langle\alpha_0, \alpha_2\rangle + \#\langle\alpha_1, \alpha_2\rangle$ can be simplified: $\chi(G) = -\#\langle\alpha_2\rangle + \#\langle\alpha_0, \alpha_1\rangle + \#\langle\alpha_0, \alpha_2\rangle + \#\langle\alpha_1, \alpha_2\rangle$. This formula can be rewritten into $\chi(G) = (\#\langle\alpha_1 \circ \alpha_0\rangle + \#\langle\alpha_2 \circ \alpha_0\rangle + \#\langle\alpha_2 \circ \alpha_1\rangle - |D|)/2$ by using properties on orbits and the fact that G has no 0 nor 1-boundary (see [165] for details).

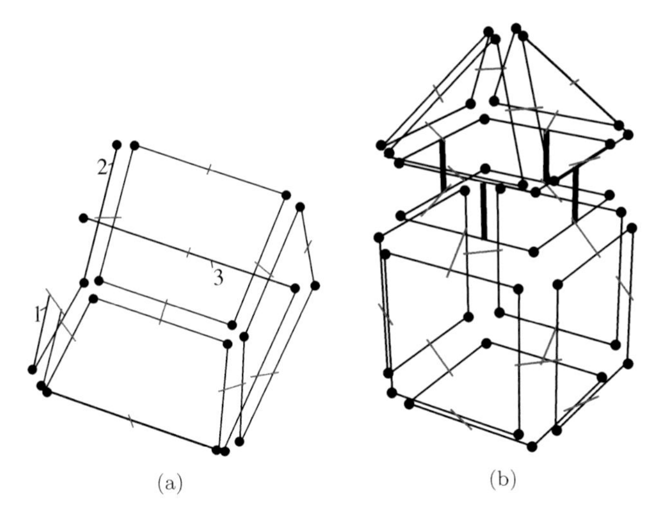

FIGURE 8.19
(a) 2-Gmap G_5, with 1-boundary and 2-boundary, describes a disk: $\chi(G_5) = 1$.
(b) 3-Gmap G_6 describes a 3D ball: $\chi(G_6) = 1$.

If the 2-Gmap has no 0 nor 1-boundary and no folded cell, $\#\langle\alpha_2\rangle = 2\#\langle\alpha_0,\alpha_2\rangle$ and the previous formula becomes $\chi(G) = \#\langle\alpha_0,\alpha_1\rangle - \#\langle\alpha_0,\alpha_2\rangle + \#\langle\alpha_1,\alpha_2\rangle$. This gives the classical definition of the Euler-Poincaré characteristics equals to the number of vertices minus the number of edges plus the number of faces.

2-Gmap G_5 depicted in Fig. 8.19(a) contains 1-boundary and 2-boundary. $\#\langle\rangle = 36$, $\#\langle\alpha_0\rangle = \#\langle\alpha_1\rangle = 19$, $\#\langle\alpha_2\rangle = 21$, $\#\langle\alpha_0,\alpha_1\rangle = 5$, $\#\langle\alpha_0,\alpha_2\rangle = 11$ and $\#\langle\alpha_1,\alpha_2\rangle = 8$: thus $\chi(G_2) = 36 - 19 - 19 - 21 + 5 + 11 + 8 = 1$: this is the Euler-Poincaré characteristic of a disk.

3-Gmap G_6 depicted in Fig. 8.19(b) is such that $\#\langle\rangle = 80$, $\#\langle\alpha_0\rangle = \#\langle\alpha_1\rangle = \#\langle\alpha_2\rangle = 40$, $\#\langle\alpha_3\rangle = 72$, $\#\langle\alpha_0,\alpha_1\rangle = 11$, $\#\langle\alpha_0,\alpha_2\rangle = 20$, $\#\langle\alpha_0,\alpha_3\rangle = 36$, $\#\langle\alpha_1,\alpha_2\rangle = 13$, $\#\langle\alpha_1,\alpha_3\rangle = 36$, $\#\langle\alpha_2,\alpha_3\rangle = 32$, $\#\langle\alpha_0,\alpha_1,\alpha_2\rangle = 2$, $\#\langle\alpha_0,\alpha_1,\alpha_3\rangle = 10$, $\#\langle\alpha_0,\alpha_2,\alpha_3\rangle = 16$, $\#\langle\alpha_1,\alpha_2,\alpha_3\rangle = 9$: thus $\chi(G_2) = -80+40+40+40+72-11-20-36-13-36-32+2+10+16+9 = 1$, it is the Euler-Poincaré characteristic of a 3D ball.

At last, the definition of the Euler-Poincaré characteristic, and thus its computation, can be simplified for n-Gmaps satisfying some properties, for instance when there is no folded cell.

8.3.2.2 For n-maps

The definition of the Euler-Poincaré characteristic of an n-map is based upon the correspondence between an n-map M and an equivalent n-Gmap G (see Section 5.5.2); each dart of the n-map is described by two darts joined by

α_0 in the corresponding n-Gmap. Let us consider here the case where M is without boundary. By definition, G is also without boundary. The Euler-Poincaré characteristic of G is:

$$\chi(G) = \sum_{i=0}^{n} (-1)^{(n-i)} \sum_{0 \leq k_0 < \ldots < k_{i-1} \leq n} \#\langle \alpha_{k_0}, \ldots, \alpha_{k_{i-1}} \rangle.$$

This sum can be decomposed into four parts:

$$\begin{aligned}\chi(G) = &(-1)^n \#\langle\rangle + \\ &(-1)^{(n-1)} \sum_{0 \leq k_0 \leq n} \#\langle \alpha_{k_0} \rangle + \\ &\sum_{i=2}^{n} (-1)^{(n-i)} \sum_{0 = k_0 < k_1 < \ldots < k_{i-1} \leq n} \#\langle \alpha_0, \alpha_{k_1}, \ldots, \alpha_{k_{i-1}} \rangle + \\ &\sum_{i=2}^{n} (-1)^{(n-i)} \sum_{0 < k_0 < \ldots < k_{i-1} \leq n} \#\langle \alpha_{k_0}, \ldots, \alpha_{k_{i-1}} \rangle.\end{aligned}$$

For the first sum, $(-1)^n \#\langle\rangle = (-1)^n |D_G|$ ($|D_G|$ is the number of darts of G which is equal to $2|D|$ the number of darts of M).

For the second sum, since G is without boundary, $\#\langle \alpha_{k_0} \rangle = \frac{|D_G|}{2}$ for any $k_0 \in \{0, \ldots, n\}$. This gives $(-1)^{(n-1)} \sum_{0 \leq k_0 \leq n} \#\langle \alpha_{k_0} \rangle = (-1)^{(n-i)}(n+1)\frac{|D_G|}{2}$.

By definition of orbits, $\langle \alpha_{k_0}, \ldots, \alpha_{k_{i-1}} \rangle = \langle \alpha_{k_1} \circ \alpha_{k_0}, \ldots, \alpha_{k_{i-1}} \circ \alpha_{k_0} \rangle$.

This property is used for the two last parts of the sum. First:

$$\begin{aligned}\langle \alpha_0, \alpha_{k_1}, \ldots, \alpha_{k_{i-1}} \rangle &= \langle \alpha_{k_1} \circ \alpha_0, \ldots, \alpha_{k_{i-1}} \circ \alpha_0 \rangle \\ &= \langle \beta_{k_1}, \ldots, \beta_{k_{i-1}} \rangle.\end{aligned}$$

Here the definition of the conversion between n-maps and n-Gmaps is used to link $\alpha_{k_j} \circ \alpha_0$ and β_{k_j}.

Second:

$$\begin{aligned}\langle \alpha_{k_0}, \ldots, \alpha_{k_{i-1}} \rangle = &\langle \alpha_{k_1} \circ \alpha_{k_0}, \ldots, \alpha_{k_{i-1}} \circ \alpha_{k_0} \rangle \\ &\langle \beta_{k_1} \circ \beta_{k_0}, \ldots, \beta_{k_{i-1}} \circ \beta_{k_0} \rangle.\end{aligned}$$

Here $\alpha_{k_j} \circ \alpha_{k_0}$ is transformed into $\alpha_{k_j} \circ \alpha_0 \circ \alpha_0 \circ \alpha_{k_0}$ and the definition of the conversion between n-maps and n-Gmaps is used to obtain $\beta_{k_j} \circ \beta_{k_0}$ (here $k_j > 1$ thus β_{k_j} is an involution).

This gives the following Euler-Poincaré characteristic of an n-map without boundary:

Definition 92 (Euler-Poincaré characteristic of an n-map) *Let $M = (D, \beta_1, \ldots, \beta_n)$ be an n-map without boundary, with $n > 0$. Its Euler-Poincaré*

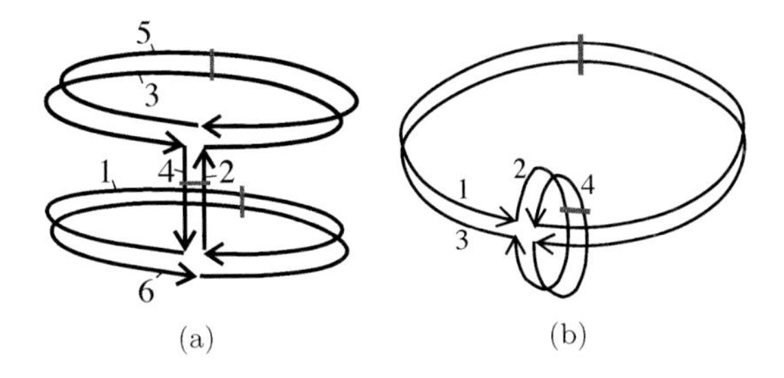

FIGURE 8.20
Examples of 2-maps and their Euler-Poincaré characteristics.
(a) M_1 describes a sphere: $\chi(M_1) = 2$.
(b) M_2 describes a torus: $\chi(M_2) = 0$.

characteristic $\chi(M)$ *is the alternated sum:*

$$\begin{aligned}\chi(M) =&(-1)^{(n-1)}(n-1)|D|+\\ &\sum_{i=2}^{n}(-1)^{(n-i)}\sum_{0<k_1<\ldots<k_{i-1}\leq n}\#\langle\beta_{k_1},\ldots,\beta_{k_{i-1}}\rangle+\\ &\sum_{i=2}^{n}(-1)^{(n-i)}\sum_{0<k_0<\ldots<k_{i-1}\leq n}\#\langle\beta_{k_1}\circ\beta_{k_0},\ldots,\beta_{k_{i-1}}\circ\beta_{k_0}\rangle.\end{aligned}$$

Note that it is possible to extend this formula to consider n-maps with boundary, but the definition becomes more complex since the case of free darts has to be taken into account.

For 1D, the previous definition becomes $\chi(M) = 0$. Any 1-map M without boundary (i.e. closed polygonal curves) has the Euler-Poincaré characteristic of a circle.

For 2D, the previous definition becomes $\chi(M) = -|D| + \#\langle\beta_1\rangle + \#\langle\beta_2\rangle + \#\langle\beta_2\circ\beta_1\rangle$. Figure 8.20 shows two examples of 2-maps and their corresponding Euler-Poincaré characteristics. For M_1 given in Fig. 8.20(a), $|D| = 6$, $\#\langle\beta_1\rangle = 3$, $\#\langle\beta_2\rangle = 3$, $\#\langle\beta_2\circ\beta_1\rangle = 2$: thus $\chi(M_1) = -6+3+3+2 = 2$: this is the Euler-Poincaré characteristic of a sphere. For M_2 given in Fig. 8.20(b), $|D| = 4$, $\#\langle\beta_1\rangle = 1$, $\#\langle\beta_2\rangle = 2$, $\#\langle\beta_2\circ\beta_1\rangle = 1$: thus $\chi(M_2) = -4+1+2+1 = 0$: this is the Euler-Poincaré characteristic of a torus.

For 3D, the previous definition becomes $\chi(M) = 2|D| - \#\langle\beta_1\rangle - \#\langle\beta_2\rangle - \#\langle\beta_3\rangle + \#\langle\beta_1,\beta_2\rangle + \#\langle\beta_1,\beta_3\rangle + \#\langle\beta_2,\beta_3\rangle - \#\langle\beta_2\circ\beta_1\rangle - \#\langle\beta_3\circ\beta_1\rangle - \#\langle\beta_3\circ\beta_2\rangle + \#\langle\beta_2\circ\beta_1,\beta_3\circ\beta_1\rangle$.

In 2D, the Euler-Poincaré characteristics of a 2-map without 1-boundary but possibly with 2-boundary can be computed by using the correspondence between a 2-map M and an equivalent 2-Gmap G. In this case, the 2-Gmap has no 0 nor 1-boundary and thus $\chi(G) = -\#\langle\alpha_2\rangle + \#\langle\alpha_0, \alpha_1\rangle + \#\langle\alpha_0, \alpha_2\rangle + \#\langle\alpha_1, \alpha_2\rangle$. Since G is without 0-boundary, $\#\langle\alpha_2\rangle = 2\#\langle\alpha_0, \alpha_2\rangle - |\{d|\alpha_0(d) = \alpha_2(d)\}|$. Indeed, for a dart d such $\alpha_0(d) = \alpha_2(d)$, there is one orbit $\langle\alpha_2\rangle$ for one orbit $\langle\alpha_0, \alpha_2\rangle$, while for other darts, there are two orbits $\langle\alpha_2\rangle$ for one orbit $\langle\alpha_0, \alpha_2\rangle$. Then the correspondence between M and G gives the Euler-Poincaré characteristics of a 2-map:

$$\chi(M) = \#\langle\beta_2 \circ \beta_0\rangle - \#\langle\beta_2\rangle + \#\langle\beta_1\rangle + |\{d|\beta_2(d) = d\}|.$$

If the 2-map has no 1-boundary and no folded cell, $|\{d|\alpha_0(d) = \alpha_2(d)\}| = 0$ and the previous formula becomes $\chi(M) = \#\langle\beta_2 \circ \beta_0\rangle - \#\langle\beta_2\rangle + \#\langle\beta_1\rangle$. This gives the classical definition of the Euler-Poincaré characteristics equal to the number of vertices minus the number of edges plus the number of faces.

Let us consider the 2-map M_3 obtained from 2-map M_1 given in Fig. 8.20(a) where we remove the two darts 5 and 6. $\#\langle\beta_1\rangle = 1$, $\#\langle\beta_2\rangle = 3$, $\#\langle\beta_2 \circ \beta_0\rangle = 2$ and $|\{d|\beta_2(d) = d\}| = 0$: thus $\chi(M_3) = 1 - 3 + 2 + 0 = 0$: this is the Euler-Poincaré characteristic of an annulus.

9

Comparison with Other Cellular Data Structures

In this chapter, n-maps and n-Gmaps are compared with related cellular data structures. Many different data structures have been proposed in order to describe subdivided objects. They can be classified according to:

- the dimension of the represented objects;
- the topological properties of the represented objects: non-manifold, pseudo-manifold, quasi-manifold, orientable, nonorientable...
- the type of cells: regular cells, as triangles (and more generally simplices), quadrangles (and more generally cubes, simploids), general cells...

Remember that n-maps allow to represent n-dimensional orientable quasi-manifolds with general cells, and n-Gmaps allow to represent n-dimensional orientable and nonorientable quasi-manifolds with general cells. Thus, we limit in this chapter the comparison with data structures corresponding to these two classes (see also for instance [164, 208]). A more general discussion is conducted in the next chapter 10.

Schematically, the first work about data structures for representing subdivided geometric objects is the definition of the *winged-edge* data structure by Baumgart at the beginning of the 1970s [17]. Then, during the 1980s, many works in geometric modeling and computational geometry dealt with the definition of data structures for representing orientable (quasi-)manifolds without boundaries, then general (quasi-)manifolds, for dimension 2, then dimension 3, at last dimension n (cf. next section). At the end of the 1980s and during the 1990s, many data structures for representing complexes have been conceived, from dimension 2 to dimension n (see for instance [102, 195, 216, 104, 159]).

In parallel, several authors studied data structures and operations for representing simplicial objects (e.g. [90]). Some ideas, as the representation of adjacency relations between n-cells, when handling n-dimensional quasi-manifolds, were applied for simplicial structures as for cellular structures (see for instance [183]).

Note that during the 1990s and 2000s, the definition of subclasses close to manifolds was a subject of great interest, for simplicial structures, incidence graphs and ordered cellular structures (see for instance [89, 91, 85, 39, 165]).

Since the beginning of the 2000s, two very interesting research directions are explored:

- the conception of hierarchical structures, for instance for representing objects at different detail levels (e.g. [88, 200, 148]); this is also a subject of interest for the image analysis community since the beginning of the 1990s (e.g. [151, 44, 197, 188]);
- the compression of the representation, for instance for the storage of huge subdivisions (e.g. [204, 190, 129]).

9.1 History of Combinatorial Maps

The first work about combinatorial maps started in the 1960s in mathematics, more precisely in the field of combinatorics, in order to represent planar graphs embedded in the plane. They were introduced in [103] to describe *polyhedral surfaces*, then formally defined in [206, 137]. In the different works, different names were used: *planar maps*, *topological graphs*, *constellations* or *rotation systems*. These concepts were used in combinatorial graph theory, for problems of compact encoding of planar graphs, enumeration of rooted planar graphs or for the map-coloring problem [54, 55, 191, 192].

When a graph is embedded in a plane, edges are ordered around vertices, for instance according to a clockwise orientation. It is then possible to define a function f, giving for each edge its next edge around one of its vertex. However, since each edge has two different successors, one for each extremity vertex, each edge has two associated edges and thus f is not a function[1]. This problem can be solved by cutting each edge in two half-edges: so, each half-edge has only one extremity, and thus only one successor by f. This cutting requires to add a relation between the two half-edges issued from the same edge. This gives the original definition of planar maps: a planar map is a set of half-edges (called "darts" according to the terminology of combinatorial maps), plus two functions defined on these half-edges: σ, which gives the next half-edge around its incident vertex, and α, which gives the other half-edge issued from the same edge. Formally, σ is a permutation and α is an involution.

Thanks to these two functions, the notions of vertices, edges and faces can be retrieved. A vertex is the cycle of half-edges obtained from a given half-edge, by applying σ until going back to the initial half-edge. An edge is the set of two half-edges linked by α. Since a planar map describes a planar graph embedded in the plane, cycles of edges describe the boundary of faces. Given a half-edge h, the next half-edge of the face is $\sigma \circ \alpha(h)$, denoted $\varphi(h)$. A face

[1]Note that this is the basis of the *winged-edge data structure*, defined in the 1970s for computer vision purposes: cf. Section 9.2.1.

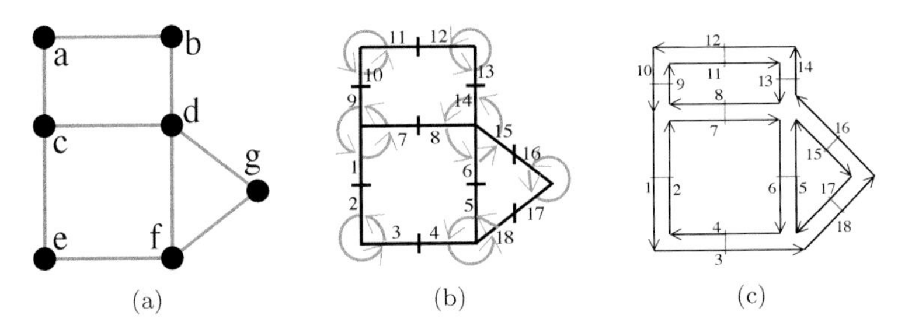

FIGURE 9.1
Planar map and link with 2-maps.
(a) Planar graph embedded in the plane.
(b) Corresponding planar map. $D = \{1, \dots, 18\}$ is the set of half-edges; σ is represented by gray arrows; α links two half-edges sharing a little perpendicular segment.
(c) Corresponding 2-map (D, φ, α).

is thus the cycle of half-edges obtained from a given half-edge, by applying φ until going back to the initial half-edge.

As illustrated in Fig. 9.1, planar maps are equivalent to 2-maps without boundary, where darts correspond to half-edges, β_1 to φ and β_2 to α. Figure 9.1(b) shows a planar map, and Fig. 9.1(c) the corresponding 2-map. For example: $\sigma(1) = 7$, $\sigma(7) = 9$ and $\sigma(9) = 1$; $\alpha(1) = 2$ and $\alpha(2) = 1$; $\varphi(2) = 7$, $\varphi(7) = 6$, $\varphi(6) = 4$ and $\varphi(4) = 2$.

Planar maps have been extended during the 1980s by several works in the field of combinatorics, leading to notions equivalent to 2-Gmaps [47] and to n-Gmaps [212].

In the 1980s, similar notions were retrieved by works related to geometric modeling, computational geometry and computer graphics. For instance, the *half-edge* data structure [214] is basically a 2-map without boundary. Several versions and variants were also defined [172, 86] (see Section 9.2.2). The *quad-edge* data structure is close to 2-Gmaps: cf. Section 9.3.1; the *facet-edge* data structure is close to 3-Gmaps: cf. Section 9.3.2. The *radial-edge* data structure [216] has been defined in order to represent 2-dimensional complexes, i.e. $2D$ *non-manifolds*. In fact, this structure is very close to 3-maps, extended to take into account several "non-manifold configurations": cf. Section 9.2.3 and Section 9.2.4.

At the end of the 1980s, 2-maps were extended in 3D [9, 10, 203] and in nD [162, 165]. In the original definition, n-maps allow to describe n-dimensional orientable quasi-manifold without boundary. The extension for representing n-dimensional orientable quasi-manifold with or without boundary was made formally much later [189, 83], even if many people use since a long time in many different softwares different implementations of modified n-maps allow-

TABLE 9.1
Winged-edge pointers, and relations with 2-maps. Each edge corresponds to two darts linked by β_2

name	vertex1	vertex2	right_face	left_face
pointer to	vertex	vertex	face	face
in 2-maps	$emb_0(d)$	$emb_0(\beta_2(d))$	$emb_2(\beta_2(d))$	$emb_2(d)$
name	prev_left_face	next_left_face	prev_right_face	next_right_face
pointer to	winged edge	winged edge	winged edge	winged edge
in 2-maps	$\beta_0(d)$	$\beta_1(d)$	$\beta_1(\beta_2(d))$	$\beta_0(\beta_2(d))$

ing boundaries. Also at the end of 1980s, n-Gmaps were introduced in the field of geometric modeling [162, 165].

Simultaneously, at the end of the 1980s, several works related to image processing started to use combinatorial maps in order to describe 2D segmented images: they extended previous works based on region adjacency graphs [36]. These works were pursued in the 1990s: several image processing algorithms have been proposed, based on 2-maps and some variants (such as the *topological graph of frontier* [109, 110] or *discrete maps* [36, 95, 40]). Mainly, image edition methods [16, 116, 37] and image segmentation algorithms [42, 41, 33, 32, 3, 144, 43, 66] were concerned. In the beginning of the 2000s, previous works were extended for processing 3D images, thanks to the use of 3-maps [34, 58, 59, 15, 98, 100, 97].

9.2 Oriented Cellular Quasi-Manifolds

9.2.1 Winged-edge

The first data structure defined in order to describe 2D subdivisions for computer vision purposes is the *winged-edge* data structure [17].

The main element of this data structure is the *winged-edge*, which is an edge having eight pointers to different incident cells. The main structural information is related to the four pointers which make it possible to access the incident edges by turning clockwise and counterclockwise around the incident vertices: cf. Fig. 9.2. The winged-edge is often used for solid modeling, i.e. for representing the surface which bounds a polyhedron. By convention, clockwise ordering corresponds to a view from outside of the polyhedron, and this ordering is used for traversing the subdivision.

In addition to the list of winged-edges, there is also a list of vertices and a list of faces. Different information is associated with vertices and faces: for

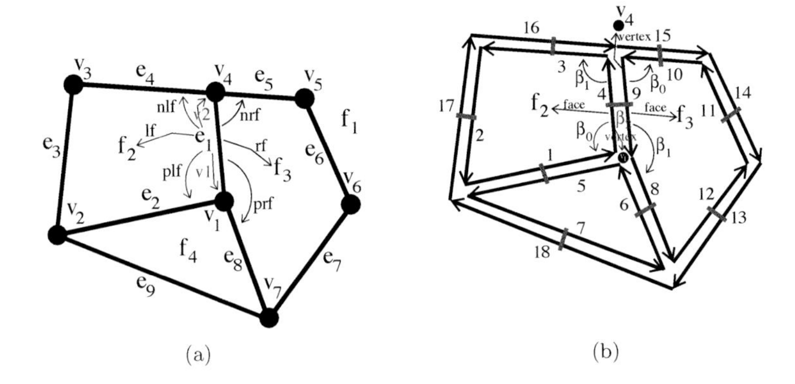

FIGURE 9.2
(a) Winged-edge representation of the 2D quasi-manifold without boundary depicted in Fig. 3.4(a) page 56. The eight pointers from edge e_1 are drawn by arrows.
(b) Corresponding 2-map.

example, a 3D point is associated with each vertex, a color is associated with each face, and a pointer is associated with each vertex (resp. face), making it possible to access one of its incident winged-edge.

Winged-edges allow to represent oriented 2D manifold without boundary: they are equivalent to 2-maps without boundary. Each winged-edge corresponds to a pair of darts linked by β_2; relations between the winged-edges can easily be translated in corresponding relations in the 2-map (see Table 9.1). Note that the lists of vertices and faces often exist in 2-maps data structures (cf. chapter 7), when 0- and 2-embeddings are defined; in this case, embeddings are often linked with one of the darts of the associated cells, in order to access the structural information from a given embedding.

A winged-edge structure is represented in Fig. 9.2(a); it corresponds to the 2D object depicted in Fig. 3.4(a) page 56; the corresponding 2-map is represented in Fig. 9.2(b). For example, vertex1$(e_1) = v_1$, vertex2$(e_1) = v_4$, left_face$(e_1) = f_2$, right_face$(e_1) = f_3$, prev_left_face$(e_1) = e_2$, next_left_face$(e_1) = e_4$, prev_right_face$(e_1) = e_8$, next_right_face$(e_1) = e_5$.

A simplified version of the winged-edge data structure was proposed, in which all pointers are not represented [179]. This version uses less memory space, but it allows only the traversal of the boundary of a face in the clockwise direction. Note that this memory optimization can also be done for 2-maps , by not explicitly representing β_0 in the dart data structure. An extension of winged-edges has also been proposed in [48], in order to describe 2-manifolds with or without boundary.

TABLE 9.2
Half-edge relations and link with 2-maps

name	vertex	face	previous	next	opposite
pointer to	vertex	face	half-edge	half-edge	half-edge
in 2-maps	$emb_0(d)$	$emb_2(d)$	$\beta_0(d)$	$\beta_1(d)$	$\beta_2(d)$

9.2.2 Half-edge

One problem related to the winged-edge data structure is the fact that pointers between edges correspond to adjacency relations between edges according to some vertex (or face). A simple way to illustrate this problem is to consider a subdivision of an orientable surface without boundary in which an edge e is a loop: so, vertex1(e) = vertex2(e). The traversal of the incident cells in a coherent way is possible, but it is more complicated for e than for an edge which is incident to two distinct vertices. Since traversals can be complicated, it can be necessary to conceive complex modification operations.

To solve this problem, the winged-edge data structure was modified by splitting each edge in two heal-edges. A half-edge is incident to exactly one face and to exactly one vertex: so, no confusion can arise, even when multi-incidence occurs.

This idea is the basis of the definition of the face-edge data structure [214], the half-edge data structure [172], and the DCEL data structure [86] (DCEL, for Doubly Connected Edge List). These three data structures are equivalent, the only difference being the names of the relations between the half-edges (e.g. *opposite* is called sometimes *twin* or *pair*, *previous* is called sometimes *prev*). This is probably the data structure which is the most used for describing surface meshes with irregular faces. See [140] for a comparison of different versions of half-edges, and for an efficient implementation.

The main element of the half-edge data structure is the *half-edge*: four pointers associate the half-edge with incident cells: cf. Fig. 9.3.

In addition to the list of half-edges, there is also a list of vertices and a list of faces. Each vertex and each face is associated with some information, and a pointer links it to one of its incident half-edges.

Half-edge data structures allow to represent oriented 2D manifolds without boundary. They are directly equivalent to 2-maps without boundary: for instance, half-edges correspond to darts (see Table 9.2). The two lists of vertices and faces exist also in 2-map data structures, when 0- and 2-embeddings are defined.

Figure 9.3(a) shows a half-edge data structure which represents the 2D object depicted in Fig. 3.4(a) page 56. Half-edges are numbered from 1 to 18. For example, previous(4) = 1, next(4) = 3, opposite(4) = 9, vertex(4) = v_1 and face(4) = f_2.

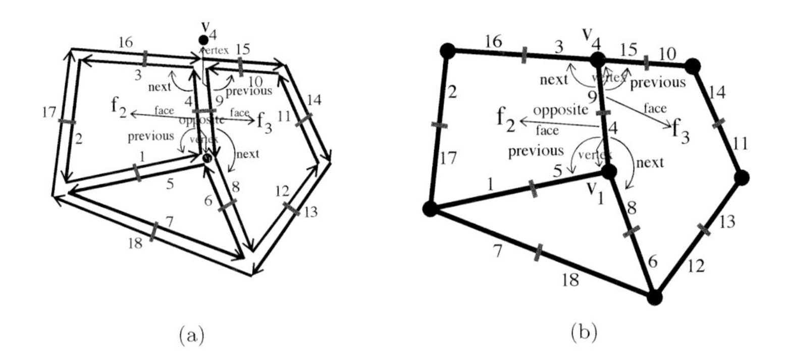

FIGURE 9.3
(a) Half-edge data structure, corresponding to the 2D quasi-manifold without boundary depicted in Fig. 3.4(a) page 56.
(b) Corresponding vertex-edge data structure.

The *vertex-edge* data structure [214] is a dual version of the face-edge data structure. Pointers *next* and *previous* are replaced in order to turn around vertices instead of turning around faces. Half-edges and opposites are the same for the two data structures. These two representations are duals: turning around vertices for one data structure consists in turning around faces for the other data structure. As seen in Section 5.5.4 page 182, it is easy to transform one representation into its dual. Let us call next_d, previous_d and opposite_d the pointers of the dual representation, the following properties hold for each half-edge h and its corresponding dual half-edge h_d:

- $\text{opposite}_d(h_d) = \text{opposite}(h)$;
- $\text{previous}_d(h_d) = \text{opposite}(\text{previous}(h))$;
- $\text{next}_d(h_d) = \text{next}(\text{opposite}(h))$.

Figure 9.3(b) shows the vertex-edge structure corresponding to the face-edge structure represented in Fig. 9.3(a). For example, $\text{opposite}_d(4) = \text{opposite}(4) = 4$, $\text{previous}_d(4) = \text{opposite}(\text{previous}(4)) = 5$, and $\text{next}_d(4) = \text{next}(\text{opposite}(4)) = 8$.

The half-edge data structure is equivalent to the *corner table* data structure [194]. Corners are equivalent to half-edges, next, previous and opposite corners relations are equivalent to next, previous and opposite relations for half-edges. Note that for triangular meshes, the corner table can be encoded in a compact way by only two arrays of integers and a set of rules. Note also that the half-edge data structure can be encoded in a similar compact way.

At last, the half-edge data structure has been extended in order to represent 2-manifolds with or without boundary [168].

TABLE 9.3
Edge-use relations, and correspondence with 3-maps

name	vertex_ptr	ccw_ptr	cw_ptr	radial_ptr	mate_ptr
pointer to	vertex-use	edge-use	edge-use	edge-use	edge-use
in 3-maps	$embo_0(d)$	$\beta_0(d)$	$\beta_1(d)$	$\beta_2(d)$	$\beta_3(d)$

9.2.3 Radial-edge

The *radial-edge* data structure was defined in [215] as an extension of the winged-edge data structure, in order to represent 2D *non-manifolds*, i.e. 2-dimensional complexes: remember that a 2-dimensional complex can be constructed by adding isolated i-cells, for $0 \leq i \leq 2$, and by identifying isomorphic j-cells, for $0 \leq j \leq 1$. So, more than two faces can be incident to the same edge; vertices or edges can exist, such that they are not incident to a face.

The main notion of a radial-edge data structure is the *edge-use* element. Let (f, e) be any pair of incident face and edge. Two edge-use elements are associated with (f, e). More precisely, remember that f can be oriented by defining a coherent orientation of its incident edges; then each edge-use associated with (f, e) corresponds to a possible coherent orientation of f and e.

Several relations are defined on edge-use elements. Let f be a face and e be an edge incident to f:

- the edge-use elements associated with f are organized into two inversely oriented cycles, according to *ccw_edge_use*; *cw_edge_use* is the inverse of *ccw_edge_use*;
- let f' be a face such that it is *directly adjacent* to f by e, i.e. f and f' share e and, starting from f, f' is reached by turning around e without traversing another face. The two coherently oriented cycles of edge-use elements corresponding to (f, e) and (f, e'), according to their direct adjacency, are linked by *radial_edge_use*;
- the two edge-use elements corresponding to (f, e) are linked by *mate_edge_use*.

Figure 9.4(b) shows an example of radial-edge data structure, which represents a 2D object embedded in $\mathbb{R}^3$, composed by three square faces incident to an edge. Only edge-uses are drawn in this figure, represented by arrows, some of them being numbered. For example ccw_ptr(1) = 7, cw_ptr(1) = 8, radial_ptr(1) = 2 and mate_ptr(1) = 6.

In addition to the list of edge-uses, ten other entities are represented and linked together. There are six "topological entities": *model* is the 3D modeling space, *region* is a volume, *shell* is a boundary surface, *face-use* is one face,

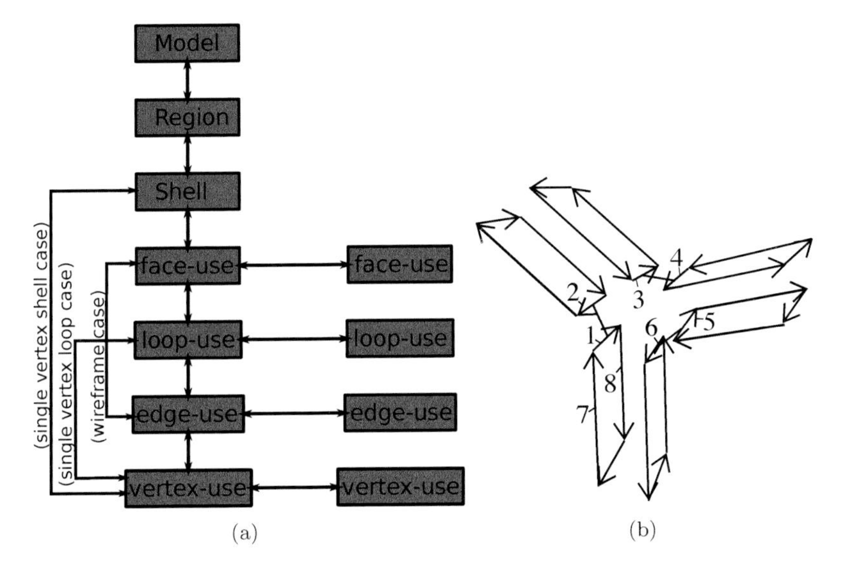

(a) (b)

FIGURE 9.4
(a) The eleven topological elements of the radial-edge data structure.
(b) Radial-edge representation of three square faces incident to a same edge.

loop-use is a boundary of a face, *vertex-use* is a vertex; and four "geometrical entities": *face*, *loop*, *edge* and *vertex*, which contain the corresponding geometric information and/or attributes associated with the "topological entities". Top-down and bottom-up hierarchical relationships are represented, i.e. each topological entity of dimension i has a list of its incident elements of dimension $i-1$ and $i+1$. *Wireframe* allows to represent paths of isolated edges which do not belong to a boundary of a face, and *single vertex* allows to represent isolated vertices, either in a shell or in a loop. Figure 9.4(a) shows these different entities and their links.

The radial-edge data structure has been designed in order to represent the topology of 2-dimensional complexes *embedded in* $\mathbb{R}^3$. Thus, *solids* can be implicitly represented by their boundary surfaces, and assemblies of such solids can be represented by the radial-edge data structure. So, this structure makes it possible to represent 3-dimensional subdivided objects. More precisely, if the topology of a 2D complex is described, there is no reason to define an order between the faces, through the different pointers acting on the edge-use elements. Take for instance three edges, choose one vertex for each edge, and identify these three vertices: you get a 1-dimensional complex. Embedded in $\mathbb{R}^3$, there is no order defined on the adjacent edges. But if the object is embedded into $\mathbb{R}^2$, an order can be defined on the edges, corresponding to the fact that *faces* exist (in this case, one face around the edges); the object *with its embedding space* is a 2-dimensional manifold. This is similar for

TABLE 9.4
Oriented-edge relations, and correspondence with 3-maps

name	svert	sedg	sfac	snxt	ssym	radial
pointer to	vertex	edge	face	oriented-edge	oriented-edge	oriented-edge
in 3-maps	$emb_0(d)$	$emb_1(d)$	$emb_2(d)$	$\beta_1(d)$	$\beta_2(d)$	$\beta_3(d)$

the radial-edge data structure. Basically, edge-use elements plus the relations defined on them, i.e. the underlying ordered structure, define 3-dimensional oriented quasi-manifolds. Note that this ordered structure is equivalent to 3-maps, as illustrated in Table 9.3. The other elements of the radial-edge data structure generalize the set of objects which can be represented; for instance, the boundary of a face can be made of several cycles of edges; the boundary of a face can be made of a single vertex; a vertex can be incident to a volume and to a path of edges which are not incident to higher dimensional cells, etc.

9.2.4 Handle-face

The *handle-face* data structure [169] is an extension of the half-edge data structure, designed in order to describe 3D orientable quasi-manifolds with or without boundary. It is very similar to the *radial-edge* data structure.

The main notion is the *oriented edge*; an oriented edge corresponds to an edge-use element in the *radial-edge* data structure, and to a dart in a 3-map: cf. Table 9.4. Oriented edges represent the boundary of half-faces, thanks to pointer *snxt*, and they are linked with their belonging cells. There are two pointers *ssym* and *radial*, which link neighbor oriented edges into pairs [2].

The handle-face data structure makes it possible to represent 3D quasi-manifolds with boundary; thus, the radial pointer is defined only for edges that do not belong to a boundary. Note also that the radial pointer defined here links two opposite half-faces; in the original radial-edge data structure, the radial pointer links two edge-use elements, i.e. two opposite half-faces along an edge.

In addition to the oriented edges, nine other entities are represented and linked together: 3-*manifold*, *surface*, *boundary surface*, *face*, *half-face*, *edge*, *surface-edge*, *vertex* and *surface vertex*. Each entity has two different versions, depending if it belongs to a boundary or to the interior. Moreover, top-down and bottom-up hierarchical relationships are represented, i.e. each topological entity of dimension i has a list of its incident elements of dimension $i-1$ and $i+1$. For this reason, this data structure is, similarly to the the radial-edge data

[2]Note that in the paper [169], these two last pointers belong to the surface edge node, and not to the oriented edge node; but the two solutions are equivalent.

structure, a mix between an incidence graph and an ordered data structure. Several constraints are defined between the different relations in order to guaranty the topological validity of the model (for example $ssym(ssym(oe)) = oe$ or $radial(radial(oe)) = oe$).

Partial Entity Structure [159] and *OpenVolumeMesh* [150] are very similar to the *radial-edge* and to the *handle-face* data structures: their basic element is also a kind of half-edge, with a relation defined in order to describe half-faces, plus a radial and an opposite pointers, and additional entities and top-down and bottom-up relations between these entities.

9.2.5 Nef Polyhedron

A n-dimensional *Nef-polyhedron* is defined as the set of points in $\mathbb{R}^n$ generated by a finite number of intersection and complement operations of open halfspaces. So, Nef polyhedra can be n-dimensional manifolds with or without boundary, or non-manifolds mixing different dimensional parts. They were first defined by the mathematician W. Nef [182, 24]. Data structures, algorithms and implementation are described in [131], and [130] provides a free available implementation in 3D.

The main interest of Nef-polyhedron is to be closed for Boolean operations, which are set union, set intersection, set difference, set complement, interior, exterior, boundary, closure. The main difference between Nef-polyhedra and other data structures introduced in this chapter is the fact that the definition of Nef-polyhedra is a geometric one (intersection of halfspaces), and not a topological one. For this reason, it is not possible here to make a clear distinction between the combinatorial part of the data structure and its embedding. Due to these intrinsic differences, a comparison is not directly possible between Nef polyhedra, and n-maps and n-Gmaps.

9.3 Orientable and Nonorientable Cellular Quasi-Manifolds

9.3.1 Quad-edge

The basic element of the *quad-edge* data structure [127] is the quad-edge, i.e. an *oriented* and *directed* edge. Let e be an edge, and let v_1, v_2, f_1, f_2 be its incident vertices and faces. Intuitively, a direction of e corresponds to the choice of v_1 or v_2 as origin vertex, the other being the destination vertex (the edge is directed from its origin to its destination). Similarly, an orientation of e corresponds intuitively to the choice of f_1 or f_2 (the edge is oriented towards a face; this corresponds to a clockwise or counterclockwise orientation, according to v_1 or v_2). Since the direction and the orientation are

TABLE 9.5
Quad-edge relations, and correspondence with 2-Gmaps

name	data	dual	flip	onext	sym
pointer to	vertex	quad-edge	quad-edge	quad-edge	quad-edge
in 2-Gmaps	$embo_0(d)$	dual(d)	$\alpha_2(d)$	$\alpha_2(\alpha_1(d))$	$\alpha_2(\alpha_0(d))$

independent, four quad-edges are associated with each edge. Moreover, the quad-edge data structure simultaneously represents the primal and the dual subdivision: so, a dual edge (i.e. four quad-edges) of the dual subdivision is associated with any edge (i.e. four quad-edges) of the primal subdivision.

Several functions are defined on the quad-edges. Given a quad-edge e, *sym* gives the quad-edge describing the same edge with the opposite direction and the same orientation; *flip* gives the quad-edge describing the same edge with the same direction and the opposite orientation; *dual* gives the dual quad-edge, and *onext* gives the next quad-edge around the origin vertex (considering the orientation of the quad-edge). Several constraints are defined between these relations in order to guaranty the topological validity (for example $flip(flip(e)) = e$, $flip(onext(flip(onext(e)))) = e$, where flip = dual $\circ$ sym $\circ$ dual). It is shown in [127] that it is possible to retrieve all topological relations for a given oriented and directed edge: its origin/destination vertex, its left/right face and its next/previous edge around a vertex/face.

The quad-edge data structure makes it possible to represent orientable and nonorientable 2D manifolds without boundary. It is equivalent to 2-Gmaps, when a 2-Gmap and its dual are handled simultaneously: cf. Table 9.5.

Figure 9.5(a) shows a quad-edge data structure representing the 2D object given in Fig. 3.4(a) page 56; Fig. 9.5(b) shows the two corresponding 2-Gmaps. For example sym(1) = 3, dual(1) = 1′, onext(1) = 8, data(1) = v_1, flip(1) = 2.

In [127], authors defined $rot(e)$ as a shortcut for $dual(flip(e))$. Thanks to this shortcut, all relations can be retrieved by a composition of a constant number of rot, $onext$ and $flip$. For example $sym(e) = rot(rot(e))$, $dual(e) = rot(flip(e))$, $oprev(e) = rot(onext(rot(e)))$ ($oprev(e)$ being the previous quad-edge around the origin vertex) and $lnext(e) = rot(onext(rot(rot(rot(e)))))$ ($lnext(e)$ being the next quad-edge around the left face).

The structure is implemented using quarter-records: a quarter-record is an array of four *nonoriented* quad-edges, corresponding to an edge. A quad-edge is represented by a triple (e, r, f), where e is a quarter-record, r is the index of the corresponding nonoriented quad-edge in this quarter-record, and f is a Boolean indicating the orientation of the quad-edge. Each nonoriented quad-edge contains a pointer to its belonging cell (a vertex for a primal quad-edge, and a face for a dual quad-edge), and two pointers to nonoriented quad-edges in order to encode *rot* and *next*. Note that this representation is an efficient way to represent the theoretical model, and such technique can be used for

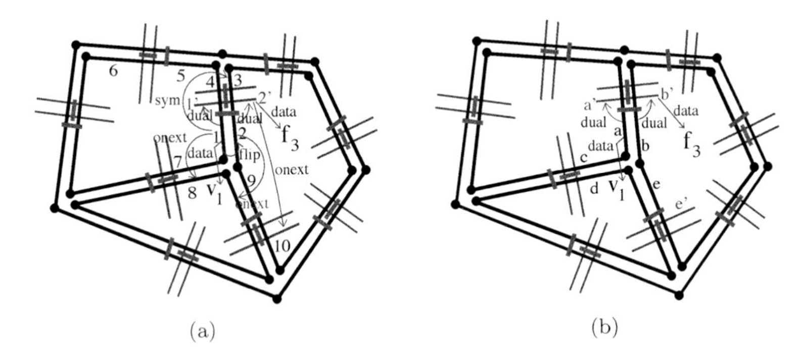

FIGURE 9.5
(a) Quad-edge representation of the 2D quasi-manifold without boundary given in Fig. 3.21(a) page 73. The four pointers from quad-edge 1 are drawn by arrows. Dual darts are drawn in grey.
(b) The two corresponding 2-Gmaps. Primal is drawn in black, and dual in grey (α_1 is not drawn for the dual).

the other structures presented above (for example 2-maps, 2-Gmaps or half-edges).

9.3.2 Facet-edge

The *facet-edge* structure [93] is the 3D extension of the quad-edge structure. Let f be a face and e be an edge of a 3D manifold. Two coherent cycles of edges can be associated with f, corresponding to its two orientations. Similarly (and by duality), two coherent cycles of faces can be associated with e, according to the chosen direction of rotation around e. Intuitively, the basic element of the facet-edge structure is the facet-edge pair $a = (e, f)$, where face f is incident to edge e. Four facet-edge elements correspond to a facet-edge pair, depending on the two directions of the rings around e and f.

Four functions are defined on the facet-edge elements: *fnext* gives the next facet-edge element in the facet-ring, i.e. the facet-edge element with the same edge and the next facet; *enext* gives the next facet-edge element in the edge-ring, i.e. the facet-edge element with the same facet and the next edge; *rev* corresponds to the same facet-edge pair, but considering the facet with its opposite direction; *clock* corresponds to the same facet-edge pair, but considering the edge and the facet with their opposite directions. Moreover the dual is also described; thus, each facet-edge element is associated with its dual, thanks to the *dual* function.

The facet-edge structure allows to represent orientable and nonorientable 3D quasi-manifold without boundary. It is equivalent to 3-Gmaps, when a 3-Gmap and its dual are simultaneously represented: cf. Table 9.6.

TABLE 9.6
Facet-edge relations, and correspondence with 3-Gmaps

name	aOrg	Sdual	eNext	fNext	Rev	Clock
pointer to	vertex	fe-pair	fe-pair	fe-pair	fe-pair	fe-pair
in 3-Gmaps	$embo_0(d)$	dual(d)	$\alpha_1(\alpha_0(d))$	$\alpha_2(\alpha_3(d))$	$\alpha_3(d)$	$\alpha_0(\alpha_3(d))$

Eight facet-edge elements correspond to each facet-edge pair: four clocked and oriented versions of the facet-edge, plus four clocked and oriented versions of its dual representation. In [93], the authors define a data structure extending the quad-edge data structure, in order to get a compact encoding: only four facet-edge pairs are represented in a 2×2 matrix, each facet-edge pair being retrieved by a pointer to the matrix plus two indices, and only two pointers are explicitly represented: *data* and *next*. Once again, such compact representation can also be adapted to encode other models.

9.3.3 Cell-tuple

Brisson [38, 39] studied the representation of subdivisions of n-dimensional manifolds without multi-incidence. He pointed out the important relations between incidence graphs and what he called "ordered models", making it possible to:

- formally define properties satisfied by incidence graphs which correspond to quasi-manifolds;
- conceive operations for converting equivalent incidence graphs and ordered models.

More precisely, let M be a n-dimensional subdivided *manifold* without boundary, such that no multi-incidence occurs. M is thus a *regular CW-complex*[3]: its structure can be described unambiguously by an incidence graph. Since M is a manifold, it satisfies particular properties studied by Brisson, mainly the *switch property* (cf. Section 8.2.4). Let (c_{i-1}, c_i, c_{i+1}) be three incident cells of respective dimensions $i-1, i, i+1$, for $0 \leq i \leq n$ (a fictive (-1)-cell (resp. $(n+1)$-cell) is added, incident to all 0-cells (resp. all n-cells)). Then $switch(c_{i-1}, c_i, c_{i+1}) = c'_i$, such that c'_i is the only i-cell different from c_i, which is also incident to both c_{i-1} and c_{i+1}. Consequently, $switch(c_{i-1}, c'_i, c_{i+1}) = c_i$. Operator *switch* is thus an *involution* on the set of triples of incident cells of consecutive dimensions.

Let C be the set of *cell-tuples*, i.e. any element $(c_0, \ldots, c_i, \ldots, c_n)$ of C is such that:

[3] *Regular* means that no multi-incidence occurs.

- the dimension of cell c_j is j, for $0 \leq j \leq n$,
- c_j is incident to c_{j+1}, for $0 \leq j \leq n-1$.

$(i+1)$ operators $switch_i : C \rightarrow C$ can be deduced from operator *switch*, by:

$$switch_i(c_0, \dots, c_{i-1}, c_i, c_{i+1}, \dots, c_n) = (c_0, \dots, c_{i-1}, c'_i, c_{i+1}, \dots, c_n)$$

where $c'_i = switch(c_{i-1}, c_i, c_{i+1})$. Any $switch_i$ operator is thus an involution on C. $(C, \{switch_i\})$ is a *cell-tuple structure*. Note that it is easy to prove that $switch_i \circ switch_j$ is an involution, for $0 \leq j \leq i-2 < i \leq n$. So, any n-dimensional cell-tuple structure is a n-Gmap: cell-tuples correspond to darts, and for $0 \leq i \leq n$, $switch_i$ corresponds to α_i.

So, starting from subdivisions of manifolds without multi-incidence, Brisson:

- pointed out and studied the properties of the associated incidence graphs;
- deduced the notion of cell-tuple structure;
- showed the link with the *barycentric triangulation* of the manifold (cf. Section 8), establishing thus the link between incidence graphs, cell-tuple structures and their simplicial analogs;
- showed the relations between the cell-tuple structure and close structures of lower dimensions, namely the quad-edge structure and the facet-edge structure.

Brisson restricted his work to subdivided manifolds without multi-incidence. In fact, the property characterizing manifolds, i.e. the neighborhood of any point is homeomorphic to a ball, is not a combinatorial property: as far as we know, it is not possible to take this property into account within a combinatorial structure. So, the results established by Brisson, which are combinatorial ones, can be generalized, namely for n-surfaces, the corresponding n-Gmaps and their simplicial analogs.

10

Concluding Remarks

Numerous data structures and operations have been proposed in order to represent and handle subdivided geometric objects, for various applications within different fields, e.g. geometric modeling, computational geometry, discrete geometry, computer graphics, image processing and analysis, etc. [17, 8, 167, 214, 127, 107, 215, 93, 102, 87, 38, 163, 162, 94, 10, 195, 128, 170, 181, 203, 108, 57, 39, 132, 183, 153, 165, 104, 184, 154, 110, 49, 19, 159, 31, 45, 90, 84, 88, 66, 175, 185, 157, 60, 50, 158].

All structures are based on the representation of the structural information (topology), the shape (embedding) being represented by attributes associated with the cells. This distinction between topology and embedding makes it possible to easily define different data structures corresponding to different embedding requirements (linear, curved embeddings); this simplifies also the definition of operations and the reuse of software parts.

Note (and this is reassuring) that the proposed solutions are often similar, whatever their original field is. More precisely, most *topological models* conceived for representing the structural information of subdivisions fit into a general frame.

General frame

Simplicial structures are the basis of this frame. Two classes can be distinguished:

- one is related to abstract simplicial complexes, and multi-incidence between simplices is not taken into account. So, such structures are usually linearly embedded, i.e. as simplicial complexes. Basic operations as identification of vertices can lead to identify other simplices, or to degenerate simplices;
- the other one is related to (semi-)simplicial sets, and multi-incidence is taken into account. So, such structures can be embedded with curved shapes, for instance triangular Bézier spaces. Basic operations modify only the simplices on which they are applied.

Operations for converting abstract simplicial complexes and semi-simplicial sets satisfying some properties have been conceived. Moreover, optimized structures have been conceived in order to represent subclasses of abstract simplicial complexes or semi-simplicial sets, mainly for representing objects close to (quasi-)manifolds (see for instance [183, 90]). For instance:

- quasi-manifolds are defined as objects one can construct by adding n-simplices and by identifying $(n-1)$-simplices, in such a way that at most two n-simplices share a $(n-1)$-simplex. This definition is consistent for semi-simplicial sets. But several problems arise when applied to abstract simplicial complexes, since the identification of simplices can lead to the degeneracy of simplices [89];
- optimized data structures have been proposed in order to represent quasi-manifolds; mainly, n-simplices are represented, together with adjacency relations between n-simplices.

Simplicial objects are well-known in combinatorial topology: several operations have been defined, as many topological properties and methods for computing these properties. The fact that the general frame for the definition of structures for representing subdivisions is based upon simplicial structures is thus very interesting, since many notions, constructions, properties and algorithms can be extended for the other structures of this frame. For instance, it has been possible to extend the definition of cartesian product from simplicial sets to combinatorial maps [166]; similarly, a topological property, namely homology, has been directly defined on combinatorial maps, inspired by the corresponding definition for semi-simplicial sets [78, 4].

Cubic and simploidal structures have been defined in order to represent subdivisions with cubic cells, and more generally with simploidal cells. They are based upon simplicial structures, and *cartesian product* [185, 186]:

- a 2-dimensional cube is the cartesian product of two 1-dimensional cubes, i.e. two edges; more generally, a n-dimensional cube is the cartesian product of n edges. It is possible to deduce cubic sets from semi-simplicial sets: cubes are the basic elements (as simplices are the basic elements of semi-simplicial sets), and they are linked by face operators;
- more generally, a simploid is the cartesian product of simplices of any dimension; for instance, a prism is the cartesian product of an edge and a triangle. Simploidal sets have been deduced from semi-simplicial sets: their basic elements are simploids, which are linked by face operators.

Such structures can be useful for image processing and analysis (since pixels, voxels, and more generally the basic elements of images, are associated with n-cubes), and also for computer graphics, geometric modeling or computational geometry, for handling assemblies of curved geometric simploids. As for simplicial structures, optimized structures for representing subclasses of cubic and simploidal sets can be defined.

A main interest of structures with regular cells is the fact that consistency constraints are taken into account within their definitions, and that they are more efficient than cellular structures for representing subdivisions with regular cells. The counterpart is the fact that subdivisions with any cells cannot be directly represented. When such a subdivision has to be represented, it is necessary to arbitrarily subdivide the cells, and to add information in order to memorize the initial cells. Moreover, several operations which naturally produce any cells (as Boolean operations, for instance) cannot be directly applied on such structures with regular cells.

Cellular structures are also based upon simplicial structures: a *numbering mechanism* makes it possible to implicitly define cells as particular subsets of simplices. Mainly two classes of cellular structures can be distinguished:

- one is related to incidence graphs; they correspond to abstract simplical complexes structured into cells, and multi-incidence between cells is not taken into account. Incidence graphs do not naturally integrate consistency constraints about cells: so, cells can be complexes, and not even quasi-manifolds. Consistency constraints have to be explicitly checked, for instance based upon the definition of n-surfaces (remember that n-surfaces are a subclass of incidence graphs corresponding to quasi-manifolds). In order to simplify, incidence graphs are often linearly embedded by associating *convex* spaces with cells; cells are thus homeomorphic to balls, and such embedded incidence graphs correspond to *regular CW-complexes*[1]. Since the structure itself does not integrate consistency constraints, operations have to take these constraints into account, but it is usually difficult. For instance, many operations, usual in Geometric Modeling, do not produce convex cells;

- the other one is related to combinatorial maps; they correspond to semi-simplicial sets structured into cells, and multi-incidence between cells is taken into account. Combinatorial maps naturally integrate consistency constraints: mainly cells are quasi-manifolds. Since constraints are taken into account within the structure definition itself, operations can be easily defined. Moreover, combinatorial maps can be embedded linearly, or by associating curved spaces with cells. Combinatorial maps can thus be used for a larger class of applications than incidence graphs. The drawback is the fact that combinatorial maps, when equivalent to incidence graphs, are generally more space consuming; note that the ratio is usually not important for applications in geometric modeling, computational geometry, discrete geometry, computer graphics, image processing and analysis; and operations for handling combinatorial maps are less time consuming than operations for handling incidence graphs, since more constraints have to be checked for these last ones.

[1] *Regular* mainly means that no multi-incidence occurs between cells.

Operations for converting incidence graphs and combinatorial maps satisfying both some properties have been conceived. Moreover, optimized structures have been defined in order to represent subclasses of subdivisions, as n-Gmaps for n-dimensional quasi-manifolds and n-maps for n-dimensional oriented quasi-manifolds. Contrary to incidence graphs, the fact that subclasses are represented is taken into account within the structure definition, leading to optimization[2].

This general frame for the definition of structures representing subdivisions is thus sound, since it is based upon simplicial structures, and safe mechanisms (cartesian product, numbering for implicit definition of cells). It is large, from structures with regular cells to structures with any cells, from complexes to oriented quasi-manifolds. Optimization mechanisms exist, for defining structures suited for subclasses of subdivisions. Given an application, it is thus possible to define a structure, well-suited to the requirements of the application.

Model choice

The choice of a topological model obviously depends on the "nature" of objects one intends to handle, and also on the operations which have to be applied, the space/time complexity and the complexity of software development [21, 96]. It is thus a classical problem: choosing a data structure, taking into account:

1. the "nature" of objects, i.e. the topological properties satisfied by the objects. It is possible to always use a general structure conceived in order to represent "any" cellular complex, but it is often space consuming, maybe also time consuming, since more information has to be handled compared to an optimized structure. It is also obvious that using a structure conceived for a subclass in order to handle subdivisions of a larger class can lead to important errors. For instance, even if it is intuitively possible to deduce an incidence graph from a subdivision in which some cells are multi-incident to other cells, the formal interpretation of the resulting incidence graph does not correspond in any way to the initial subdivision;

2. the operations applied to the objects. It is impossible to define some operations for some structures, since these operations naturally create objects which do not satisfy the implicit properties of the structures (e.g. Boolean operations for simplicial objects);

[2] n-surfaces are incidence graphs satisfying some structural property. This property does not lead to an optimized representation, and it has to be checked in order to know if an incidence graph represents a quasi-manifold. Conversely, there is a bijection between n-Gmaps and quasi-manifolds, and n-Gmaps is an optimized representation, compared to chains of maps, which have been designed for representing cellular complexes.

3. the space/time costs of structures and operations. Given a class C of subdivisions (e.g. quasi-manifolds), a structure S_L corresponding to a larger class L (e.g. cellular complexes) uses generally more memory space than an optimized structure S_C corresponding to C: this corresponds to the fact that some information is explicit within S_L, and implicit within S_C.

   It is clear that the time complexity of operations has also to be taken into account: when some operations often need to make explicit some implicit information, a less efficient model (according to space complexity) could be a better choice. For instance, it is possible to represent a polyhedron by a list of faces, but adjacency and incidence relations are thus not explicitly represented; so, this representation is often not efficient when constructing objects, since adjacency and incidence information are used by many construction operations. Conversely, many algorithms do not need all information contained in the whole geometric data structures, and specialized structures can be a better choice [190] (e.g. a list of faces for some rendering algorithms).

4. the cost of operation conception. For instance, the definitions of several structures do not take into account the constraints of consistency which have to be satisfied by the represented objects (e.g. an edge incident to three vertices can be represented by an incidence graph). The construction process has thus to control the object validity. For instance, *Euler operators* have been defined in order to construct any subdivision of any orientable surface without boundary [173, 171]: each operator simultaneously creates or removes several cells; so, the conception of Euler operators for handling incidence graphs, even if not complicated, is not so easy than the conception of other basic operations defined for handling structures in which the constraints of consistency are directly taken into account, e.g. inserting a dart and sewing for n-Gmaps. Moreover, Euler operators have never been generalized for constructing n-dimensional quasi-manifolds; is it inefficient, difficult or both?

So, it is clear for us that "one best structure" does not exist, but some structures can be well suited for particular processes. Thus, conversion algorithms which can be deduced from the general frame presented in this book are very important for practical purposes.

Combinatorial maps

When cellular objects have to be handled for some reason ("nature" of represented objects, or specificities of operations), we think that combinatorial maps are a good starting point for the design of a data structure:

1. consistency properties are taken into account in the very definition

of combinatorial maps. For instance, any n-Gmap represents a n-dimensional quasi-manifold. So, any operation constructing an n-Gmap construct a valid quasi-manifold; it is not necessary to control its validity through Euler operators, for instance;

2. few basic operations make it possible to construct any combinatorial map. It is possible to conceive many low- or high-level operations, in order to satisfy different construction needs for various applications, and to prove the validity of these operations;

3. local operations can be defined, for carefully controlling the modifications of the object during its construction. Global operations have also been defined, based on these local operations, for instance for different processes related to geometric modeling, computational geometry, discrete geometry, computer graphics, image processing and analysis;

4. optimized structures can be defined for the efficient representation of subclasses of cellular subdivisions. Since the structures are close to each other, it is possible to:

    - either reuse for a structure parts of a software conceived for another structure;
    - either conceive the software in order to efficiently implement several structures;

5. combinatorial maps are based on a single type of elements, i.e. the darts, and functions between darts satisfying clear and simple consistency constraints. This simplifies the implementation of related data structures and algorithms.

For representing subdivisions of the plane, and more generally surface subdivisions, 2-maps and 2-Gmaps correspond to the kernel of other well-known data structures. This makes it possible to clearly distinguish the structure information from the shape information or redundant information added for optimizing some processes. Based on combinatorial maps, several implementations have been proposed, including more or less explicit information, in order to get efficient data structures.

Advantages of combinatorial maps increase for higher dimensions, since defining specific data structures becomes more difficult. It is thus possible to safely conceive many local and global construction operations, operations for computing topological properties, and to deduce efficient implementations for different types of (linear or curved) embeddings.

More specifically, both n-maps and n-Gmaps correspond to quasi-manifolds with or without boundary, but n-maps represent orientable quasi-manifolds while n-Gmaps can also represent nonorientable ones. However the two structures have their one advantages, and the choice of one of them depends on the needs of your specific application. For instance, the definition of n-Gmaps is "homogeneous", since all functions are involutions. This simplifies the definitions of related notions and the conception of algorithms. The definition of n-maps is more complex since β_1 is a partial permutation while other β_i are partial involutions. However n-maps use twice less darts than n-Gmaps for representing an orientable object. Thus if memory footprint is a crucial criterion for your application, and if only orientable objects are handled, using n-maps is perhaps a better solution. But if nonorientable objects are handled, or if you want to simplify the implementation of operations by avoiding special cases, which are sometimes complex, using n-Gmaps is perhaps a better solution.

Some references about n-Gmaps and n-maps

n-maps and n-Gmaps are used for many different works in geometric modeling, computational geometry, discrete geometry, computer graphics, image processing and analysis, for instance:

- 2D and 3D image segmentation and processing: segmentation [36, 16, 95, 40, 110, 42, 41, 33, 32, 35, 19, 34, 20, 26, 144, 31, 43, 63, 66, 25, 175, 67, 59, 99, 98, 14, 73], discrete operations [81, 82, 100], robust region filling [65], contour approximation [68, 69], deformable model [101, 75, 74];
- definition of hierarchical models based on combinatorial maps, and use of these hierarchical models for different applications: [44, 46, 45, 198, 199, 72, 200, 197, 147, 112, 118, 121, 148, 113, 122, 120, 119, 210, 209];
- definition of comparison tools of combinatorial maps; isomorphism and subisomorphism [71, 70, 83, 202], combinatorial map signature [123, 124], frequent submap discovery [124], edit distance [51, 52, 213, 53];
- definition of generic operations: removal and contraction [77, 72, 112, 113], insertion and expansion [13], chamfering [156], compression [190, 106], thickening [28], cartesian product [166];
- computation of topological invariants: incremental Euler-Poincaré characteristic [80], canonical polygonal schema [64], Betti numbers and homologie groups [78, 187, 79, 6, 188, 76];
- geometric modeling for geology, architecture, etc. [21, 7, 29, 30, 114, 126, 135, 136, 11];
- animation and simulation for computer graphics: [205, 27, 201, 160, 161, 177, 176, 22, 138, 23, 111].

Several libraries implement some combinatorial maps and high level operations, for instance:

- the "Combinatorial maps" package in CGAL [61] is a generic implementation of n-maps, the "Linear cell complexes" package [62] is the embedding layer;
- CGoGN [149] is a library implementing n-maps and n-Gmaps, and hierarchical structures deduced from them, for $n = 1, 2, 3$;
- Girl [117] is a 2D image processing software based on 2-maps;
- MOKA [211] is a geometrical modeler based on 3-Gmaps.

Note that these libraries implement n-maps or n-Gmaps for different n and various uses, depending on the specific needs of the addressed applications. This illustrates once again the interest of the generic definitions of n-maps and n-Gmaps, which are versatile enough for various applications of geometric modeling, computational geometry, discrete geometry, computer graphics, image processing and analysis, etc.

Bibliography

[1] M. K. Agoston. *Algebraic Topology, a first course.* Pure and applied mathematics. Marcel Dekker Ed., 1976.

[2] M. K. Agoston. *Computer Graphics and Geometric Modeling.* Springer, 2005.

[3] E. Ahronovitz, C. Fiorio, and S. Glaize. Topological operators on the topological graph of frontiers. In *Proc. of International Conference Discrete Geometry for Computer Imagery*, volume 1568 of *LNCS*, pages 207–217, Marne-la-Vallée, France, 1999.

[4] S. Alayrangues, G. Damiand, P. Lienhardt, and S. Peltier. Homology of cellular structures allowing multi-incidence. *Submitted*, 2013.

[5] S. Alayrangues, X. Daragon, J.-O. Lachaud, and P. Lienhardt. Equivalence between closed connected n-G-maps without multi-incidence and n-surfaces. *Journal of Mathematical Imaging and Vision*, 32(1):1–22, 2008.

[6] S. Alayrangues, S. Peltier, G. Damiand, and P. Lienhardt. Border operator for generalized maps. In *Proc. of International Conference on Discrete Geometry for Computer Imagery*, volume 5810 of *LNCS*, pages 300–312, Montréal, Canada, September 2009. Springer Berlin/Heidelberg.

[7] E. Andres, R. Breton, and P. Lienhardt. SpaMod: design of a spatial modeler. In *Proc. of Digital and Image Geometry*, volume 2243 of *LNCS*, pages 90–107. Springer Verlag, 2001.

[8] S. Ansaldi, L. de Floriani, and B. Falcidieno. Geometric modeling of solid objects by using a face adjacency graph representation. *Computer Graphics*, 19(3):131–139, 1985.

[9] D. Arques and P. Koch. Définition et implémentation de pavages dans l'espace. Technical Report 46, Laboratoire d'Informatique, UFR Sciences et Techniques, Besançon, France, August 1988.

[10] D. Arques and P. Koch. Modélisation de solides par les pavages. In *Proc. of Pixim*, pages 47–61, Paris, France, 1989.

[11] K. Arroyo Ohori, G. Damiand, and H. Ledoux. Constructing an n-dimensional cell complex from a soup of (n-1)-dimensional faces. In *Proc. of International Conference on Applied Algorithms*, volume 8321 of *LNCS*, page to appear, Kolkata, India, January 2014.

[12] M. H. Austern. *Generic Programming and the STL*. Addison Wesley, 1999.

[13] M. Baba-Ali, G. Damiand, X. Skapin, and D. Marcheix. Insertion and expansion operations for n-dimensional generalized maps. In *Proc. of International Conference on Discrete Geometry for Computer Imagery*, volume 4992 of *LNCS*, pages 141–152, Lyon, France, April 2008. Springer Berlin/Heidelberg.

[14] F. Baldacci, A. Braquelaire, and G. Damiand. 3D topological map extraction from oriented boundary graph. In *Proc. of International Workshop on Graph-Based Representations in Pattern Recognition*, volume 5534 of *LNCS*, pages 283–292, Venice, Italy, May 2009. Springer Berlin/Heidelberg.

[15] F. Baldacci, A. Braquelaire, P. Desbarats, and J.P. Domenger. 3D image topological structuring with an oriented boundary graph for split and merge segmentation. In *Proc. of International Conference on Discrete Geometry for Computer Imagery*, volume 4992 of *LNCS*, pages 541–552, Lyon, France, April 2008. Springer Berlin/Heidelberg.

[16] P. Baudelaire and M. Gangnet. Planar maps: an interaction paradigm for graphic design. In *Proc. of SIGCHI Conference on Human Factors in Computing Systems*, volume 20, pages 313–318. ACM, May 1989.

[17] B. Baumgart. A polyhedron representation for computer vision. In *Proc. of AFIPS National Computer Conference*, volume 44, pages 589–596, 1975.

[18] G. Bertrand. New notions for discrete topology. In *Proc. of International Conference on Discrete Geometry for Computer Imagery*, volume 1568 of *LNCS*, pages 218–228, Marne-la-Vallée, France, 1999. Springer.

[19] Y. Bertrand, G. Damiand, and C. Fiorio. Topological encoding of 3D segmented images. In *Proc. of International Conference on Discrete Geometry for Computer Imagery*, volume 1953 of *LNCS*, pages 311–324, Uppsala, Sweden, December 2000. Springer Berlin/Heidelberg.

[20] Y. Bertrand, G. Damiand, and C. Fiorio. Topological map: Minimal encoding of 3D segmented images. In *Proc. of International Workshop on Graph-Based Representations in Pattern Recognition*, pages 64–73, Ischia, Italy, May 2001.

[21] Y. Bertrand and J-F. Dufourd. Algebraic specification of a 3D-modeler based on hypermaps. *CVGIP: Graphical Models and Image Processing*, 56(1):29–60, January 1994.

[22] R. Bézin, B. Crespin, X. Skapin, O. Terraz, and P. Meseure. Topological operations for geomorphological evolution. In *Proc. of Workshop on Virtual Reality Interaction and Physical Simulation*, pages 139–148, Lyon, France, 2011.

[23] R. Bézin, A. Peyrat, B. Crespin, O. Terraz, X. Skapin, and P. Meseure. Interactive hydraulic erosion using CUDA. *Machine Graphics and Vision*, 20(2):157–172, 2011.

[24] H. Bieri and W. Nef. Elementary set operations with d-dimensional polyhedra. In *Computational Geometry and its Applications*, volume 333 of *LNCS*, pages 97–112. Springer Berlin Heidelberg, 1988.

[25] K. Bouchefra, P. Bonnin, and A. De Cabrol. Data image fusion using combinatorial maps. In *Proc. of SPIE*, volume 5909, pages 59091I–59091I–8, 2005.

[26] P. Bourdon, O. Alata, G. Damiand, C. Olivier, and Y. Bertrand. Geometrical and topological informations for image segmentation with monte carlo markov chain implementation. In *Proc. of International Conference on Vision Interface*, pages 413–420, Calgary, Alberta, Canada, May 2002.

[27] S. Brandel, D. Bechmann, and Y. Bertrand. STIGMA: a 4-dimensional modeller for animation. In *Proc. of Computer Animation and Simulation*, Eurographics, pages 103–126, Lisbonne, September 1999. Springer Vienna.

[28] S. Brandel, D. Bechmann, and Y. Bertrand. Thickening: an operation for animation. *The Journal of Vision and Computer Animation*, 11:261–277, 2000.

[29] S. Brandel, S. Schneider, M. Perrin, N. Guiard, J.-F. Rainaud, P. Lienhardt, and Y. Bertrand. Automatic building of structured geological models. In *Proc. of ACM Symposium on Solid Modeling and Applications*, pages 59–69. Eurographics Association, June 2004.

[30] S. Brandel, S. Schneider, M. Perrin, N. Guiard, J.-F. Rainaud, P. Lienhardt, and Y. Bertrand. Automatic building of structured geological models. *Journal of Computing and Information Science in Ingeneering*, 5(2), 2005.

[31] A. Braquelaire, G. Damiand, J.-P. Domenger, and F. Vidil. Comparison and convergence of two topological models for 3D image segmentation. In *Proc. of International Workshop on Graph-Based Representations in*

Pattern Recognition, volume 2726 of *LNCS*, pages 59–70, York, England, July 2003. Springer Berlin/Heidelberg.

[32] J.-P. Braquelaire and J.-P. Domenger. Representation of segmented images with discrete geometric maps. *Image and Vision Computing*, 17(10):715–735, 1999.

[33] J.P. Braquelaire and L. Brun. Image segmentation with topological maps and inter-pixel representation. *Journal of Visual Communication and Image Representation*, 9(1):62–79, March 1998.

[34] J.P. Braquelaire, P. Desbarats, and J.P. Domenger. 3D split and merge with 3-maps. In *Proc. of International Workshop on Graph-Based Representations in Pattern Recognition*, pages 32–43, Ischia, Italy, May 2001.

[35] J.P. Braquelaire, P. Desbarats, J.P. Domenger, and C.A. Wüthrich. A topological structuring for aggregates of 3D discrete objects. In *Proc. of International Workshop on Graph-Based Representations in Pattern Recognition*, pages 193–202, Austria, May 1999.

[36] J.P. Braquelaire and P. Guitton. A model for image structuration. In *Proc. of Computer Graphics International*, pages 426–435, Genève, Switzerland, May 1988.

[37] J.P. Braquelaire and P. Guitton. 2D1/2 scene update by insertion of contour. *Computers & Graphics*, 15(1):41–48, 1991.

[38] E. Brisson. Representing geometric structures in d dimensions: topology and order. In *Proc. of ACM Symposium Computational Geometry*, pages 218–227, Saarbrücken, Germany, June 1989.

[39] E. Brisson. Representing geometric structures in d dimensions: topology and order. *Discrete & Computational Geometry*, 9(1):387–426, 1993.

[40] L. Brun. *Segmentation d'images couleur à base topologique.* Thèse de doctorat, Université Bordeaux 1, December 1996.

[41] L. Brun and J.P. Domenger. A new split and merge algorithm with topological maps and inter-pixel boundaries. In *Proc. of International Conference in Central Europe on Computer Graphics and Visualization*, pages 21–30, February 1997.

[42] L. Brun, J.P. Domenger, and J.P. Braquelaire. Discrete maps: a framework for region segmentation algorithms. In *Proc. of International Workshop on Graph-Based Representations in Pattern Recognition*, pages 83–92, Lyon, April 1997. published in Advances in Computing (Springer).

[43] L. Brun, J.P. Domenger, and M. Mokhtari. Incremental modifications of segmented image defined by discrete maps. *Journal of Visual Communication and Image Representation*, 14(3):251–290, 2003.

[44] L. Brun and W.G. Kropatsch. Introduction to combinatorial pyramids. In *Proc. of Digital and Image Geometry*, number 2243 in LNCS, pages 108–127, 2001.

[45] L. Brun and W.G. Kropatsch. Combinatorial pyramids. In *Proc. of IEEE International Conference on Image Processing*, volume 2, pages 33–37, Barcelona, Spain, 2003.

[46] L. Brun and W.G. Kropatsch. Contraction kernels and combinatorial maps. *Pattern Recognition Letters*, 24(8):1051–1057, 2003.

[47] R. Bryant and D. Singerman. Foundations of the theory of maps on surfaces with boundaries. *Quarterly Journal of Mathematics Oxford*, 2(36):17–41, 1985.

[48] A. Castelo, H.C.V. Lopes, and G. Tavares. Handlebody representation for surfaces and morse operators. In *Proc. of SPIE*, volume 1830, pages 270–283, 1992.

[49] P.R. Cavalcanti, P.C.P Carvalho, and L.F. Martha. Non-manifold modeling: an approach based on spatial subdivisions. *Computer-Aided Design*, 29(3):299–320, 1997.

[50] D. Cazier and P. Kraemer. X-maps: an efficient model for non-manifold modeling. In *Proc. of Shape Modeling International*, pages 226–230. IEEE Conference Publishing Services, June 2010. Short paper.

[51] C. Combier, G. Damiand, and C. Solnon. Measuring the distance of generalized maps. In *Proc. of International Workshop on Graph-Based Representations in Pattern Recognition*, volume 6658 of *LNCS*, pages 82–91, Münster, Germany, May 2011. Springer Berlin/Heidelberg.

[52] C. Combier, G. Damiand, and C. Solnon. From maximum common submaps to edit distances of generalized maps. *Pattern Recognition Letters*, 33(15):2020–2028, November 2012.

[53] C. Combier, G. Damiand, and C. Solnon. Map edit distance vs graph edit distance for matching images. In *Proc. of International Workshop on Graph-Based Representations in Pattern Recognition*, volume 7877 of *LNCS*, pages 152–161, Vienna, Austria, May 2013. Springer Berlin/Heidelberg.

[54] R. Cori. *Un code pour les graphes planaires et ses applications.* PhD thesis, Université Paris VII, 1973.

[55] R. Cori. Un code pour les graphes planaires et ses applications. In *Astérisque*, volume 27. Soc. Math. de France, Paris, France, 1975.

[56] T.H. Cormen, C.E. Leiserson, R.L. Rivest, and C. Stein. *Introduction to Algorithms*. The MIT Press, Cambridge, London, 3rd edition, 2009.

[57] G. Crocker and W. Reinke. An editable non-manifold boundary representation. *Computer Graphics and Applications*, 11(2), 1991.

[58] G. Damiand. *Définition et étude d'un modèle topologique minimal de représentation d'images 2d et 3d*. Thèse de doctorat, Université Montpellier II, December 2001.

[59] G. Damiand. Topological model for 3D image representation: Definition and incremental extraction algorithm. *Computer Vision and Image Understanding*, 109(3):260–289, March 2008.

[60] G. Damiand. *Contributions aux Cartes Combinatoires et Cartes Généralisées : Simplification, Modèles, Invariants Topologiques et Applications*. Habilitation diriger des recherches, Université Lyon 1, September 2010.

[61] G. Damiand. Combinatorial maps. In *CGAL User and Reference Manual*. CGAL Editorial Board, 3.9 edition, 2011. http://www.cgal.org/Pkg/CombinatorialMaps.

[62] G. Damiand. Linear cell complex. In *CGAL User and Reference Manual*. CGAL Editorial Board, 4.0 edition, 2012. http://www.cgal.org/Pkg/LinearCellComplex.

[63] G. Damiand, O. Alata, and C. Bihoreau. Using 2D topological map information in a markovian image segmentation. In *Proc. of International Conference on Discrete Geometry for Computer Imagery*, volume 2886 of *LNCS*, pages 288–297, Naples, Italy, November 2003. Springer Berlin/Heidelberg.

[64] G. Damiand and S. Alayrangues. Computing canonical polygonal schemata with generalized maps. In *Proc. of International Conference on Topological & Geometric Graph Theory*, volume 31 of *Electronic Notes in Discrete Mathematics*, pages 287–292, Paris, France, August 2008. Elsevier.

[65] G. Damiand and D. Arrivault. A new contour filling algorithm based on 2D topological map. In *Proc. of International Workshop on Graph-Based Representations in Pattern Recognition*, volume 4538 of *LNCS*, pages 319–329, Alicante, Spain, June 2007. Springer Berlin/Heidelberg.

[66] G. Damiand, Y. Bertrand, and C. Fiorio. Topological model for two-dimensional image representation: Definition and optimal extraction algorithm. *Computer Vision and Image Understanding*, 93(2):111–154, February 2004.

[67] G. Damiand and L. Brun. *Géométrie discrète et images numériques, chapitre 4. Cartes combinatoires pour l'analyse d'images*, chapter 4, pages 103–120. Traité IC2, Signal et Image. Hermès Paris, France, August 2007.

[68] G. Damiand and D. Coeurjolly. A generic and parallel algorithm for 2D image discrete contour reconstruction. In *Proc. of International Symposium on Visual Computing*, volume 5359 of *LNCS*, pages 792–801, Las Vegas, Nevada, USA, December 2008. Springer Berlin/Heidelberg.

[69] G. Damiand and D. Coeurjolly. A generic and parallel algorithm for 2D digital curve polygonal approximation. *Journal of Real-Time Image Processing*, 6(3):145–157, September 2011.

[70] G. Damiand, C. de la Higuera, J.-C. Janodet, E. Samuel, and C. Solnon. A polynomial algorithm for subisomorphism of holey plane graphs. In *Proc. of International Workshop on Mining and Learning with Graphs*, Leuven, Belgium, July 2009.

[71] G. Damiand, C. de la Higuera, J.-C. Janodet, E. Samuel, and C. Solnon. Polynomial algorithm for submap isomorphism: Application to searching patterns in images. In *Proc. of International Workshop on Graph-Based Representations in Pattern Recognition*, volume 5534 of *LNCS*, pages 102–112, Venice, Italy, May 2009. Springer Berlin/Heidelberg.

[72] G. Damiand, M. Dexet-Guiard, P. Lienhardt, and E. Andres. Removal and contraction operations to define combinatorial pyramids: Application to the design of a spatial modeler. *Image and Vision Computing*, 23(2):259–269, February 2005.

[73] G. Damiand and A. Dupas. Combinatorial maps for 2D and 3D image segmentation. In *Digital Geometry Algorithms*, volume 2 of *LNCVB*, pages 359–393. Springer Netherlands, 2012.

[74] G. Damiand, A. Dupas, and J.-O. Lachaud. Combining topological maps, multi-label simple points, and minimum-length polygons for efficient digital partition model. In *Proc. of International Workshop on Combinatorial Image Analysis*, volume 6636 of *LNCS*, pages 56–69, Madrid, Spain, May 2011. Springer Berlin/Heidelberg.

[75] G. Damiand, A. Dupas, and J.-O. Lachaud. Fully deformable 3D digital partition model with topological control. *Pattern Recognition Letters*, 32(9):1374–1383, July 2011.

[76] G. Damiand, R. Gonzalez-Diaz, and S. Peltier. Removal operations in nD generalized maps for efficient homology computation. In *Proc. of International Workshop on Computational Topology in Image Context*, volume 7309 of *LNCS*, pages 20–29, Bertinoro, Italy, May 2012. Springer.

[77] G. Damiand and P. Lienhardt. Removal and contraction for n-dimensional generalized maps. In *Proc. of International Conference on Discrete Geometry for Computer Imagery*, volume 2886 of *LNCS*, pages 408–419, Naples, Italy, November 2003. Springer Berlin/Heidelberg.

[78] G. Damiand, S. Peltier, and L. Fuchs. Computing homology for surfaces with generalized maps: Application to 3D images. In *Proc. of International Symposium on Visual Computing*, volume 4292 of *LNCS*, pages 235–244, Lake Tahoe, Nevada, USA, November 2006. Springer Berlin/Heidelberg.

[79] G. Damiand, S. Peltier, and L. Fuchs. Computing homology generators for volumes using minimal generalized maps. In *Proc. of International Workshop on Combinatorial Image Analysis*, volume 4958 of *LNCS*, pages 63–74, Buffalo, NY, USA, April 2008. Springer Berlin/Heidelberg.

[80] G. Damiand, S. Peltier, L. Fuchs, and P. Lienhardt. Topological map: An efficient tool to compute incrementally topological features on 3D images. In *Proc. of International Workshop on Combinatorial Image Analysis*, volume 4040 of *LNCS*, pages 1–15, Berlin, Germany, June 2006. Springer Berlin/Heidelberg.

[81] G. Damiand and P. Resch. Topological map based algorithms for 3D image segmentation. In *Proc. of International Conference on Discrete Geometry for Computer Imagery*, volume 2301 of *LNCS*, pages 220–231, Bordeaux, France, April 2002. Springer Berlin/Heidelberg.

[82] G. Damiand and P. Resch. Split and merge algorithms defined on topological maps for 3D image segmentation. *Graphical Models*, 65(1-3):149–167, May 2003.

[83] G. Damiand, C. Solnon, C. de la Higuera, J.-C. Janodet, and E. Samuel. Polynomial algorithms for subisomorphism of nD open combinatorial maps. *Computer Vision and Image Understanding*, 115(7):996–1010, July 2011.

[84] X. Daragon, M. Couprie, and G. Bertrand. Discrete frontiers. In *Proc. of International Conference on Discrete Geometry for Computer Imagery*, volume 2886 of *LNCS*, pages 236–245, Naples, Italy, 2003. Springer.

[85] X. Daragon, M. Couprie, and G. Bertrand. Discrete surfaces and frontier orders. *Journal of Mathematical Imaging and Vision*, 23:379–399, 2005.

[86] M. de Berg, M. van Kreveld, M. Overmars, and O. Schwarzkopf. *Computational Geometry: Algorithms and Applications.* Springer-Verlag, January 2000.

[87] L. De Floriani and B. Falcidieno. A hierarchical boundary model for solid object representation. *ACM Transactions on Graphics*, 7(1):42–60, 1988.

[88] L. De Floriani, P. Magillo, and E. Puppo. A multiresolution topological representation for non-manifold meshes. *Computer-Aided Design*, 36(2):141–159, February 2004.

[89] L. De Floriani, M. Mesmoudi, F. Morando, and E. Puppo. Decomposing non-manifold objects in arbitrary dimensions. *Graphical Models*, 65(1-3):2–22, 2003.

[90] L. De Floriani, F. Morando, and E. Puppo. A representation for abstract simplicial complexes: An analysis and a comparison. In *Proc. of International Conference on Discrete Geometry for Computer Imagery*, volume 2886 of *LNCS*, pages 454–464, Naples, Italy, November 2003. Springer.

[91] L. De Floriani, F. Morando, and E. Puppo. Representation of non-manifold objects through decomposition into nearly manifold parts. In *Proc. of ACM Symposium on Solid Modeling and Applications*, pages 304–309, Seattle, WA, USA, June 2003. ACM Press.

[92] M. Dexet. *Architecture d'un modeleur géométrique à base topologique d'objets discrets et méthodes de reconstruction en dimensions 2 et 3.* PhD thesis, Université de Poitiers, December 2006.

[93] D. Dobkin and M. Laszlo. Primitives for the manipulation of three-dimensional subdivisions. In *Proc. of Symposium on Computational Geometry*, pages 86–99, Waterloo, Canada, June 1987.

[94] D. Dobkin and M. Laszlo. Primitives for the manipulation of three-dimensional subdivisions. *Algorithmica*, 5(4):3–32, 1989.

[95] J.-P. Domenger. *Conception et implémentation du noyau graphique d'un environnement 2D1/2 d'édition d'images discrètes.* Thèse de doctorat, Université Bordeaux 1, April 1992.

[96] J.-F. Dufourd. Algebras and formal specifications in geometric modelling. *The Visual Computer*, 13:131–154, 1997.

[97] A. Dupas. *Opérations et Algorithmes pour la Segmentation Topologique d'Images 3D.* Thèse de doctorat, Université de Poitiers, November 2009.

[98] A. Dupas and G. Damiand. Comparison of local and global region merging in the topological map. In *Proc. of International Workshop on Combinatorial Image Analysis*, volume 4958 of *LNCS*, pages 420–431, Buffalo, NY, USA, April 2008. Springer Berlin/Heidelberg.

[99] A. Dupas and G. Damiand. First results for 3D image segmentation with topological map. In *Proc. of International Conference on Discrete Geometry for Computer Imagery*, volume 4992 of *LNCS*, pages 507–518, Lyon, France, April 2008. Springer Berlin/Heidelberg.

[100] A. Dupas and G. Damiand. Region merging with topological control. *Discrete Applied Mathematics*, 157(16):3435–3446, August 2009.

[101] A. Dupas, G. Damiand, and J.-O. Lachaud. Multi-label simple points definition for 3D images digital deformable model. In *Proc. of International Conference on Discrete Geometry for Computer Imagery*, volume 5810 of *LNCS*, pages 156–167, Montréal, Canada, September 2009. Springer Berlin/Heidelberg.

[102] H. Edelsbrunner. *Algorithms in Computational Geometry.* Springer, New-York, 1987.

[103] J. Edmonds. A combinatorial representation for polyhedral surfaces. *Notices American Mathematical Society*, 7, 1960.

[104] H. Elter and P. Lienhardt. Cellular complexes as structured semi-simplicial sets. *International Journal of Shape Modeling*, 1(2):191–217, 1994.

[105] P.F. Felzenszwalb and D.P. Huttenlocher. Efficient graph-based image segmentation. *International Journal of Computer Vision*, 59(2):167–181, 2004.

[106] X. Feng, Y. Wang, Y. Weng, and Y. Tong. Compact combinatorial maps in 3d. In *Proc. of the First International Conference on Computational Visual Media*, CVM'12, pages 194–201, Berlin, Heidelberg, 2012. Springer-Verlag.

[107] M. Ferri, C. Gagliardi, and L. Grasselli. A graph-theoretical representation of pl-manifolds: a survey on crystallizations. *Aequationes Mathematicae*, 31:121–141, 1986.

[108] V. Ferruci and A. Paoluzzi. Extrusion and boundary evaluation for multidimensional polyhedra. *Computer-Aided Design*, 23(1):40–50, 1991.

[109] C. Fiorio. *Approche interpixel en analyse d'images : une topologie et des algorithmes de segmentation.* Thèse de doctorat, Université Montpellier II, November 1995.

[110] C. Fiorio. A topologically consistent representation for image analysis: the frontiers topological graph. In *Proc. of International Conference Discrete Geometry for Computer Imagery*, volume 1176 of *LNCS*, pages 151–162, Lyon, France, November 1996.

[111] E. Fléchon, F. Zara, G. Damiand, and F. Jaillet. A generic topological framework for physical simulation. In *Proc. of International Conference in Central Europe on Computer Graphics and Visualization*, WSCG Full papers proceedings, pages 104–113, Plzen, Czech Republic, June 2013.

[112] S. Fourey and L. Brun. A first step toward combinatorial pyramids in n-D spaces. In *Proc. of International Workshop on Graph-Based Representations in Pattern Recognition*, volume 5534 of *LNCS*, pages 304–313, Venice, Italy, May 2009. Springer.

[113] S. Fourey and L. Brun. Efficient encoding of n-d combinatorial pyramids. In *Proc. of International Conference on Pattern Recognition*, pages 1036–1039. IEEE Computer Society, 2010.

[114] D. Fradin, D. Meneveaux, and P. Lienhardt. A hierarchical topology-based model for handling complex indoor scenes. *Computer Graphics Forum*, 25(2):149–162, June 2006.

[115] R. Fritsch and R.A. Piccinini. *Cellular Structures in Topology*. Cambridge University Press, 1990.

[116] M. Gangnet, J.C. Herve, T. Pudet, and J.M. Van Thong. Incremental computation of planar maps. *ACM Computer Graphics*, 23(3):345–354, 1989.

[117] GIRL: **G**eneral **I**mage **R**epresentation **L**ibrary. `http://girl.labri.fr`.

[118] R. Goffe, L. Brun, and G. Damiand. A top-down construction scheme for irregular pyramids. In *Proc. of International Conference On Computer Vision Theory and Applications*, pages 163–170, Lisboa, Portugal, February 2009.

[119] R. Goffe, L. Brun, and G. Damiand. Tiled top-down combinatorial pyramids for large images representation. *International Journal of Imaging Systems and Technology*, 21(1):28–36, March 2011.

[120] R. Goffe, L. Brun, and G. Damiand. Tiled top-down pyramids and segmentation of large histological images. In *Proc. of International Workshop on Graph-Based Representations in Pattern Recognition*, volume 6658 of *LNCS*, pages 255–264, Münster, Germany, May 2011. Springer Berlin/Heidelberg.

[121] R. Goffe, G. Damiand, and L. Brun. Extraction of tiled top-down irregular pyramids from large images. In *Proc. of International Workshop*

on Combinatorial Image Analysis, Research Publishing Services, pages 123–137, Cancun, Mexico, November 2009. RPS, Singapore.

[122] R. Goffe, G. Damiand, and L. Brun. A causal extraction scheme in top-down pyramids for large images segmentation. In *Proc. of International Workshop on Structural and Syntactic Pattern Recognition*, volume 6218 of *LNCS*, pages 264–274, Cesme, Izmir, Turkey, August 2010. Springer Berlin/Heidelberg.

[123] S. Gosselin, G. Damiand, and C. Solnon. Signatures of combinatorial maps. In *Proc. of International Workshop on Combinatorial Image Analysis*, volume 5852 of *LNCS*, pages 370–382, Cancun, Mexico, November 2009. Springer Berlin/Heidelberg.

[124] S. Gosselin, G. Damiand, and C. Solnon. Frequent submap discovery. In *Proc. of Symposium on Combinatorial Pattern Matching*, volume 6661 of *LNCS*, pages 429–440, Palermo, Italy, June 2011. Springer Berlin/Heidelberg.

[125] H.-B. Griffiths. *Surfaces*. Cambridge University Press, Cambridge, 2nd edition, 1981.

[126] N. Guiard. *Construction de modéles géologiques 3D par co-raffinement de surfaces*. Thèse de doctorat, École des Mines de Paris, May 2006.

[127] L. Guibas and J. Stolfi. Primitives for the manipulation of general subdivisions and the computation of voronoi diagrams. *ACM Transaction on Graphics*, 4(2):74–123, 1985.

[128] E.L. Gursoz, Y. Choi, and F.B. Prinz. Vertex-based representation of non-manifolds boundaries. In *Proc. of Geometric Modeling for Product Engineering*, pages 107–130. North-Holland, 1990.

[129] T. Gurung and J. Rossignac. SOT: compact representation for tetrahedral meshes. In *Proc. of SIAM/ACM Joint Conference on Geometric and Physical Modeling*, SPM, pages 79–88, New York, NY, USA, 2009. ACM.

[130] P. Hachenberger and L. Kettner. 3D Boolean operations on Nef polyhedra. In *CGAL User and Reference Manual*. CGAL Editorial Board, 4.3 edition, 2013.

[131] P. Hachenberger, L. Kettner, and K. Mehlhorn. Boolean operations on 3d selective nef complexes: Data structure, algorithms, optimized implementation and experiments. *Computational Geometry*, 38(1-2):64–99, 2007. Special Issue on {CGAL}.

[132] O.H. Hansen and N.J. Christiansen. A model for n-dimensional boundary topology. In *Proc. of ACM Symposium on Solid Modeling Foundations and CAD/CAM Applications*, pages 65–73, Montréal, Canada, May 1993.

[133] A. Hatcher. *Algebraic Topology.* Cambridge University Press, 2002.

[134] S. Horna. *Reconstruction géométrique et topologique de complexes architecturaux 3D à partir de plans numériques 2D.* Thèse de doctorat, Université de Poitiers, November 2008.

[135] S. Horna, G. Damiand, D. Meneveaux, and Y. Bertrand. Building 3D indoor scenes topology from 2D architectural plans. In *Proc. of International Conference on Computer Graphics Theory and Applications*, pages 37–44, Barcelona, Spain, March 2007.

[136] S. Horna, D. Meneveaux, G. Damiand, and Y. Bertrand. Consistency constraints and 3D building reconstruction. *Computer-Aided Design*, 41(1):13–27, January 2009.

[137] A. Jacques. Constellations et graphes topologiques. In *Proc. of Combinatorial Theory and Applications*, volume 2, pages 657–673, Budapest, Hungary, 1970.

[138] T. Jund, P. Kraemer, and D. Cazier. A unified structure for crowd simulation. *Computer Animation and Virtual Worlds*, 23(3):311–320, 2012.

[139] T. Kaczynski, K. Mischaikow, and M. Mrozek. *Computational Homology.* Springer, 2004.

[140] L. Kettner. Using generic programming for designing a data structure for polyhedral surfaces. *Computational Geometry*, 13(1):65–90, 1999.

[141] E. Khalimsky, R. Kopperman, and P.R. Meyer. Boundaries in digital planes. *Journal of Applied Mathematics and Stochastic Analysis*, 3(1):27–55, 1990.

[142] T.Y. Kong, R. Kopperman, and P.R. Meyer. A topological approach to digital topology. *American Mathematical Monthly*, 98(10):901–917, 1991.

[143] T.Y. Kong and A. Rosenfeld. Digital topology: introduction and survey. *Computer Vision, Graphics, and Image Processing*, 48(3):357–393, 1989.

[144] U. Köthe. Xpmaps and topological segmentation - a unified approach to finite topologies in the plane. In *Proc. of International Conference Discrete Geometry for Computer Imagery*, volume 2301 of *LNCS*, pages 22–33, Bordeaux, France, April 2002.

[145] V.A. Kovalevsky. Discrete topology and contour definition. *Pattern Recognition Letters*, 2(5):281–288, 1984.

[146] V.A. Kovalevsky. Finite topology as applied to image analysis. *Computer Vision, Graphics, and Image Processing*, 46:141–161, 1989.

[147] P. Kraemer, D. Cazier, and D. Bechmann. Multiresolution half-edges. In *Proc. of Spring Conference on Computer Graphics*, pages 242–249. ACM Press, April 2007.

[148] P. Kraemer, D. Cazier, and D. Bechmann. Extension of half-edges for the representation of multiresolution subdivision surfaces. *The Visual Computer*, 25(2):149–163, 2009.

[149] P. Kraemer, L. Untereiner, T. Jund, S. Thery, and D. Cazier. CGoGN: N-dimensional meshes with combinatorial maps. In *Proc. of International Meshing Roundtable*, pages 485–503. Springer International Publishing, october 2013. `http://cgogn.unistra.fr/`.

[150] M. Kremer, D. Bommes, and L. Kobbelt. OpenVolumeMesh - A versatile index-based data structure for 3D polytopal complexes. In *Proc. of International Meshing Roundtable*, pages 531–548, 2012.

[151] W.G. Kropatsch. Building irregular pyramids by dual-graph contraction. *Vision, image and signal processing*, 142(6):366–374, 1995.

[152] W.G. Kropatsch and H. Macho. Finding the structure of connected components using dual irregular pyramids. In *Proc. of International Conference on Discrete Geometry for Computer Imagery*, pages 147–158, September 1995.

[153] T.L. Kunii and S. Takahashi. Area guide map modeling by manifolds and CW-complexes. In *Proc. of IFIP TC 5/WG 2 Working Conf. on Geometric Modeling in Computer Graphics*, pages 5–20, Genoa, Italy, 1993.

[154] V. Lang and P. Lienhardt. Geometric modeling with simplicial sets. In *Proc. of Pacific Graphics*, Seoul, Korea, 1995.

[155] V. Lang and P. Lienhardt. Simplicial sets and triangular patches. In *Proc. of Computer Graphics International*, Pohang, Korea, 1996.

[156] F. Ledoux, J.-M. Mota, A. Arnould, C. Dubois, P. Le Gall, and Y. Bertrand. Spécifications formelles du chanfreinage. *Techniques et Sciences Informatiquess*, 21(8):1–26, 2002.

[157] F. Ledoux, J.-C. Weill, and Y. Bertrand. GMDS: A generic mesh data structure. In *Proc. of International Meshing Roundtable*, USA, 2008.

[158] F. Ledoux, J.-C. Weill, and Y. Bertrand. Definition of a generic mesh data structure in the high performance computing context. *Developments and applications in engineering computational technology*, 26:49–80, 2010.

[159] S.H. Lee and K. Lee. Partial entity structure: a fast and compact non-manifold boundary representation based on partial topological entities. In *Proc. of ACM Symposium on Solid Modeling and Applications*, Ann Arbor, USA, 2001.

[160] P.-F. Léon, X. Skapin, and P. Meseure. Topologically-based animation for describing geological evolution. *Machine Graphics & Vision*, 15(3):481–491, January 2006.

[161] P.-F. Léon, X. Skapin, and P. Meseure. A topology-based animation model for the description of 2D models with a dynamic structure. In *Proc. of Workshop on Virtual Reality Interaction and Physical Simulation*, pages 67–76, Grenoble, France, 2008. Eurographics Association.

[162] P. Lienhardt. Subdivisions of n-dimensional spaces and n-dimensional generalized maps. In *Proc. of ACM Symposium on Computational Geometry*, pages 228–236, Saarbrücken, Germany, June 1989.

[163] P. Lienhardt. Subdivisions of surfaces and generalized maps. In *Proc. of Eurographics*, pages 439–452, Hamburg, Germany, 1989.

[164] P. Lienhardt. Subdivisions de surfaces et cartes généralisées de dimension 2. *RAIRO: Theoretical Informatics and Applications*, 25(2):171–202, 1991.

[165] P. Lienhardt. N-dimensional generalized combinatorial maps and cellular quasi-manifolds. *International Journal of Computational Geometry and Applications*, 4(3):275–324, 1994.

[166] P. Lienhardt, X. Skapin, and A. Bergey. Cartesian product of simplicial and cellular structures. *International Journal on Computational Geometry and Aplications*, 14(3):115–159, June 2004.

[167] S. Lins and A. Mandel. Graph-encoded 3-manifolds. *Discrete Mathematics*, 57:261–284, 1985.

[168] H. Lopes, S. Pesco, G. Tavares, M. Maia, and A. Xavier. Handlebody representation for surfaces and its applications to terrain modeling. *International Journal of Shape Modeling*, 09(01):61–77, 2003.

[169] H. Lopes and G. Tavares. Structural operators for modeling 3-manifolds. In *Proc. of ACM Symposium on Solid Modeling and Applications*, SMA '97, pages 10–18, New York, NY, USA, 1997. ACM.

[170] Y. Luo and G.A. Lukacs. A boundary representation for form-features and non-manifold solid objects. In *Proc. of ACM/Siggraph Symposium on Solid Modeling Foundations and CAD/CAM Applications*, Austin, Texas, USA, 1990.

[171] M. Mäntylä. Computational topology: a study of topological manipulations and interrogations in computer graphics and geometric modeling. *Acta Polytechnica Scandinavia*, 37, 1983.

[172] M. Mäntylä. *An Introduction to Solid Modeling.* Computer Science Press, 1988.

[173] M. Mäntylä and R. Sulonen. GWB: A solid modeler with euler operators. *Computer Graphics and Applications*, 2(7):17–31, 1982.

[174] J. P. May. *Simplicial objects in algebraic topology.* Van Nostrand, Princeton, 1967.

[175] H. Meine and U. Köthe. The GeoMap: a unified representation for topology and geometry. In *Proc. of International Workshop on Graph-Based Representations in Pattern Recognition*, volume 3434 of *LNCS*, pages 132–141, Poitiers, France, April 2005. Springer Berlin Heidelberg.

[176] P. Meseure, E. Darles, and X. Skapin. A topology-based mass/spring system. In *Proc. of Computer Animation and Social Agents (short papers)*, St Malo, France, June 2010.

[177] P. Meseure, E. Darles, and X. Skapin. Topology-based physical simulation. In *Proc. of Workshop on Virtual Reality Interaction and Physical Simulation*, pages 1–10, Copenhagen, Danemark, November 2010.

[178] M. E. Mortenson. *Geometric Modeling.* John Wiley & Sons, Inc., New York, NY, USA, 1985.

[179] D.E. Muller and F.P. Preparata. Finding the intersection of two convex polyhedra. *Theoretical Computer Science*, 7(2):217 – 236, 1978.

[180] J. R. Munkres. *Elements of Algebraic Topology.* Perseus Books, 1984.

[181] S. Murabata and M. Higashi. Non-manifold geometric modeling for set operations and surface operations. In *Proc. of IFIP/RPI Geometric Modeling Conference*, Rensselaerville, NY, USA, 1990.

[182] W. Nef. *Beiträge zur Theorie der Polyeder, mit Anwendungen in der Computergraphik.* Herbert Lang, Bern, 1978.

[183] A. Paoluzzi, F. Bernardini, C. Cattani, and V. Ferrucci. Dimension-independent modeling with simplicial complexes. *ACM Transactions on Graphics*, 12(1):56–102, 1993.

[184] V. Pascucci, V. Ferrucci, and A. Paoluzzi. Dimension-independent convex-cell based hierarchical polyhedral complex: representation scheme and implementation issues. In *Proc. of Symposium on Solid Modeling and Applications*, pages 163–174, 1995.

[185] S. Peltier, L. Fuchs, and P. Lienhardt. Homology of simploidal sets. In *Proc. of International Conference on Discrete Geometry for Computer Imagery*, volume 4245 of *LNCS*, pages 235–246, Szeged, Hungary, 2006. Springer Berlin Heidelberg.

[186] S. Peltier, L. Fuchs, and P. Lienhardt. Simploidal sets: definitions, operations and comparison with simplicial sets. *Discrete Applied Mathematics*, 157:542–557, 2009.

[187] S. Peltier, A. Ion, Y. Haxhimusa, W.G. Kropatsch, and G. Damiand. Computing homology group generators of images using irregular graph pyramids. In *Proc. of International Workshop on Graph-Based Representations in Pattern Recognition*, volume 4538 of *LNCS*, pages 283–294, Alicante, Spain, June 2007. Springer Berlin/Heidelberg.

[188] S. Peltier, A. Ion, W.G. Kropatsch, G. Damiand, and Y. Haxhimusa. Directly computing the generators of image homology using graph pyramids. *Image and Vision Computing*, 27(7):846–853, June 2009.

[189] M. Poudret, A. Arnould, Y. Bertrand, and P. Lienhardt. Cartes combinatoires ouvertes. Research Notes 2007-1, Laboratoire SIC E.A. 4103, F-86962 Futuroscope Cedex, France, october 2007.

[190] S. Prat, P. Gioia, Y. Bertrand, and D. Meneveaux. Connectivity compression in arbitrary dimension. *The Visual Computer*, 21(8):876–885, 2005.

[191] G. Ringel. *Map Color Theorem.* Springer-Verlag, 1974.

[192] G. Ringel. The combinatorial map color theorem. *Journal of Graph Theory*, 1(2):141–155, 1977.

[193] A. Rosenfeld. Adjacency in digital pictures. *Information and Control*, 26(1):24–33, 1974.

[194] J. Rossignac. 3D compression made simple: Edgebreaker with zipandwrap on a corner-table. In *Proc. of International Conference on Shape Modeling and Applications*, pages 278–283, 2001.

[195] J. Rossignac and M. O'Connor. SGC: A dimension-independant model for pointsets with internal structures and incomplete boundaries. *Geometric modeling for Product Engineering*, pages 145–180, 1989.

[196] H. Seifert and W. Threlfall. *A textbook of topology.* Academic Press, New York, 1980.

[197] C. Simon and G. Damiand. Generalized map pyramid for multi-level 3D image segmentation. In *Proc. of International Conference on Discrete Geometry for Computer Imagery*, volume 4245 of *LNCS*, pages 530–541, Szeged, Hungary, october 2006. Springer Berlin/Heidelberg.

[198] C. Simon, G. Damiand, and P. Lienhardt. Pyramids of n-dimensional generalized maps. In *Proc. of International Workshop on Graph-Based Representations in Pattern Recognition*, volume 3434 of *LNCS*, pages 142–152, Poitiers, France, April 2005. Springer Berlin/Heidelberg.

[199] C. Simon, G. Damiand, and P. Lienhardt. Receptive fields for generalized map pyramids: The notion of generalized orbit. In *Proc. of International Conference on Discrete Geometry for Computer Imagery*, volume 3429 of *LNCS*, pages 56–67, Poitiers, France, April 2005. Springer Berlin/Heidelberg.

[200] C. Simon, G. Damiand, and P. Lienhardt. nD generalized map pyramids: Definition, representations and basic operations. *Pattern Recognition*, 39(4):527–538, April 2006.

[201] X. Skapin and P. Lienhardt. Using cartesian product for animation. In *Proc. of Computer Animation and Simulation*, Eurographics, pages 187–201. Springer Vienna, 2000.

[202] C. Solnon, G. Damiand, C. de la Higuera, and J.-C. Janodet. On the complexity of submap isomorphism. In *Proc. of International Workshop on Graph-Based Representations in Pattern Recognition*, volume 7877 of *LNCS*, pages 21–30, Vienna, Austria, May 2013. Springer Berlin/Heidelberg.

[203] J.-C. Spehner. Merging in maps and pavings. *Theoretical Computer Science*, 86:205–232, 1991.

[204] G. Taubin and J. Rossignac. Geometric compression through topological surgery. *ACM Trans. Graph.*, 17(2):84–115, 1998.

[205] O. Terraz and P. Lienhardt. A study of basic tools for simulating metamorphoses of subdivided 2D and 3D objects. Application to the internal growing of wood and to the simulation of the growing of fishes. In *Proc. of Computer Animation and Simulation*, 6th Eurographics Workshop on Animation and Simulation, pages 104–129, Maastricht, Pays-Bas, 1995. Springer.

[206] W.T. Tutte. A census of planar maps. *Canad. J. Math.*, 15:249–271, 1963.

[207] W.T. Tutte. Combinatorial oriented maps. *Canad. J. Math.*, 5(31):986–1004, 1979.

[208] L. Čomič and L. de Floriani. Modeling and manipulating cell complexes in two, three and higher dimensions. In *Digital Geometry Algorithms*, volume 2 of *LNCVB*, pages 109–144. Springer Netherlands, 2012.

[209] L. Untereiner. *Représentation des maillages multirésolutions: application aux volumes de subdivision.* PhD thesis, Université de Strasbourg, November 2013.

[210] L. Untereiner, D. Cazier, and D. Bechmann. n-dimensional multiresolution representation of subdivision meshes with arbitrary topology. *Graphical Models*, 75(5):231–246, September 2013.

[211] F. Vidil and G. Damiand. Moka. http://moka-modeller.sourceforge.net/, 2003.

[212] A. Vince. Combinatorial maps. *Journal of Combinatorial Theory Series B*, 34:1–21, 1983.

[213] T. Wang, G. Dai, B. Ni, D. Xu, and F. Siewe. A distance measure between labeled combinatorial maps. *Computer Vision and Image Understanding*, 116(12):1168–1177, 2012.

[214] K. Weiler. Edge-based data structures for solid modelling in curved-surface environments. *Computer Graphics and Applications*, 5(1):21–40, 1985.

[215] K. Weiler. The radial-edge data structure: A topological representation for non-manifold geometry boundary modeling. In *Proc. of IFIP WG 5.2 Working Conference*, Rensselaerville, USA, 1986.

[216] K. Weiler. The radial edge structure: a topological representation for non-manifold geometric boundary modeling. In *Geometric Modeling for CAD Applications*, pages 217–254. Elsevier Science, 1988.

[217] D. Willersinn and W.G. Kropatsch. Dual graph contraction for irregular pyramids. In *Proc. of International Conference on Signal Processing*, volume 3, pages 251–256, Jerusalem, Israel, 1994.

Index